FIFTH EDITION

A LABORATORY MANUAL FOR GENERAL BOTANY

HOLT, RINEHART and WINSTON

New York

Chicago

San Francisco

Atlanta

Dallas

Montreal

Toronto

London

Sydney

MARGARET K. BALBACH
Illinois State University

LAWRENCE C. BLISS
University of Alberta

HARRY J. FULLER
(deceased)   Univerity of Illinois

FIFTH EDITION

A LABORATORY MANUAL FOR GENERAL BOTANY

Figure Credits

Figures 2.2, 2.3, 6.1, 16.2, 24.1, 24.5, 24.6, 26.1, 26.2, 28.1, 28.1, 28.3, 28.4, 29.5, 30.2, 30.3, 30.4, 32.2 come in part or in their entirety from *The Plant World*, Fifth Edition by Harry J. Fuller, Zane B. Carothers, Willard W. Payne and Margaret K. Balbach. Copyright ©1941, 1951, 1955, 1963, 1972 by Holt, Rinehart and Winston, Inc. Reprinted by permission of Holt, Rinehart and Winston.

Cover photo: Grant Heilman

Cataloging in Publication Data

Balbach, Margaret K
    A laboratory manual for general botany.

    Includes index.
    1. Botany—Laboratory manuals.   I. Bliss, Lawrence C., joint author.   II. Fuller, Harry James, 1907–   joint author.   III. Title.
QK53.B22 1977   581'.028   76-30542
ISBN 0-03-089749-1

Printed in the United States of America
789   140   98765432

CONTENTS

PREFACE

We live today in a world of relevancy. This fifth edition of *A Laboratory Manual for General Botany* is dedicated to presenting basic botany as a useful, biologically meaningful, and scientifically pertinent topic. The manual has been totally revised and rewritten. Special thanks are due to the youth of our day whose lively interests in plants have sparked our efforts to make botany of real human interest as well as a scientific study.

The philosophy has been to present the basic, classical botanical principles and topics in such a way as to relate them not only to the general concepts of our botanical heritage but also to the practical, applied interests in plants. Applied botany is uppermost in the minds of people, both young and old, and we must meet the challenge of both educating our young college people in the science of botany and also, hopefully, making knowledgeable plantsmen of them all.

The exercises fulfill all the needs of a traditional introduction to the principles of laboratory botany. Their organization also makes them directly usable in the individualized-instruction self-learning laboratory format.

Further the introduction to each exercise gives more background and perspective than is typical for a laboratory manual. This makes the manual especially recommended for those circumstances in which the lecture topics are not closely aligned to the specific needs of the laboratory coverage.

Part I concerns the structure and function of the flowering plant. Structure is always interpreted as regards its selective advantage to the plant. Economic importance is stressed where it is applicable. Nothing is presented simply as a vacuous fact.

Totally new chapters have been introduced on seed germination, the factors affecting seed germination and dendrochronology, and also an algal key that has been successfully used for several years on an individual classroom basis.

Part II concerns the survey of the plant kindgom. Life cycles, often so boring to beginning students, are presented as each being an efficient solution to the universal problems of survival. Each successive plant group is evaluated as to its progressive features over those of the preceding group.

In Part III the chapters on ecology have been consolidated and revamped so as to provide more participatory involvement on the part of the students in the very difficult laboratory presentation of ecological principles.

Many new diagrams have been introduced and old ones revised. Each exercise has summary questions to be handed in and graded, as well as internal questions to liven the students' awareness of their progress through an exercise. Many of these internal questions are to be answered simply by circling YES or NO.

Users of the manual should note that the tables and figures may not always appear in the order they are mentioned in the exercise. This is because, for the sake of space, certain tables and figures have been positioned at the end of the chapters on special tear-out sheets.

An appendix has been added, with information on the needs and sources of materials for each exercise.

Appreciation goes mainly to Dr. Harold E. Balbach, Environmental Biologist, U.S. Army Construction Engineering Research Laboratory. Thanks are due to Dr. Balbach specifically for his help in the revision of the ecology exercise, and generally for his consultation and advice in resolving many of the questions and decisions that came with the writing and reorganization of the manual.

Particular thanks go to Dr. Richard L. Phipps, Research Botanist, U.S. Geological Survey, for his suggestions and advice regarding the exercise on dendrochronology; to Dr. Wesley Whiteside of Eastern Illinois University for his review of the chapters on fungi; to Dr. Richard L. Smith of Eastern Illinois University for his "Key to Some of the Fresh Water Algae"; to Professor Laurence E. Crofutt of Eastern Illinois University for his recommendations for the exercise on bacteria; and to M. Diana Webb for her excellent drawings in the algae key.

Grateful acknowledgement is given to the diligent reviewers of the manuscript, Drs. Jordan J. Choper, Montgomery College; Harold E. Eversmeyer, Murray State University; Hugo A. Ferchau, Western State College of Colorado; James F. Matthews, University of North Carolina; William H. Miller, William Rainey Harper College; B. John Syrocki, State University College at Brockport; James J. Tobolski, Indiana University–Purdue University; and T. J. Watson, Jr., University of Montana, whose critiques and examinations were thoughtful, accurate, and helpful in many ways.

Finally, appreciation is expressed to Hilda Phipps for her forbearance and conscientious and accurate typing of the manuscript and to the editors of Holt, Rinehart and Winston whose skill and professional production of the manual has been first class.

Illinois State University   M.B.
University of Alberta   L.C.B.
January, 1977

PART I

Structure and Function of Seed Plants

EXERCISE

1

THE WHOLE PLANT

We live in a world dominated by flowering plants. Flowering plants are the ones most commonly known and utilized by today's consumer, as they were by our prehistoric ancestors. They occur as woody-stemmed trees and shrubs and as soft-stemmed herbs. Herbs are a broad category which includes weeds, grains, lawn grasses, garden flowers and vegetables, certain fruits, and most house plants. All these plants go through the life stages shown in Figure 1.1.

There are two major groups of flowering plants, the dicots and the monocots (or, more formally, the dicotyledons and the monocotyledons). Dicots are those plants having two food-storage organs (cotyledons) in the seed, monocots have one.

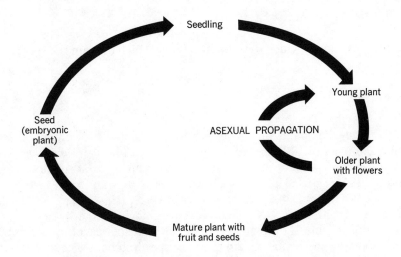

Figure 1.1   Stages in Life Cycle of Flowering Plant

I. THE MATURE DICOT PLANT

A. THE VEGETATIVE PLANT BODY

Vegetative parts of a plant are those concerned with photosynthesis, growth, and storage. In photosynthesis, the green plant absorbs light and converts it to a usable chemical form of energy. Essentially the plant traps sun energy and puts it in storage in sugar molecules. To do this, the plant needs water, CO_2, and light. To get enough of these three essentials, the plant needs big absorptive surfaces to absorb water from the soil, CO_2 from the air, and energy from the sun. Leaves are the absorptive surface for light and CO_2, although in many plants, stems also function this way. Roots are the absorptive surfaces for water.

The following are suggested examples of dicots to study: two-month-old bean plant (*Phaseolus vulgaris*); two-month-old cucumber plant (*Cucumis sativus*); "Tiny Tim" or any dwarf variety of tomato (*Lycopersicon* sp.); *Coleus*; ornamental pepper (*Capsicum annuum*); Christmas cherry (*Solanum pseudocapsicum*); rubber plant (*Ficus elastica*); or any suitable species of house plant or potted woody dicot; or any leafy shrub, such as lilac (*Syringa vulgaris*), privet (*Ligustrum* sp.), etc.

► ACTIVITY 1   **Examine** the plants available and **identify**

1. LEAF   Each is composed of two parts, (a) the BLADE—the expanded flat portion, and (b) the PETIOLE—a stalk that attaches the blade to the stem. In some plants the blade is divided into a number of segments, called LEAFLETS.
2. STIPULES   Green leaf-like blades attached near the base of the petiole. These do not occur in all plants.
3. LEAF VEINS   The veins form a meshwork throughout the blade. The meshwork of a dicot leaf is described as NET VENATION.
4. AXIL   The angle formed between the stem and the upper side of the petiole.
5. AXILLARY or LATERAL BUD   A bud in the axil. It will produce a leafy branch or a flower, or both.
6. TERMINAL or APICAL BUD   The bud at the top of the main stem or at the end of each branch. Its growth produces more stem and leaves.
7. NODE   The place on the stem where a leaf occurs.
8. INTERNODE   The segment of the stem between any two nodes. Stem elongation is due primarily to elongation of the internodes.

Label Figure 1.2, using the capitalized terms defined in the preceding description.

B. FLOWERS, FRUIT, AND SEED

The plant forms flowers which develop into fruits which contain the seeds. Seeds grow into plants.

The Dicot Flower

► ACTIVITY 2   **Study** the flowers of any number of representative dicot plants. Suggested plants are three-month-old bean plant (*Phaseolus vulgaris*) or cucumber (*Cucumis sativus*); any dwarf tomato variety (*Lycopersicon* sp.); house plants, such as *Oxalis* sp., *Fuchsia* sp., flowering maple (*Abutilon* sp.); or any available garden plants, such as *Petunia* sp., snapdragon (*Antirrhinum* sp.), etc. **Find** (1) the colored PETALS, (2) green leafy-looking SEPALS. (In some plants sepals look like petals.) Color attracts birds and insects. Not all flowers are colorful and aromatic.

Count and **record** in Table 1.1 the number of petals in the flowers of the dicots available for your study. Have the laboratory instructor explain to you or show you how to count them if some of these plants have double flowers.

Table 1.1   Number of Petals in Flowers of Dicots

Plant		Number of Petals
common name	*botanical name*	

The Fruit—The Structure that Protects and Disperses Seeds

► ACTIVITY 3   **Examine** plants bearing fruit, such as a three-month-old bean plant (*Phaseolus vulgaris*); dwarf varieties of tomato, such as "Tiny Tim" (*Lycopersicon* sp.); ornamental pepper (*Capsicum annuum*); or Christmas cherry (*Solanum pseudocapsicum*). Fruits such as most of these are colorful. Of what value is the color? _____.

Examine fruits of other plants which are not colorful and/or fleshy but are dull colored and hard or dry, such as those of elm (*Ulmus* sp.); tree of heaven (*Ailanthus altissima*); hickory (*Carya* sp.); beggar tick (*Bidens* sp.); cocklebur (*Xanthium* sp.); milkweed (*Asclepias syriaca*); peanut (*Arachis hypogaea*); ragweed (*Ambrosia* sp.); velvet leaf (*Abutilon theophrasti*); or other available types. These are variously adapted for dispersal by wind, gravity, burrowing rodents, and water or by clinging to the fur of animals or to clothing.

The Dicot Seedling

Seeds germinate into young plants, called SEEDLINGS.

► ACTIVITY 4   **Examine** seedlings of the bean plant (*Phaseolus vulgaris*); cucumber (*Cucumis sativus*); or tomato plant (*Lycopersicon* sp.) that have been grown in a glass or plastic tumbler, jar, or beaker with only water and some sort of physical support supplied to them. Since only water has been supplied, from where have the seedlings obtained the energy and nutrients for this growth?

II. THE MATURE MONOCOT PLANT

A. THE VEGETATIVE PLANT BODY

► ACTIVITY 5   **Examine** representative monocot plants, such as species of wandering Jew (*Zebrina pendula*); spiderwort (*Tradescantia virginiana*); purple heart (*Setcreasia pallida*); day-flower (*Commelina* sp.); *Philodendron*; dumbcane (*Dieffenbachia* sp.) water hyacinth (*Eichornia* sp.); or any lawn or pasture grass; or commercial grain such as corn (*Zea mays*) or oats (*Avena sativa*). **Find**

1. The leaf BLADE. It is relatively narrow, parallel-sided, and flat.
2. SHEATH   The base of the blade. It partially or completely encloses the stem.

3. NODES   The swollen-looking joints in the stem. Swollen nodes are characteristic of most monocots.

4. VEINS which run parallel to the leaf margin.

Note: Some monocots, such as *Dieffenbachia* and *Philodendron*, have venation like that of many dicots.

B. FLOWERS AND SEED

The Monocot Flower

► ACTIVITY 6   **Examine** the flowers of representative monocots, such as spiderwort (*Tradescantia* sp.); wandering Jew (*Zebrina* sp.); or *Rhoeo* sp. If in season, the following could also be studied: lily (*Lilium* sp.); *Iris* sp.; *Gladiolus* sp.; tulip (*Tulipa* sp.); or *Crocus* sp.

Count the number of petals in the flowers of the monocots available and **record** in Table 1.2.

Table 1.2   Number of Petals in Flowers of Monocots

Plant		Number of Petals
common name	*botanical name*	
rododendron		ten

The Monocot Seedling

► ACTIVITY 7   **Examine** seedlings of corn, wheat, or oats that have been grown in glass or plastic tumblers, jars, or beakers with only water and some physical support supplied to them. (1) Does the seedling show the same leaf and stem characteristics as the mature plant?   YES   NO. (2) Is there a leaf sheath?   YES   NO. (3) Is the leaf parallel-sided?   YES   NO.

► ACTIVITY 8   **Fill** in Table 1.3.

EXERCISE 2

SEED STRUCTURE AND GERMINATION

Seeds are the great staple food of the world. They feed more people and livestock than any other type of food. Of all the world's harvested acreage, 70 percent is devoted to growing the seeds of cereals. The most important of these are wheat, rice, and corn. Other valuable seeds for food are the soybean, the common bean, rye, oats, sorghum, and millet. All seeds are embryonic plants supplied with the nutrients needed for germination. Seed nutrients ensure the growth of the new plant during its seedling stage when it is most vulnerable to the hazards of weather and other factors. All peoples have exploited seeds for these nutrients.

Nutrients are supplied to the embryo of the ripening seed by the parent plant in a special tissue, called the *endosperm*. It supplies nutrients and growth-promoting hormones. As the seed is ripening on the parent plant, the embryo of the seed digests some or all of the endosperm. It then restocks the nutrients in its own storage organs, called *cotyledons*.

In some dicot flowering plants all the endosperm is digested during seed ripening, and two plump cotyledons occupy most of the ripe seed. In other dicots only part of the endosperm is digested, so the ripe seed contains some endosperm as well as two cotyledons. In monocots the ripe seed contains an embryo with one cotyledon and variable amounts of endosperm, depending on the species.

Note: Seed ripening is not the same as germination. Ripening is the growth that produces the seed; germination is the growth of the seed that produces the seedling.

I. THE DICOT SEED

A. THE BEAN SEED—A DICOT SEED WITHOUT AN ENDOSPERM

► ACTIVITY 1   **Examine** seeds of the pinto, kidney, or lima bean (*Phaseolus* sp.) that have been soaked in water overnight or for several hours. **Find**

1. SEED COAT   The skin-like protective covering of the seed.
2. HILUM   An elliptical scar on the concave side of the seed. It marks the point of attachment of a stalk that attached the seed to the inside of the fruit (pod).
3. RAPHE   A small ridge at one end of the hilum. It is a remnant of the stalk.
4. MICROPYLE   A small circular scar at the other end of the hilum. It marks the hole from the entry of the pollen tube just prior to fertilization.

Label the external parts of the bean seed in Figure 2.1, using the capitalized terms defined in the preceding description.
Peel off the seed coat. **Find**

1. COTYLEDONS   The two halves of the seed. These are modified storage leaves. Because cotyledons are already part of the living embryo, the food they contain can be quickly digested during germination.
2. EMBRYONIC AXIS   You will find this attached to one of the cotyledons. It is easily identified by its two miniature leaves. These are attached near the COTYLEDONARY NODE to the shoot of the axis, the EPICOTYL. The epicotyl and leaves together form the PLUMULE. The remainder of the axis is the HYPOCOTYL, the tip of which is the RADICLE, or embryonic root. This tiny embryonic axis utilizes the foods of its two cotyledons during germination and grows into the seedling. The epicotyl forms the shoot and the radicle forms the root.

Label the internal parts of the bean seed shown in Figure 2.2, using the capitalized terms defined in the preceding description.

B. GERMINATION OF THE DICOT SEED

Imbibition—Water Absorption and Swelling of the Seed

As long as the seed is dry, it will not germinate. But when moisture is available, and the seed coat is porous, the seed will imbibe water. This is IMBIBITION, the first phase of germination. If the seed is not dormant, it will then germinate.

► ACTIVITY 2   **Compare** seeds of beans, peas, and other dicots in the dry state versus those which have been kept moistened for several days in wet rolled paper towels or on moist paper in petri dishes, kept in a warm (21°C) place.

Although small, the dry seed contains large amounts of starches, fats, and oils, the proportions varying with different species. The dehydrated state of these foods permits the storage of these large quantities. Imbibition builds up tremendous water pressure, which is usually sufficient to split the coat and allow the emergence of the growing embryo.

► ACTIVITY 3   **Observe** paper cups containing plaster of paris into which dry seeds were mixed before it hardened. Describe the condition of the plaster of paris. _____

Seedling Stages of the Bean—Epigean Type Germination

► ACTIVITY 4   **Examine** bean seedlings of different ages that have been grown in sand or soil flats. Refer to Figure 2.3 as a guide.

three-day-old seedling   The growth of the radicle has formed the PRIMARY ROOT. Although the shoot (epicotyl) has enlarged, it has not yet left its position between the COTYLEDONS. What are two advantages of the root being the first organ to grow? _____

_____

Label Figure 2.3 (A), using the capitalized terms defined in the preceding description.

six-to-eight-day-old seedlings   These have just broken through the soil surface. **Identify**

1. HYPOCOTYL   It grows in an arched fashion and pulls the cotyledons and plumule up. In this way the delicate growing tip is protected from damage during emergence.
2. COTYLEDONS   These may now be green and are still close together. They may or may not yet have emerged from the split SEED COAT.
3. PLUMULE   The two LEAVES of the plumule may or may not extend from between the cotyledons.
4. PRIMARY ROOT   This now has LATERAL ROOTS.

Label Figure 2.3 (B) and (C), using the capitalized terms defined in the preceding description.

ten-day-old-seedlings   In this stage, the hypocotyl has straightened up and the cotyledons are spread apart and well above ground. This type of germination, which elevates the cotyledons above ground, is called EPIGEAN GERMINATION. Most dicots exhibit this type.

Find young plants with just one pair of leaves. **Identify**

1. COTYLEDONS   These fleshy organs may now be wrinkled and somewhat shriveled. Why? _____.
2. HYPOCOTYL   The stem portion below the cotyledons.
3. FIRST TRUE LEAVES   These are just above the cotyledons.
4. TERMINAL BUD   This is the cluster of very tiny leaves at the tip of the stem which is between the first true leaves.
5. ROOTS   There are many BRANCH or LATERAL ROOTS now, as well as the original PRIMARY ROOT.

Label Figure 2.3 (D), using the capitalized terms defined in the preceding description.

Seedling Stages of the Pea—Hypogean Type Germination

► ACTIVITY 5   **Examine** successive stages of seedlings of the garden pea (*Pisum sativum*) and refer to Figure 2.4 as a guide. Are the cotyledons above the ground in the later stages? YES   NO. (Circle one.) In the younger stages is there a loop in the axis that is growing upward?   YES   NO. This arched axis is the EPICOTYL. It functions like the hypocotyl of the bean seedling in that it elongates and tunnels a path through the soil dragging the plumule (but not the cotyledons) upward. The hypocotyl does not elongate and remains below ground. This type of seedling growth, in which the cotyledons remain below ground, is called HYPOGEAN GERMINATION.

Label the different seedling stages of the pea in Figure 2.4, using the same terms as those used for the bean seedling.

C. A DICOT SEED WITH AN ENDOSPERM

Seeds of the honey locust tree (*Gleditisia triacanthos*), like those of many other dicots, contain an endosperm as well as two cotyledons. These honey locust seeds have a very hard coat which must be perforated with a file or knife to allow imbibition to occur.

► ACTIVITY 6   **Study** seeds of honey locust that have been perforated and allowed to soak in water for a couple of days. Cut the seed in half across its long axis. Note the two different colored zones in the cut surface. The outer, clear, light yellowish brown zone is the endosperm. The two cotyledons make up the central, yellowish green zone.

Dissect another seed so that you can separate the endosperm from the cotyledons intact. Note the size and shape of the cotyledons. Are they as large and plump as those of the bean seed? YES   NO. Much of the food for germination of the seedling is stored where?   _____ _____ Is there any food also stored in the cotyledons?   YES   NO. Is there an embryonic axis?   YES   NO.

II. THE MONOCOT SEED

A. THE CORN GRAIN

Although the corn grain is really a fruit, there is barely any fruit development, and the bulk of the grain is the seed.

▶ ACTIVITY 7   **Examine** a corn grain that has been soaked for several hours. **Find**

    1. SCUTELLUM   The white, oval-shaped depression. Basically it is a modified cotyledon and functions to digest the endosperm just as cotyledons do in dicot plants.
    2. EMBRYONIC AXIS   Faintly visible as a linear depression in the center of the scutellum.
    3. ENDOSPERM   The outer, yellow portion of the grain.

Label the external features of the corn grain in Figure 2.5 (A). **Obtain** a soaked corn grain. Hold it flat on the table and make a lengthwise cut through it with a razor blade [see Figure 2.5 (A) for line of cut]. **Find**

    1. SCUTELLUM   The white portion of the seed.
    2. ENDOSPERM   The yellow portion of the seed.

Use a hand lens or dissecting microscope to identify the rest of the seed parts. Refer to Figure 2.5 (B) as a guide. **Find**

    3. COLEOPTILE   A tubular sheath pointed at one end. It is a modified part of the cotyledon and protects the plumule during germination.
    4. PLUMULE   The growing shoot tip of the embryonic axis.
    5. RADICLE   The rudimentary root. It is the root tip of the axis.
    6. COLEORHIZA   A tubular sheath enclosing the radicle. It protects the radicle in the early stages of germination.
    7. MESOCOTYL   The part of the embryonic axis that is intermediate between the radicle and the plumule.

Label the internal parts of the corn grain in Figure 2.5 (B), using the capitalized terms defined in the preceding description.

B. SEEDLING STAGES OF THE CORN GRAIN

▶ ACTIVITY 8   **Examine** germinating corn grains and the successive stages of seedling development. Refer to Figure 2.6 as a guide. **Find**

    1. ROOT   This is the first organ to emerge. In its early stages it is enclosed and protected by the coleorhiza which you probably won't see.
    2. COLEOPTILE   A colorless tubular sheath which encloses the mesocotyl. It is seen growing up from the seed, or directly opposite the root.
    3. MESOCOTYL   The stem portion enclosed by the coleoptile and bearing at its tip the plumule.
    4. PLUMULE   The folded first leaf that emerges from the coleoptile sheath.

When the coleoptile reaches the surface, it splits open and the leaves of the plumule emerge. Later, new crown roots form at the base (first node) of the stem. In monocot seeds the cotyledon functions mainly in the digestion and translocation of the foods in the endosperm to the growing axis.

Is the cotyledon elevated above ground?   YES   NO. What type of seed germination is this called? _____

Label the stages of the corn seedling development in Figure 2.6 using the capitalized terms defined in the preceding description.

EXERCISE 2   **STUDENT NAME** _____

QUESTIONS

1. What is the selective advantage of the abundant supply of foods in the seed? _____

_____

2. There are at least two advantages to the storage of foods in the dehydrated state in seeds. What are these?

(a) _____

(b) _____

3. Primitive land plants like the ferns don't supply their embryos with any food reserves. What problems

does such an embryo face compared to that of a seed plant? _____

_____

4. How do grains lend themselves so readily to worldwide trade? _____

_____

Why are they in such demand anyway? _____

_____

5. Epigean germination is characteristic of the majority of which group of flowering plants? _____

6. What does the radicle of the seed form? _____

7. Dicots and monocots differ in the way by which the delicate plumule is protected as it is being elevated

through the soil. The dicot process involves _____

_____

The monocot process involves _____

_____

8. What phase of seed germination ruptured the plaster of paris? _____

9. Is the endosperm part of the embryo? _____

10. Is a cotyledon part of the embryo? _____

11. Do cotyledons develop during seed ripening or during seed germination? _____

12. Distinguish between seed ripening and seed germination on the following bases:

a. Ripening takes place usually when the seed is located where? _____

b. Germination takes place when the seed is where? _____

c. During ripening of the bean seed, the endosperm is being formed or digested? _____

d. During ripening of any type of seed, what three components of the seed are being formed? _____

_____

13. Do all dicot seeds germinate like the bean seed? _____

14. Which do you think is more protective of the seedling's food supply, (a) epigean germination or (b)

hypogean germination? Circle (a) or (b). _____

15. Which type of germination do you think, therefore, is considered to be the more highly evolved?

_____

Figure 2.1   External Features of the Bean Seed

Figure 2.2   Internal Parts of the Bean Seed

Figure 2.3   Germination of the Bean Seed

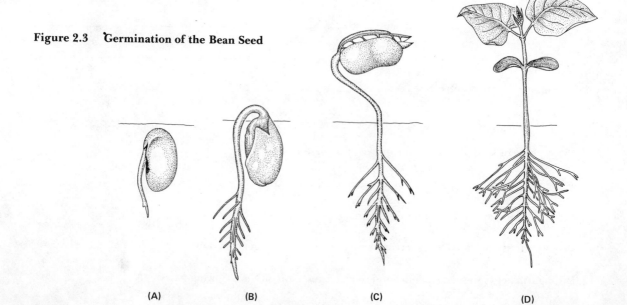

(A)   (B)   (C)   (D)

EXERCISE 2   **STUDENT NAME** _____

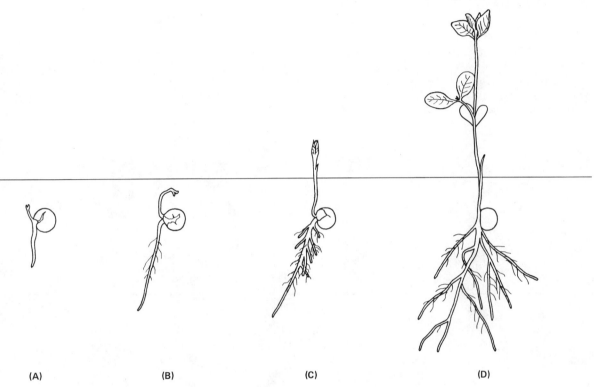

(A)   (B)   (C)   (D)

Figure 2.4   Germination of the Pea Seed

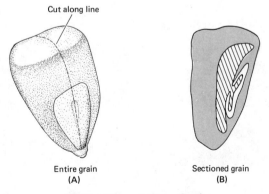

Cut along line

Entire grain   Sectioned grain
(A)   (B)

Figure 2.5   A Corn Grain

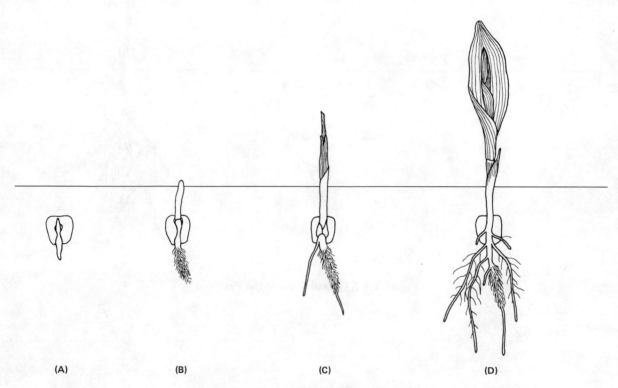

(A)   (B)   (C)   (D)

Figure 2.6   Germination of the Corn Seed

EXERCISE 3

FACTORS THAT INFLUENCE SEED GERMINATION

Three major factors influence seed germination. They are (1) seed viability; (2) seed dormancy; and (3) environmental conditions. If germination is to occur, certain conditions must be met as regards all three. (1) The seed must be viable, that is, the embryo must be alive and capable of growth; (2) any dormancy conditions within the seed (which may chemically or physically prevent germination) must have disappeared or been eliminated; and (3) the seed must be exposed to favorable conditions of moisture, temperature, and oxygen. (Some seeds also require light or darkness.)

I. SEED VIABILITY

A viable seed is one that contains a living embryo capable of germinating. The U.S. Federal Seed Act requires that all lots of seed entered in trade must be tested for their viability beforehand. The most commonly used test is the Germination Percentage Test. Another one is the Tetrazolium Test. Both can be done in the classroom laboratory.

A. GERMINATION PERCENTAGE TEST

The germination percentage is the percentage of seed tested that produces normal seedlings under optimal conditions. When you buy a package or a lot of seed, the information on the label will usually include the minimum germination percentage.

In licensed laboratories the germination percentage test is done using 400 seeds. For classroom accuracy, you can use only 100 seeds of any one type divided into lots of 25. The total germination of the four lots is the germination percentage.

► ACTIVITY 1   **Select** seeds of different plants, such as lettuce (*Lactuca sativa*); radish (*Raphanus sativus*); spinach (*Spinacia oleracea*); corn (*Zea mays*); beans (*Phaseolus vulgaris*); pot marigold

17

(*Calendula officinalis*); celosia (*Celosia argentea* v. *cristata*); dahlia (*Dahlia pinnata*); nasturtium (*Tropaeolum majus*); or morning glory (*Ipomoea purpurea*).

Work in groups of four students. Any four students will test 100 seeds of one type of plant. All materials and the work area must be as clean as possible to reduce contamination with fungi. Each group of students will be given a beaker containing a bleach type disinfectant made of 9 parts water and 1 part commercial bleach (5 percent sodium hypochlorite). This gives a solution of about 0.5 percent sodium hypochlorite.

Soak the seeds in the bleach for 5-10 minutes. While the seeds are soaking, each student should soak two paper towels in some bleach. Reserve one for preparation of the seeds and use the second one to wipe clean the table work area.

Each student prepares 25 seeds as follows. (1) Moisten a paper towel and spread it out. (2) Space five seeds along one side of the towel and fold the edge over the row of seeds. (3) Space another row of five seeds and roll the first fold over them. Continue adding rows. Do not make tight rows. Five rows are needed. (4) Tie the rolls at each end with a string or twist tape. You have now made a "rag-doll tester." (5) Put the rag-doll testers in plastic bags and store in a warm place (about 21° C). Make certain that the testers do not dry out during the germination period.

After ten days unfold the rag-doll testers and **count** the number of seeds that have germinated. **Compute** the germination percentage for each seed species tested. Record in Table 3.1. Some will unavoidedly be lost to fungal contamination.

Table 3.1   Germination Percentage Test Results

Plant Name	Number of Seeds Tested	Number of Seeds Germinated	Germination Percentage

B. TETRAZOLIUM TEST

This is a quick test for germinability. It simply determines if the embryo in the seed is alive or dead. It is done by testing the respiration of the seed.

Viable seeds respire and can change colorless tetrazolium dyes into a red color. The test is not foolproof and, therefore, is not legal for germination labeling on seed tags. In research, however, it is valuable because it can determine if a dormant seed is viable whereas germination tests cannot.

► ACTIVITY 2   **Select** seeds of corn (*Zea mays*) that have been soaked in water overnight in the dark. (1) With a razor blade cut each seed in half and place the cut surface on a filter paper in a petri dish. (2) Moisten the paper with 0.1 percent tetrazolium chloride. (3) Set the dish aside for 15-30 minutes in the dark. (4) Prepare and test soybeans or pinto beans in the same way. (5) Examine the seeds for any red coloration. Any of the following results indicates a viable seed: embryo entirely red; embryo mostly red; endosperm also red. Dead seeds are indicated by pink embryo or embryo remaining white.

II. PRACTICAL CONSEQUENCE OF DIFFERENCES IN SEEDS AND SEED GERMINATION

Commonly, large discount stores in the north temperate zones of the United States display inexpensive grass seed in three- to five-pound bags sold at the same price as one-pound bags of other grass seed mixes. These less expensive mixes are usually composed mostly of annual ryegrass, whereas the more expensive mixes will contain more of the bluegrass varieties and fescues. In the Gulf states of the United States, Bahiagrass is less expensive than Bermuda grass as a lawn grass.

► ACTIVITY 3   Ignoring the relative merit of one grass over the other, and given the information in Table 3.2, **calculate** for each grass (1) the approximate number of seeds in a pound (or kilogram) that will germinate; and (2) the number of pounds of seed needed per 1000 square feet (or the number of kilograms per 100 square meters), that is, the seeding rate.

Results and conclusions of your calculations.

1. If the price of Kentucky bluegrass is $2.50 per pound and the price of annual ryegrass is $0.65 per pound and you have a lawn of 5000 square feet to seed, what would be the cost to you to seed your lawn with the bluegrass? _____ with the ryegrass? _____
2. If you saw a sign in the store advertising the Bahiagrass at 4 pounds for $2.40 and Bermuda grass at $3.50 per pound, which would be the better buy? _____
3. Do "sales" of seed really always mean less outlay of your money?   YES   NO.
4. What is one important fact about different grass seeds that is worth knowing? _____
_____
5. Do differences in germination percentages make up for differences in seeding rates? YES   NO.
6. Seed tags and packages of seeds are required by law to show the germination percentage. What other information would be helpful, especially for turfgrass seed? _____
_____

III. SEED DORMANCY

Seed dormancy is the failure of a seed to germinate even though optimal environmental conditions exist. It is due to chemical or physical barriers within the seed. In most wild (and even cultivated) plants this is of value. For example, it is very common for seeds to be produced at the end of the summer or early fall. Conditions at this time are often perfect for germination, but if germination took place, there would not be time for much seedling growth before the first killing frost. It is, therefore, a survival advantage to the species if its seeds remain dormant at this time.

Seed dormancy is caused by many factors, but three very common ones are (1) dormant embryo; (2) impermeable seed coat; and (3) chemical inhibitors within the seed. The easiest of these to demonstrate is dormancy due to an impermeable seed coat.

Dormancy Due to an Impermeable Seed Coat

Impermeable seed coats prevent water absorption. Without water, the seed's reserve foods are not hydrolyzed (made soluble) and the digestive enzymes are not activated, so the embryo cannot grow.

► ACTIVITY 4   **Set** up rag-doll testers as you did for the germination percentage test. Use seeds of the honey locust tree (*Gleditsia triacanthos*) collected in the autumn.

Proceed as follows: (1) Scarify the seeds by scraping each with a file until an opening is visible. (2) Do this for several seeds and wrap them in a rag-doll tester. (3) Leave an equal number of seeds intact and wrap in a separate tester. (4) Label the testers and store them in plastic bags in a warm (21° C) place for five to seven days. Make certain the testers do not dry out during the germination period.

After five to seven days, unfold the testers and count the number of seeds that have germinated. **Compute** the germination percentage. **Record** in Table 3.3.

IV. ENVIRONMENTAL CONDITIONS INFLUENCING SEED GERMINATION

If a seed is viable and not dormant, it should germinate when exposed to favorable environmental conditions. These are four: (1) adequate and available moisture; (2) adequate and available oxygen; (3) suitable temperature; and (4) (for some seeds) light or darkness. We will consider only oxygen and temperature requirements.

A. OXYGEN REQUIREMENT FOR GERMINATION

► ACTIVITY 5   **Prepare** a few *dry* corn seeds in two rag-doll testers. Get two jars (baby- or junior-food jars are just right) and put one rag-doll tester in each. Add enough water to each jar so that the testers are thoroughly wet and there is about 6 mm of water in the bottom of the jars. Immediately close one jar with a metal lid or some impervious stopper. Leave the second jar open to the air, or cover with filter paper (to prevent mold growth). Set both jars in a warm place for five to ten days. Make certain that the unsealed jar doesn't dry out during this time. After the period is up, examine each jar. Has germination occurred in the sealed jar?   YES   NO. In the unsealed jar?   YES   NO. This is a very simple demonstration of the need for aeration for germination to occur. What is the gas in the air that promotes germination? _____

B. TEMPERATURE AS A FACTOR IN SEED GERMINATION

The experienced gardener living in the north temperate zone knows that you don't have to wait until the really warm days to set out the seeds of all the different vegetables you plan to grow. Some can be planted earlier than others. This is because different plants have different minimum temperatures at which their seeds will germinate. For instance, peas (*Pisum sativum*) and turnips (*Brassica rapa*) can be planted outdoors six weeks *before* the average last freeze date of winter; lettuce (*Lactuca sativa*) can be planted five weeks *before* the average last freeze date of winter; and spinach (*Spinacia oleracea*) and beets (*Beta vulgaris*), three weeks. However, for cucumbers (*Cucumis sativus*), green peppers (*Capsicum frutescens* v. *grossum*), pumpkins (*Cucurbita pepo*), and watermelon (*Citrullus vulgaris*) you must wait to plant two weeks *after* the average last freeze date, and for sweet corn (*Zea mays*), about three weeks *after* the last average freeze date.

► ACTIVITY 6   **Obtain** seeds of peas, turnips, or cabbage and prepare them for germination in rag-doll testers. Put seeds of only one seed type in each tester, label, and date it. Do the same for seeds of cucumber, green pepper, eggplant (*Solanum melongena*), and sweet corn. Store all these in plastic bags in a refrigerator maintained about 10° C (somewhat warmer than a typical household refrigerator). Store another set of each of these types in some warm place or a germination chamber if one is available.

Between one and two weeks later, check the rag-doll testers to observe whether or not germination has occurred. The days required for germination under optimum conditions are listed in Table 3.4, and these should be consulted and used as a time guide for making your observations. **Record** your findings in Table 3.4.

EXERCISE 3   **STUDENT NAME** _____

QUESTIONS

1. Give two advantages to delay in seed germination.

    (a) _____

    (b) _____

2. What are the three major factors that influence seed germination? (a) _____ ;

    (b) _____ ;   (c) _____

3. What is a viable seed? _____

4. What is the standard, legally recognized test for the determination of viable seeds? _____

5. Would you expect the seeding rate to be high or low for seeds with low germination percentages? _____

6. What else, besides germination percentage, determines the seeding rate for plants such as turfgrasses

    or crops such as rice and wheat? _____

7. Suppose a tetrazolium test on a test batch of oak (*Quercus* sp.) seeds is positive. You then plant the seeds

    of this lot, but you get zero germination. What is the most probable explanation? _____

    _____

8. Give two specific uses of water imbibed by the seed. (a) _____ ;

    (b) _____

9. Why is reseeding necessary in a field flooded by rains? Do the seeds burst from imbibing too much wa-

    ter? _____ What is the limiting or missing factor in this situation? _____

10. Is the minimum germination temperature the same for all plants in any one geographic area? _____

11. In the laboratory, you filed a hole in the seed coat of the honey locust seed in order to make it permeable
    to water. In nature, how do you suppose the seed coat is made permeable by the time spring arrives?

    _____

Table 3.2   Comparison of Price and Cost of "Expensive" versus "Inexpensive" Grasses

Turfgrass	Approximate Number of Seeds per Pound (per Kilogram)	Germi-nation (%)	Approximate Number of Seeds in a Pound (Kilogram) that Will Germinate	Number of Seeds Needed per 1000 ft² (per 100 m²)	Number of Pounds (Kilograms) of Seed Needed per 1000 ft² (per 100 m²)
"Expensive" grasses Kentucky bluegrass (*Poa pratensis*)	2,177,000/lb (4,799,000/kg)	75		2,736,000/1000 ft² (2,944,000/100 m²)	
Bermuda grass (*Cynodon* sp.)	1,787,000/lb (3,921,000/kg)	80		2,304,000/1000 ft² (2,470,000/100 m²)	
"Inexpensive" grasses Annual ryegrass (*Lolium multiflorum*)	227,000/lb (500,000/kg)	95		1,736,000/1000 ft² (1,868,000/100 m²)	
Bahiagrass (*Paspalum notatum*)	166,000/lb (366,000/kg)	75		1,152,000/1000 ft² (1,240,000/100 m²)	

Table 3.3   Germination Test of Seeds with Intact Impermeable Seed Coats Versus Identical Seeds with Scarified Seed Coats

	Number of Seeds Tested	Number of Seeds Germinated	Germination Percentage
Scarified Seeds			
Intact Seeds			

Table 3.4   The Effect of Temperature on Seed Germination and the Germination Response of Different Plants to Low Temperature

Plant	Standard Germination Time (days at optimum temperature)	Observed Germination Time (number of days at low (___° C) temperature)	Germination Percentage	Observed Germination Time (number of days at higher (___° C) temperature)	Germination Percentage
Cabbage	6–9				
Pea	7–10				
Turnip	5–10				
Cucumber	7–10				
Green pepper	10–14				
Eggplant	10–21				
Sweet corn	5–12				

EXERCISE 4

USE OF THE MICROSCOPE
FOR STUDY OF PLANT CELLS

Plants, like most living things, are made up of cells. The *cell* is the basic building unit. To study most cells you need a microscope. So, today you will look at a few cells and learn how to use the microscope.

I. THE MICROSCOPE

A. MICROSCOPE PARTS

Figure 4.1 is a diagram of one type of microscope in general use for elementary botany courses. **Refer** to this diagram as an aid when following directions for using the microscope. Not all microscopes are exactly like the illustration. Differences that you may encounter are a built-in substage light and therefore, no mirror; the fine adjustment knob on the arm instead of down near the base; a substage condenser lens system instead of a diaphragm disk; ocular lens and tube may slant toward or away from the arm; three objectives on the nosepiece rather than two.

B. SETTING UP THE MICROSCOPE

► ACTIVITY 1

1. Use two hands to carry the microscope vertically.
2. Set the microscope down with the arm toward you and the stage away from you (arm may be away from you if your microscope has a reverse-angled tube).
3. Turn the nosepiece so that the low-power $10\times$ magnification objective (the shorter one) is in line with the body tube. It should snap into position.
4. Position the microscope so that the mirror faces some light source (unless you have a built-in light).

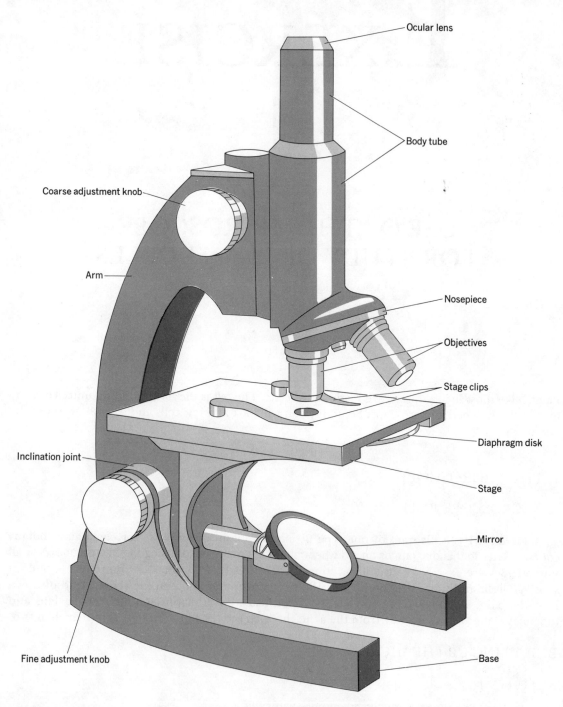

Figure 4.1   Use of Microscope

5. Look through the ocular and adjust the mirror (and diaphragm) until you see a round field of view evenly illuminated.
6. If the lenses or mirror are cloudy or dusty, wipe them gently with lens paper. Never use tissue paper or cloth.

C. FOCUSING THE MICROSCOPE

► ACTIVITY 2   **Obtain** a slide that has three or four silk, wool, or cotton fibers of different colors wetted and crossed over each other and held down by a cover glass.

1. Place the slide on the stage of the microscope so that the specimen material is centered over the hole in the stage.
2. Clip the slide in place with the stage clips.
3. While looking at the microscope *from one side,* use the coarse adjustment knob to lower the body tube or raise the stage (depending on how your microscope is designed) until the low-power objective almost touches the slide. In most microscopes there is an automatic stop that limits how close the objective can come to the stage.
4. Look into the ocular and slowly, with the coarse adjustment, increase the distance between stage and low-power objective, and soon the specimen will come into view.
5. If you find that your objective is more than 10 or 15 mm above the slide and you still haven't seen anything in the field of view, then start over. Decrease the distance between the objective and the stage as in step 3.
6. Repeat step 4, and this time *very very* slowly turn the knob.
7. Focus the specimen as sharply as possible with the coarse adjustment knob.
8. If you need to see more minute differences between various parts of the fibers, turn the fine adjustment knob very, very slightly. Turn it no more than 1 mm at a time! Turn it backwards (toward you) or forwards (away from you) until the item is clearly in focus.

Under low power, are *all* the fibers in focus at one time, that is, are they all in *one focal plane?*   YES   NO.   Are all sections of any one fiber in focus at any one time (that is, in any one focal plane)?   YES   NO.

Focus carefully with the fine adjustment knob. Does this help to show you which is the top fiber where two cross?   YES   NO.   **Admit** more light by rotating the diaphragm disk to a much larger hole (or open the substage condenser unit). Are all fibers clearly seen?   YES   NO.   **Reduce** the light by rotating the diaphragm to a smaller hole. Are the fibers more sharply visible or less?

—————————

Rotate the nosepiece to put the high-power objective (43× magnification) in line with the body tube. Can you see an entire fiber?   YES   NO.   Do you see more detail in each fiber?   YES   NO.   Are all the fibers in focus at one time (that is, in one focal plane)?   YES   NO.

The *greater the number of different objects* in focus at one time (that is, in one focal plane), the *greater the depth of focus.* The *fewer the number of, or portions of, objects* in focus at one time, the *less the depth of focus.* At which magnification is there greater depth of focus, low or high power? ———

D. PREPARING A WET MOUNT

► ACTIVITY 3   **Make** a wet mount. Use elodea (*Anacharis canadensis*) or other appropriate material.

1. Use forceps (tweezers) and pick off one or two of the youngest leaves near the tip of the elodea plant.
2. Place the leaf on the center of a clean glass microscope slide.
3. With a pipette (dropper) put one drop of water over the leaf.
4. Balance a cover glass at one edge of the drop of water at 45 degrees. Wait until the water spreads across the edge of the cover glass. Then, with a dissecting needle slowly lower the cover glass on the water. If air bubbles are trapped, tap the cover glass gently.
5. Use this slide for your study of the cell—coming up next.

II. THE PLANT CELL

Refer to the following list of terms as a guide while you work.

A. SOME TERMS AND DEFINITIONS FOR THE PLANT CELL

PROTOPLAST	The organized, living contents of the cell.
CELL WALL	A nonliving, porous, semirigid "casing" for the living protoplast. Its thickness varies with the different functions of different cells.
CYTOPLASM	The viscous, granular ground-mass part of the protoplast. Usually looks gray or colorless. In mature cells occurs as a thin layer lining the wall and extending as fine STRANDS OF CYTOPLASM in a meshwork throughout the cell. In young cells it fills the cell.
VACUOLE	A more or less clear area in the cell not occupied by cytoplasm; typically not visible in young cells.
CELL SAP	Watery solution filling the vacuoles.
ORGANELLE	A discrete body in the cytoplasm with a specialized structure and function.
NUCLEUS	The organelle that contains the chromosomes. Usually visible as a round body, darker colored than the cytoplasm.
CHLOROPLAST	An organelle that contains chlorophyll and functions in photosynthesis. Appears as a bright green, oval body and in varying numbers depending on the cell type.
PLASMA MEMBRANE AND VACUOLAR MEMBRANE	A vital, living membrane system enclosing the cytoplasm and regulating passage of material in and out of the cytoplasm. Called the *plasma membrane* where it covers the cytoplasm adjacent to the cell wall; called the *vacuolar membrane* where it covers the cytoplasm adjacent to vacuoles.
CYTOPLASMIC STREAMING	A flowing motion of the cytoplasmic granules. Chloroplasts and other organelles are pushed around by it.

B. A LIVING GREEN CELL—ELODEA LEAF CELL

➤ ACTIVITY 4   **Use** the wet mount of elodea leaf that you have just prepared (in Section I.D). This leaf isn't very thick or complicated. **Note** the midrib of the leaf. It has long, slender cells. On the margin of the leaf, **find** the thorn-like spur cells. **Focus** on or near the margin of the leaf. Here there are few layers of cells and you can see cellular details better.

Study the leaf cells and answer these questions.

1. What is the most abundant organelle in the cells?_____
2. Are these organelles moving or stationary? (You'll have to look at more than one cell to answer this.) _____
3. What force is causing this? _____
4. Do the four sides (walls) of a cell always remain visible as you focus up and down, or do they disappear in different planes of focus? _____
5. **Focus** up and down carefully with the fine adjustment. Are the cells of the different layers exactly aligned one over the other?   YES   NO.
6. If a wall suddenly goes out of view (as you turn the fine adjustment), this means that your focal plane is either above or below the wall. Do you agree?   YES   NO.
7. **Turn** the fine adjustment very slightly in *one direction* as you look at one small area. Do you see the walls come in and out of focus?   YES   NO.
8. What does this tell you about the number of layers of cells? Is there one layer or more than one? _____

Focus on a spur cell on the leaf margin. **Reduce** your light by rotating the diaphragm disk to a smaller hole or by moving the lever on the substage condenser. The nucleus should be fairly visible in the spur cell. Can you see nuclei in the green cells of the leaf?   YES   NO.

Label the drawing of the leaf cell of elodea (Figure 4.2), using the capitalized terms defined in Section II. A.

C. A LIVING NONGREEN CELL—EPIDERMIS CELL OF AN ONION BULB

► ACTIVITY 5

1. **Prepare** a wet mount of onion epidermis tissue. **Obtain** from the instructor a piece of an onion bulb scale. The scale has a thin epidermal layer which is easily removed. **Use** forceps to peel off a portion of the epidermis from the *concave* surface of the scale. Put the tissue on a clean glass slide. Add a drop of water, and spread out the tissue so that there are no folds in it. Cover with a cover glass and examine.

2. **Staining the cell.**   For better definition of the nuclei (and only after you have examined the cell in its living condition), place one or two drops of iodine at the edge of the cover glass. Draw the iodine under the cover glass by holding a small piece of paper toweling at the opposite edge of the cover glass. Your instructor will demonstrate this procedure if necessary. The iodine will kill the cells but stain the nuclei.

Examine the onion cells and answer these questions.

1. Are these cells arranged the same way as in the elodea leaf?   YES   NO.
2. Are there any chloroplasts?   YES   NO.
3. Give a reason why you would, or would not, expect to find chloroplasts in these cells.
   _____
4. Are nuclei visible?   YES   NO.
5. Is the cytoplasm visible?   YES   NO.
6. Which has the darker color, the nucleus or the cytoplasm? _____
7. Is the cytoplasm abundant or is it in fine strands? _____
8. Can you see vacuoles?   YES   NO. What is the clue that tells you if these cells are mature or young? _____
9. How many layers compose this tissue, one or several? _____
10. Are the cytoplasmic granules in motion?   YES   NO. Do they jiggle back and forth or show a directional flow? _____ If the granules simply jiggle, you are seeing what is called "Brownian movement." This is a motion of the granules resulting from their collisions with water molecules. Brownian movement occurs in any aqueous medium, living and nonliving.

D. MORE EXAMPLES OF CELL TYPES

Up to now you have seen simple, unmodified cells. These are in no way representative of all plant cells. But they do show you the basic major components of the plant cell—wall, nucleus, cytoplasm, and vacuoles. Animal cells never have a wall or vacuoles. Plants have many kinds of cells and you will study these in later exercises.

► ACTIVITY 6   **Examine** slides set up on demonstration microscopes. These will give you an idea of the variety of cell types. Examples you might see are

1. Cells specialized for support in mature woody stems—for example, SCLERENCHYMA FIBER CELLS from macerated basswood (*Tilia* sp.).
2. Cells specialized for support in young soft stems—for example, COLLENCHYMA FIBER CELLS from a freshly made thin section of a *Coleus* stem.
3. Cells specialized to change shape—for example, GUARD CELLS from the epidermis of the leaf of wandering Jew (*Zebrina pendula*) or *Rhoeo discolor*.
4. Cells specialized to conduct water—for example, TRACHEIDS and VESSEL ELEMENTS from macerated wood of pine (*Pinus* sp.) and oak (*Quercus* sp.) or basswood (*Tilia* sp.).

EXERCISE 4   STUDENT NAME _____

QUESTIONS

1. What is the name of the magnifying lens that your eye looks into? _____

2. Which adjustment knob should you use when you are first trying to simply locate the material on the slide and bring it into the field of view? _____

3. When first locating material on the slide, it is bad practice to look into the ocular as you decrease the distance between stage and objective. Circle one: (a) true; (b) false.

4. What are the two components of plant cells that are not found in animal cells? (a) _____;

   (b) _____

5. Ordinarily you *never* need to give the fine adjustment knob more than a millimeter turn. Circle one: (a) true; (b) false.

6. Define an organelle: _____

   _____

7. What is the name of the part of the microscope that is used to rotate the objectives? _____

8. What is the name of the part of the microscope through which the image passes from the objective to the eyepiece? _____

9. Do young cells have vacuoles? _____

10. How is the cytoplasm distributed in a mature cell? _____

11. Do all plant cells have chloroplasts? _____

12. Is the cell wall alive? _____

13. Is the plasma membrane alive? _____

14. Is the vacuolar membrane alive? _____

15. In a living cell, could the nucleus occur in the vacuole? _____

16. If cytoplasmic granules show a jiggling, jerking motion, it is referred to as _____

   _____

17. The high-power objective magnifies the specimen $43\times$, and this magnified image passes up the body tube to the ocular lens. If your ocular lens has a magnification power of $5\times$, what is the final magnification of the image that the eye sees? _____; what would be the final magnification with an ocular lens of $10\times$ magnification? _____

18. What material fills the vacuole? _____

19. Are the vacuolar and plasma membranes (a) separate, distinct structures or (b) one continuous membrane bounding all portions of the cytoplasm? Circle (a) or (b).

20. Which can occur in a living cell, (a) only cytoplasmic streaming; (b) only Brownian movement; (c) both (a) and (b)? Circle one.

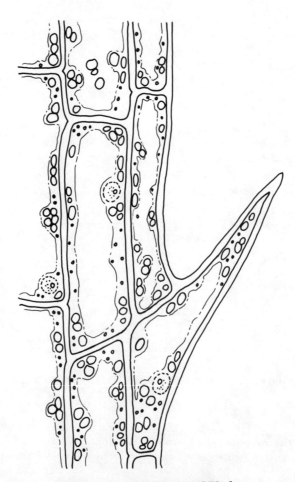

Figure 4.2   Leaf Cells of Elodea

EXERCISE 5

CELL DIVISION

Cells are unique. Not only are they the basic building units of an organism, but they also carry within their nuclei the "blueprints," the directions for building the organism. These directions are stored in the nucleus in a code composed of DNA (deoxyribonucleic acid). DNA forms most of the chromosomes.

When a cell divides and produces two daughter cells by *cytokinesis*, it is essential that this information be accurately duplicated and none be lost. Herein lies the importance of the process known as *mitosis*. Mitosis is the orderly and equal distribution of nuclear material (chromosomes) to the two daughter nuclei formed by division of a cell.

Cell division is a three-stage process, as follows:

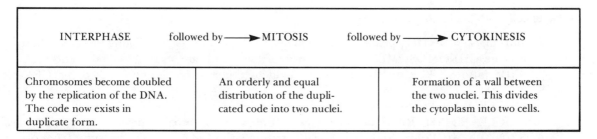

INTERPHASE	followed by ⟶ MITOSIS	followed by ⟶ CYTOKINESIS
Chromosomes become doubled by the replication of the DNA. The code now exists in duplicate form.	An orderly and equal distribution of the duplicated code into two nuclei.	Formation of a wall between the two nuclei. This divides the cytoplasm into two cells.

Active cell division occurs (usually) seasonally in certain parts of the plant. These parts are the tips of roots and shoots, lateral buds, and a tissue called the *cambium*, which causes the stem and root to grow in diameter. Cell division also occurs in any growing organ such as leaves, flowers, fruit, and seeds. Tissue in which cell division occurs on a temporary or permanent basis during the life of the plant is known as a *meristem*. Meristems, thus, are located in stem and root tips, in the cambium, in buds, etc. Today's work consists of observing the meristem of a root tip and learning to recognize the stages of cell division.

I. LOCATING THE MERISTEM

▶ ACTIVITY 1   **Obtain** a prepared slide of a longitudinal section through the root tip of an onion *(Allium cepa).* **Use** low power on the microscope and locate the pointed end of the section. **Orient** the slide so this pointed end is at "six o'clock" (the bottom center) of your field of view. The extreme tip is not the meristem. It is the root cap, a zone of somewhat loosely arranged corky cells. These root cap cells are a shield for the meristem. They protect its delicate dividing cells from the damage encountered by growth between abrasive soil particles.

The MERISTEM is behind the root cap. Its cells are more densely packed and arranged in orderly rows parallel to the long axis of the root. **Note** that many of these cells are almost square in outline. Some have very large nuclei, almost completely filling the cell (as in Figure 5.1). Some have nuclei in a mitotic stage (as in Figures 5.2 to 5.5). These are the cells to study. They were undergoing division when the root tissue was sectioned and stained to make this slide. Hence the cells are now "caught dead" in some phase of cell division.

II. LOCATING THE STAGES OF CELL DIVISION

A. INTERPHASE

▶ ACTIVITY 2   **Look** for cells with very large nuclei that almost fill the cell. The features of interphase are

1. Dark-stained CHROMATIN GRANULES irregularly scattered throughout the nucleus. These granules are all that can be seen of the duplicating chromosomes.
2. NUCLEAR MEMBRANE   Encircles the nuclear material.
3. NUCLEOLI   One or more small, dark-stained bodies, larger than the chromatin granules. They are centers of protein synthesis.

Label Figure 5.1, using the capitalized terms defined in the preceding description. **Title** the figure as to the stage it represents.

B. MITOSIS

▶ ACTIVITY 3   **Find** the four stages of mitosis, (1) prophase, (2) metaphase, (3) anaphase, and (4) telophase.

Prophase

Look for cells with the following features:

1. Chromatin material appears as a tangled loose mass of dark-stained, long threads. These are the CHROMOSOMES. Each consists of two duplicate strands, called CHROMATIDS (also SISTER CHROMATIDS), held together at one region, called the CENTROMERE.
2. NUCLEOLI still present but decreasing in size.
3. NUCLEAR MEMBRANE becoming less distinct as it now breaks down.
4. SPINDLE MICROTUBULES (FIBERS) faintly visible as fine lines diverging from opposite poles of the nucleus and radiating toward the midline (equatorial region).

Label Figure 5.2, using capitalized terms in the preceding description. **Title** the figure as to the stage it represents.

Metaphase

Look for cells in which the dark-stained chromosomes are clumped in the midregion of the cell and fine lines extend through them from pole to pole. The configurations that you see are

1. SISTER CHROMATIDS   These have condensed into tightly coiled, thick, finger-like objects.
2. EQUATORIAL PLANE   This is not an object in the cell. It is the midregion of the cell where CENTROMERES of the sister chromatid pairs are all aligned in one plane, the EQUATORIAL PLANE. Here each centromere is attached to four or more SPINDLE MICROTUBULES (FIBERS) which run perpendicular to the plane. Because conditions become very crowded in this equatorial plane, the arms of the CHROMATIDS themselves extend off this plane in all directions.

Label Figure 5.3, using the capitalized terms in the preceding description. **Title** the figure as to the stage it represents.

Anaphase

Look for cells in which there are two clusters of chromatids lying close to, but not at, the poles of the cell. Fine lines extend across the cell between the chromatid clusters. The features you see here are

1. CHROMATIDS clumped together. Many appear V-shaped, with the point of the V toward the pole of the cell. By use of fine focusing, you will see that most of the chromatids have this shape. This is because they are physically being dragged to the poles by the spindle microtubules.
2. SPINDLE MICROTUBULES (FIBERS)   The shortening of these tubules pulls the sister chromatids apart and carries each to an opposite pole of the cell.
3. CENTROMERE   This is the point of attachment for the microtubules. The centromere thus is the "leading" section of the dragged chromatid.
4. ARMS of the CHROMATID   These are the chromatid parts being pulled along behind the centromere.

Here in anaphase occurs the equal distribution of genetic material to daughter nuclei. The duplicate sister chromatids have been split apart at their centromere, and each is drawn to an opposite pole by contracting microtubules. Once the chromatids have become so separated, they are essentially chromosomes.

Label Figure 5.4, using the capitalized terms in the preceding description. **Title** the figure as to the stage it represents.

Telophase

Look for cells in which there are two dense, dark clumps and between them a faint black line. The features you see here are

1. NEWLY FORMED NUCLEI   Each composed of the same number and the same kind of chromosomes as the original parent cell. The CHROMOSOMES are beginning to return to their long, uncoiled, linear form.
2. A CELL PLATE is forming between the nuclei. It serves as the cementing middle layer between the cell walls which will be laid down later.
3. NUCLEOLI reappear.
4. NUCLEAR MEMBRANE reappears.
5. The REMNANT SPINDLE MICROTUBULES will soon disappear.

Label Figure 5.5, using the capitalized terms in the preceding description. **Title** the figure as to the stage it represents.

C. CYTOKINESIS

Cytokinesis is the stage that follows nuclear division. Walls form between the nuclei, and the cytoplasm of the parent cell is divided.

▶ ACTIVITY 4   **Look** for pairs of cells which are brick-shaped with large, round nuclei filling about one third to one half the cell. These cells and other features that you see are

1. DAUGHTER CELLS.
2. NEW CELL WALL, which separates the daughter nuclei.
3. DUPLICATE NUCLEAR MATERIAL, which occupies each nucleus.

Label Figure 5.6, using the capitalized terms in the preceding description. **Title** the figure as to the stage it represents.

III. PREPARING AND STAINING A ROOT MERISTEM

▶ ACTIVITY 5   If time permits, you might like to prepare and stain a slide comparable to the commercially made one you have just examined. There are two stains that can be used, aceto-carmine or Magenta XX. Try one or both. Magenta tends to give better results and requires fewer steps.

Staining with Aceto-Carmine	Staining with Magenta XX
1. With a razor blade, cut off about 5 mm of the tip of an onion root.	1. With a razor blade, cut off about 5 mm of the tip of an onion root.
2. Place tip in a Syracuse glass containing a weak (1M) solution of hydrochloric acid.	2. Place tip in a Syracuse glass containing a 3M solution of hydrochloric acid.
3. Leave in acid for 10 minutes.	3. Leave in acid for 5 minutes.
4. Draw off the acid with a pipette and rinse two or three times with water.	4. Blot the root dry with filter paper.
5. Draw off all the water and add a few drops of aceto-carmine (or aceto-orcein) stain to the root. Leave it on 5-10 minutes. With a dissecting needle, tease the root tissues apart at some time during the period.	5. Place root on a slide and cover with a cover glass.
	6. Press the cover glass gently to flatten the tissue.
	7. Stain by running Magenta XX dye under the cover slip. (The tissue will be yellow at first.)
6. Place the root on a slide, add a drop of water, and cover with a cover glass.	8. Draw off the stain at the opposite edge of the cover glass. Use a rough torn bit of paper towel.
7. Press cover glass gently to flatten the tissue.	9. Add more stain at one edge.
8. Heat the slide gently by passing it slowly and closely over the flame of an alcohol lamp.	10. Continue this flushing with the dye until the tissue turns red.
9. Examine the slide under the microscope and look for the various stages of cell division.	11. Examine the slide under the microscope and look for the various stages of cell division.

EXERCISE
6

ROOTS I:
TYPES OF ROOT SYSTEMS AND
GROWTH OF ROOT TIP

There is no life without soil and there is no soil without life. Roots play an important link between the two. Secretions of roots, as well as their binding action and aerating effect on the soil, help preserve a good soil and the variety of organisms in it. Components of soil sustain the organisms.

I. THE MAJOR TYPES OF ROOT SYSTEMS

1. Tap root system
   a. One predominant root commonly growing straight downward, with smaller lateral roots.
   b. Originates from the seed radicle.
2. Fibrous root system
   a. Numerous equal-sized roots growing in many directions.
   b. Originates from the seed radicle.
   c. In general, but not always, more shallow-growing than tap root systems.
   d. Most trees have fibrous root systems within 1 meter of the soil surface.
3. Adventitious root system
   a. Fibrous-like roots growing from stems and leaves.
   b. Originates from the stem or leaf and not the seed radicle.
   c. In most monocots (like the mature corn plant and all turfgrasses) it forms the major root system; in many others it supplements the primary root system.
   d. The ability of stems to form adventitious roots is the key to the highly successful method of propagating ornamental plants by stem cuttings.

The root system of most plants is not exclusively one or the other of these three types, but some modification of them. This is because root growth is influenced by many variables: temperature and available moisture; soil type (sands, clays, loams, and silt); competition with other plants; and cultural practices such as plowing and root pruning.

► ACTIVITY 1   **Examine** the plants available for study and determine the root system of each. **Record** in Table 6.1 the names of the plants exhibiting the root systems as described.

Suggested plants for study: dandelion *(Taraxacum officinale)*; any lawn grass or grassy weed; a mature corn plant *(Zea mays)*; mature barley *(Hordeum vulgare)*; turnip *(Brassica rapa)*; beet *(Beta vulgaris)*; carrot *(Daucus carota* v. *sativa)*; bean *(Phaseolus vulgaris)*; *Coleus*; English ivy *(Hedera helix)*; nasturtium *(Tropaeolum majus)*; African marigold *(Tagetes erecta)*; *Rhoeo* sp.; wandering Jew *(Zebrina pendula)*; *Kalanchoe* sp.

Suggested cuttings of plants that have been growing for several weeks in water, sand, or rooted in commercially prepared peat cubes or blocks: stems of geranium *(Pelargonium* sp.); *Col-*

Table 6.1   Some Common Plants Illustrating the Different Types of Root Systems that Occur in Plants

Plants with Tap Roots	Plants with Fibrous Roots	Plants with Adventitious Roots
		1. Whole Plants
		2. Stem Cuttings geranium

eus blumei; Kalanchoe sp.; bloodleaf *(Iresine* sp.); *Impatiens* sp.; *Peperomia* sp.; petiole of African violet *(Saintpaulia ionantha)*; leaf of snake plant *(Sansevieria* sp.).

II. GROWTH OF THE ROOT TIP

A. THE ROOT CAP — MACROSCOPIC APPEARANCE

In your studies of seed germination you saw how the tip of the shoot is protected as it emerges through the soil either by a coleoptile (as in corn) or by the tunneling action of the arched growth of the hypocotyl (as in bean). The root tip, too, needs protection as it progresses through the soil. A thimble-like mantle of cells, the ROOT CAP, provides this protection.

The root cap consists of large cells that bear the brunt of friction with the soil particles. These cells rupture and exude a slimy fluid that lubricates the root tip so that it passes more easily through the soil.

► ACTIVITY 2   **Examine** the tips of the roots of water hyacinth *(Eichornia crassipes),* water lettuce *(Pistia* sp.), or other suitable material. Use a hand lens or dissecting microscope. If root samples of American green alder *(Alnus crispa* v. *mollis)* are available, they show interesting red-colored root caps. Can you give an explanation for the presence of root caps on water hyacinth? Aren't they unnecessary? _____

B. THE ROOT CAP AND THE ROOT APICAL MERISTEM—
MICROSCOPIC APPEARANCE

► ACTIVITY 3   **Examine** a prepared slide of a longitudinal section of the root tip of onion *(Allium cepa).* Locate the pointed end of the section and find the root cap. The ROOT CAP is a zone of loosely arranged cells at the extreme tip of the root. You have already studied the onion root tip in Exercise 5 which concerned cell division. What is the name of the region of densely packed cells immediately behind the root cap cells? ~~meristematic region~~ What nuclear configurations can you recognize that identify this zone? _____
If root cap cells are gradually worn away all the time, how do you suppose the root cap is maintained? _____
Label the root cap and the apical meristem of Figure 6.1.

C. REGION OF ELONGATION

This is the region just behind the meristem. The cells have enlarged chiefly in length. Most of the elongation is due to internal pressures built up by an increase of water within the large vacuoles that occupy most of the cell. The elongation of these cells increases root length and is the force pushing the tip of the root forward.
Label the region of elongation in Figure 6.1.

D. REGION OF MATURATION

In this region (which may not be present on the slide) the enlarged cells have now developed certain structural features and you can begin to see some differences among them. The cells are organized into three tissue systems: (1) The VASCULAR CYLINDER, which usually appears as the darker central portion. Here you may be able to see some XYLEM (water-conducting) CELLS that look like vertical rows of loose spirals (because their walls have spiral thickenings). (2) The EPIDERMIS, the layer of cells at the surface and from which ROOT

HAIRS arise. In the process of preparing the slide, the root hairs are destroyed, so you will not find them. (3) The CORTEX, the zone between the epidermis and the vascular cylinder.

Label the region of maturation and its component parts in Figure 6.1, using the capitalized terms defined in the preceding description.

III. DEMONSTRATION OF THE REGIONS OF PRIMARY GROWTH IN THE ROOT

► ACTIVITY 4   **Examine** ink markings on the root of a germinating seed. Either you or the instructor could have prepared these seeds a day or two earlier. The procedure is as follows.

Germinate seeds of bean (*Phaseolus vulgaris*) or pea *(Pisum sativum)* in moist vermiculite or sand until the primary root is about 3 to 4 cm long. Then gently blot the root dry. Lay the seed flat and place a ruler alongside the root. Starting with the root tip as your zero point, mark off intervals on the root 2 mm apart with a thread wetted with India ink. Gently touch the inked thread to the root at the measured intervals. Refer to Table 6.2. The region between any two marks can be recorded as regions 1, 2, and so on. Record the date on which the marks were made. Draw your marks on the root of the seed shown in Figure 6.2 (A), and label the regions between the marks as 1, 2, and so on. Pin the germinating seed to the underside of a cork and insert it into the top of a jar with some water in the bottom. An air hole in the cork would ensure adequate oxygen. The root should be above the water. After a day or two, inspect the marks on the root. Measure the distances between the marks and record in Table 6.2. Record the date also. In Figure 6.2 (B) draw the positions of the marks as they now appear. Label the intervening regions 1, 2, and so on. Draw dotted lines between Figure 6.2 (A) and (B), connecting the corresponding marks.

Has the distance increased between all the marks or just between some? _____. In which numbered region or regions do you see any change? _____. Which numbered region shows the most increase? _____. Is it just behind the root tip or much farther up the root? _____ What is the name of this region? _____

IV. ROOT HAIRS

Although old parts of roots are absorptive to some degree, most water absorption is by the root hairs in the region of cell maturation. Minerals are absorbed mainly by the apical meristem and the region of cell elongation. This young absorptive tip of the root is known as a feeder root. Feeder roots of most trees lie within 20 to 25 cm of the soil surface. At this depth oxygen and moisture can readily reach the roots.

► ACTIVITY 5   **Examine** the roots of young seedlings of radish (*Raphanus sativus*) that have been germinated on moist filter paper in closed petri dishes. Note the white "fuzz"-like growth on the roots. This "fuzz" is really a mass of ROOT HAIRS. Each root hair is a thin, tube-like extension of an epidermal cell. These hair-like cells extend out into the soil and adhere to soil particles from which they absorb moisture. All together, they greatly increase the area of absorption of the root.

Examine for root hairs the roots of other plants or stem cuttings of plants that have been rooted and grown in peat cubes. Suggested plants might be coleus (*Coleus blumei*); geranium (*Pelargonium* sp.); common garden bean (*Phaseolus vulgaris*); cucumber (*Cucumis sativus*); or soybean (*Glycine max*).

V. MYCORRHIZA

► ACTIVITY 6   **Examine** slides of roots of an orchid or other plant species that show mycorrhiza. A MYCORRHIZA is the association of a root and fungus growing together in a physiological union that is frequently mutualistic. The fungus derives nutrients from the root cells and the root depends to some extent on the fungus for absorption of soil moisture and minerals. Mycorrhizae probably occur in the majority of woody plants and in many herbaceous plants. In these plants, root hairs may be absent and their role in absorption is in large part carried on by the fungus, which in some cases has absorptive filaments extending far out into the soil. There are two types of mycorrhiza, ectotrophic and endotrophic. In an ectotrophic mycorrhiza the fungus grows over the root as a sheath and many fungal filaments grow between the cells of the cortex of the root. In an endotrophic type mycorrhiza, there is no fungal sheath formed over the root and many of the fungal filaments extend into living cells of the root cortex as well as between them. **Study** the slides showing the mycorrhizae and determine which type is present. Does the root have an outer mass of intertwining tubular filaments completely enveloping it?   YES   NO. If so, do these filaments also extend between the cells of the root?   YES   NO. If you have answered YES to both questions, then the mycorrhiza type present is _____. If you have answered NO to both questions, then look again closely at the root cells. Each cell has a very large nucleus. In the remainder of the cell are there many thread-like, intermingled filaments?   YES   NO. What are these filaments? _____ What type of mycorrhiza is this? _____

EXERCISE 6   **STUDENT NAME** _____

Table 6.2   Record of Root Tip Growth

Markings Made Between	Region	Date Markings Made	Amount of Increase in Length of Region (mm)	Date of Second Measure- ment
Tip–2 mm	1			
2 mm–4 mm	2			
4 mm–6 mm	3			
6 mm–8 mm	4			
8 mm–10 mm	5			

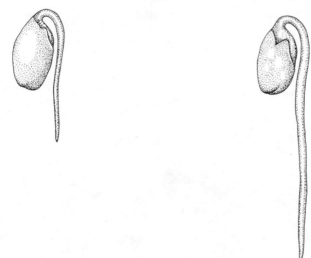

(A)   (B)

Figure 6.2   Growth of the Root Tip

EXERCISE 7

ROOTS II:
INTERNAL ANATOMY OF ROOTS

Most absorption in roots occurs in the youngest cells near the tip. Thus an old tree with a massive, woody root system each year produces new root tips. This new growth makes up its "feeder roots." These are within 20 to 25 cm of the soil surface. Here air and moisture are optimal.

Feeder roots of a tree or shrub resemble roots of soft-stemmed plants like the radish seedling roots you've seen. Their internal makeup is similar to the buttercup *(Ranunculus* sp.) which you will study today. We have you studying buttercup roots because it's easier than obtaining feeder roots of trees.

The young root is a double cylinder: (1) an inner vascular cylinder "sealed off" from the soil and (2) an outer cylinder permeated with air, water, and mineral ions from the soil. The outer cylinder controls entry of these into the vascular cylinder.

I. TISSUES OF THE ROOT OF A HERBACEOUS DICOT

➤ ACTIVITY 1   **Obtain** a prepared slide of a cross section of a young root of a dicot plant such as buttercup *(Ranunculus* sp.) and study the following parts.

A. THE OUTER CYLINDER—COMPOSED OF EPIDERMIS, CORTEX, AND ENDODERMIS

Epidermis

Focus on the outermost layer of cells, the EPIDERMIS. Some epidermal cells grow far out into the soil. This increases the root's area of contact with the soil. What do you call such epidermal cells?

In more mature root regions, a thin CUTICLE may cover the epidermis. If one is present, you will see a clear, pink-stained coating on the epidermis. **Note** the walls of the epidermal cells. These are highly permeable to the soil solution. Water of the soil solution moves into the cell interior, as well as through and along the walls. But the mineral ions are thought to move mainly through the walls of the epidermal cells. **Label** the epidermis in Figure 7.1.

Cortex

The CORTEX is the very large zone of rounded PARENCHYMA cells situated between the epidermis and the vascular cylinder. The latter is the star-shaped central axis of (red-stained) cells. **Look** at the cortex. Note that it is a big zone with big, loosely arranged PARENCHYMA cells and much free space between them. This construction serves three general functions: (1) it provides space for a high volume of soil solution to quickly move into the root; (2) the plasma membranes of these big cells all together form a large surface area for absorption; and (3) their respiration supplies the energy needed to move mineral ions into the cytoplasm of the cells.

Ions move through the cytoplasm of cortex cells across to the vascular cylinder. Water moves mostly through and along the walls of the cortex cells and not through the cytoplasm. Thus water absorption and transfer are independent of that of mineral ions. What advantages does this have to the plant? _____

_____

Does the cortex have an outer zone of smaller cells with thicker walls?   YES   NO. If so, then your root section was cut through an older part of the root where absorption has more or less ceased.

Note if there are any (blue-stained) particles in the cortex cells. These particles are STARCH GRANULES, storage products translocated into the root from the shoot system.

Label the cortex in Figure 7.1, using the capitalized terms defined in the preceding description.

Endodermis

Look at the inner boundary of the cortex and **focus** on the distinct ring of smaller, rectangular-looking cells surrounding the central vascular cylinder. This ring of cells is the ENDODERMIS, the innermost layer of the cortex. If your root section was cut from a *young* portion of the root, most of the endodermal cell walls will be thin. If the walls are thick, then you have an *older* root section which will be considered further on in this description.

On a young root section, **look** at the walls between any two adjacent endodermal cells. These are the radial walls. They are thick because they have a layer of suberin—an impermeable material—concentrated in bands, called CASPARIAN STRIPS. These strips are thought to form a barrier to the passage of anything moving through the cell walls. They occur in all the walls except the tangential ones (those that face the cortex and vascular cylinder). Materials are thus routed through the porous tangential walls and into the cytoplasm of the endodermal cells. The living cytoplasm thus controls what passes through it. In this way, anything approaching entry into the vascular cylinder is first "screened" by the endodermis.

If your slide contains an older section of the root where less absorption is taking place, most of the endodermal cells will have thick walls. A few thin-walled PASSAGE CELLS may allow some movement of materials into the vascular cylinder. **Label** the endodermis in Figure 7.1, indicating either those cell walls having the casparian strip, or if an older section, indicating the passage cells.

B. THE CENTRAL VASCULAR CYLINDER

The VASCULAR CYLINDER is the central core of the root. It functions in lengthwise conduction and (in many plants) in growth too. It is made up of three tissues, and sometimes a fourth. The three tissues always present are the PERICYCLE, the PRIMARY XYLEM, and the PRIMARY PHLOEM. In a mature buttercup root, and in those plants destined to have some secondary root growth, the fourth tissue is the CAMBIUM.

Pericycle

Find the PERICYCLE. It is a narrow zone, commonly one cell wide, just within the endodermis. It is composed of meristematic parenchyma cells which at times resume active division to form *branch roots*. In some species the pericycle also gives rise to adventitious shoots known as *root sprouts* or *suckers*. Many perennials, including some trees and shrubs, send up new growth this way. This makes them somewhat of a nuisance because the sprouts may appear in your lawn or yard where you don't want them. **Label** the pericycle in Figure 7.1.

Primary Xylem

Three-dimensionally the PRIMARY XYLEM is a fluted cylinder extending the length of the root. In cross section it looks like a star-shaped mass of (red-stained) cells. It has three, four, or five ridges reaching to the pericycle. **Note** the PRIMARY XYLEM, composed of large, thick-walled cells, called VESSEL MEMBERS. Vessel members are open-ended, barrel-shaped, dead cells devoid of all cytoplasm. They are joined end to end to form water tubes, called VESSELS. **Note** that vessel members (or cells) are different sizes. The smaller cells are called PROTOXYLEM. They developed lignified walls early in their growth, and this prevented further cell enlargement. The large cells are called METAXYLEM. They grew for a longer time before their walls became lignified. **Label** the PRIMARY XYLEM in Figure 7.1, using the capitalized terms defined in the preceding description.

Primary Phloem

Look for small semicircular clusters of cells nested between the ridges of xylem. These are the PRIMARY PHLOEM. Phloem consists of two cell types, SIEVE TUBE MEMBERS and COMPANION CELLS. These cell types are difficult to distinguish from each other in this section. Companion cells are the smaller cells, more dense looking, and may have a nucleus which fills most of the cell. Sieve tube members are larger in diameter, lack nuclei, and are vertically elongate, conducting cells. They are stacked one above the other, forming SIEVE TUBES. Strands of cytoplasm extend between them through their porous end walls, called SIEVE PLATES. Sucrose and other materials synthesized by the plant move through the tubes in the connecting cytoplasmic strands. Foods may move up or down, depending on the changing needs of the plant. **Label** at least two of the clusters of primary phloem in Figure 7.1, using the capitalized terms defined in the preceding description.

Vascular Cambium

Cambial cells may not be easy to identify. **Look** in the region between a cluster of phloem cells and the concave part of the xylem. **Find** a band of flat, rectangular-shaped cells. This band is the VASCULAR CAMBIUM. In the buttercup root these cambial cells may produce only very few secondary xylem and phloem cells. But in roots that become woody, the cambium produces much secondary tissues, and the root comes to resemble a shoot in cross section. **Label** the vascular cambium in Figure 7.1.

II. THE ORIGIN AND GROWTH
OF BRANCH OR LATERAL ROOTS

► ACTIVITY 2   **Examine** a slide of a young willow (*Salix* sp.) root showing the origin of a branch root from a primary root. Which tissue of the root forms branch roots? _____ **Look** at the pattern of the cells in the branch root. In what other study of the root did you see this particular pattern? _____ The newly formed branch root cells elongate and push the root through the cortex, out into the soil.

III. TISSUES OF THE ROOT OF AN HERBACEOUS MONOCOT

Monocot anatomy is more complex than that of dicots. There are more variations. This makes it difficult to choose any one plant as typical of all.

► ACTIVITY 3   **Examine** at least one monocot root so you can see some of the differences from a dicot. **Examine** a prepared slide showing a cross section of the root of a monocot, such as wheat (*Triticum* sp.); *Iris* sp.; corn *(Zea mays)*; or greenbrier *(Smilax* sp.).

Keeping this slide on your microscope, **compare** this monocot root with your dicot root (Figure 7.1) and **complete** Table 7.1.

Is this root a double cylinder as in dicots?   YES   NO. **Note** the circle of large METAXYLEM cells, and tapering radially outward from them the smaller PROTOXYLEM cells. All together they form a multirayed ring of xylem. What are the small clusters of cells nested between the protoxylem points of the xylem ridges? _____

The numerous ridges of xylem are characteristic of monocot roots. A large PITH occupies the center of the VASCULAR CYLINDER. Was there a pith in the dicot root?   YES   NO. **Look** at the ENDODERMIS. Its cell walls have developed more suberin and more cellulose than is found in dicot root sections of similar age. This is typical in monocots where roots last a long time but never develop secondary growth. Just inside the endodermis is the PERICYCLE which, in monocots, is several layers thick. In mature roots it will be composed of thick-walled cells. In most dicots this is not the case. How many layers of pericycle did you see in the dicot root? _____

The CORTEX in *Zea* or *Smilax* is smaller than the VASCULAR CYLINDER, but in other monocots it may be larger. These size differences are related to the number of xylem ridges. The more numerous the xylem ridges, the larger the pith and vascular cylinder and the smaller the cortex. Conversely, the fewer the xylem ridges, the smaller the pith and vascular cylinder and the larger the cortex. To check this, **look** at slides of roots of other monocots, such as wheat (*Triticum* sp.) or *Iris* sp. Are the number of xylem ridges as numerous as in corn or greenbrier? YES   NO. Which is the more massive, the cortex or the vascular cylinder? _____

IV. TISSUES OF THE WOODY ROOT OF A DICOT

► ACTIVITY 4   **Study** a prepared slide of a cross section of a woody root of a dicot plant, such as basswood (*Tilia* sp.) or tulip tree (*Liriodendron tulipifera*). This sample section is a young woody root only a few years old. When this root was less than one year old, it had the same structure as the buttercup root. Here, though, SECONDARY XYLEM has been laid down by the cambium in concentric annual rings similar to those in stems. The regularity of the rings in these very young woody roots is, however, not very clear cut and may be difficult for you to recognize. SECONDARY PHLOEM lies outside the xylem and is not as distinct as the clusters of primary phloem you saw in younger roots. Is the endodermis present?   YES   NO. Is the epidermis still intact? YES   NO. Is the cortex present?   YES   NO. Your answer to this will depend on the particular plant you are examining. Cortex is eventually replaced by cork and/or wood, but how soon this happens varies with different species. In a section of root this old, is absorption taking place to the same degree as in young root sections?   YES   NO. In some root sections you may find CORK already present.

EXERCISE 7   **STUDENT NAME** _Nicolette Pawlowski_

QUESTIONS

1. Does the woody portion of a tree's root system play the major role in active absorption? _____

2. How deep are the feeder roots of most trees and shrubs? _____

3. You are having a swimming pool built in your backyard and the contractor piles the excavated soil on the ground under some of your trees. Why is this bad for the health of most trees? _because the_ _roots will be farther from a source of air and minerals_ _from top soil_

4. What part of the root is concerned with lengthwise conduction? _sieve tubes_ ,
   with lateral conduction? _tracheids & vessel members_

5. What layer of cells regulates the entry of water and other materials into the vascular cylinder? _____

6. The absorption and movement of water in the root is independent of that of mineral ions. Circle one:
   (a) true; (b) false.

7. What two features of the cortex make it a most suitable tissue for the *influx* of the soil solution?
   (a) _____ ; (b) _____

8. What is the passage route of water across the cortex? _Through the living protoplasts_ _of the endodermal cells._

9. What is the passage route of mineral ions across the cortex? _____

10. Are mineral ions passively absorbed or do the cortex cells have to work to "pull" them in? _____

11. In what two ways does the large size of the cortex cells promote the absorption of mineral ions?
    (a) _____ ;
    (b) _____

12. Besides absorption and lateral conduction, what other function does the cortex serve? _____

13. What prevents the water that has freely flowed through the cortex from simply moving into the vascular cylinder? _suberin_

14. Buttercups aren't such important plants. Why did we study buttercup roots? _____

15. Sieve tube members are alive, but vessel members are dead. Circle one: (a) true; (b) false.

16. Which tissue in the root gives rise to secondary (lateral) roots? _pericycle_

17. How does water get past the endodermis when most of the endodermal cell walls are "waterproof"?

    _____

18. In general, the vascular cylinder of monocot roots has many fewer xylem ridges than that of dicots.

    Circle one: (a) true: (b) false.

19. Other than by seeds, how do some plants increase their numbers in an area? _____

    _____

20. What role does the cytoplasm of the endodermal cells play? _____

    _____

21. The cambium is a meristematic tissue. What other tissue in the root is meristematic? _____

22. Dicot roots have a pith. Circle one: (a) true; (b) false.

23. All monocot roots look very similar. Circle one: (a) true; (b) false.

24. Is a root sprout a shoot or a root? _____

25. When a dicot root grows old and woody, its internal structure is essentially similar to that of a stem.

    Circle one: (a) true; (b) false.

Table 7.1   Comparison of the Internal Anatomy of Herbaceous Monocot and Dicot Roots

Feature	Dicot		Monocot		
	buttercup	*corn*	*greenbrier*	*wheat*	*Iris sp.*
Number of cylinders					
Number of primary xylem ridges (few or numerous)					
Pith (present or absent)					
Endodermis—indicate in which group it is thicker, monocots or dicots					
Vascular cylinder size —larger or smaller than cortex					

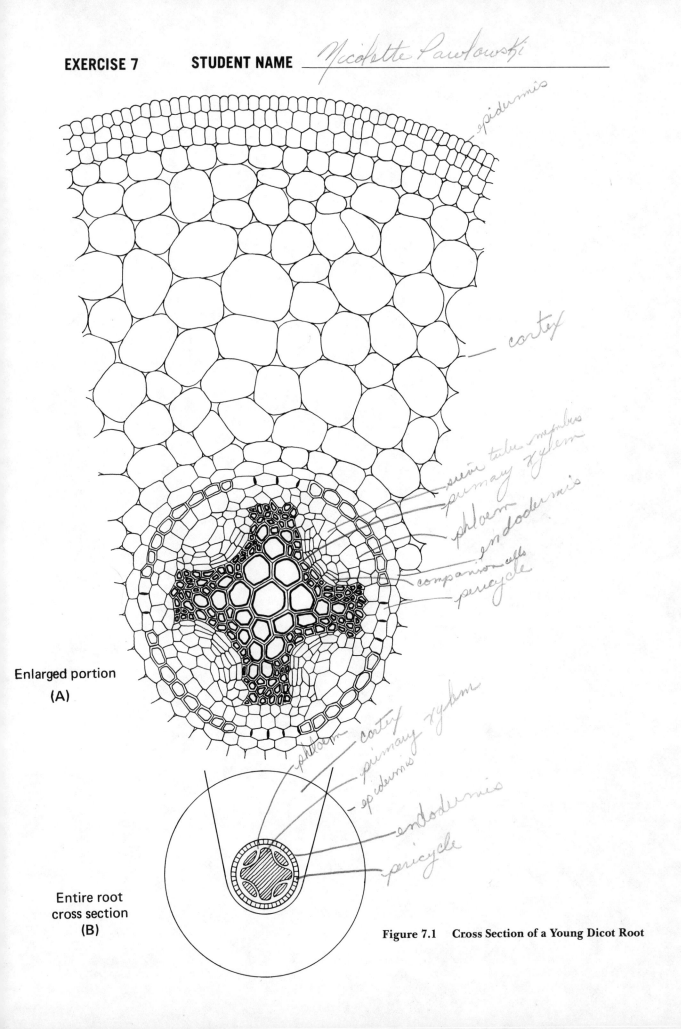

epidermis

cortex

sieve tube member
primary xylem
phloem
endodermis
companion cells
pericycle

Enlarged portion
(A)

phloem
cortex
primary xylem
epidermis
endodermis
pericycle

Entire root
cross section
(B)

Figure 7.1   Cross Section of a Young Dicot Root

EXERCISE 8

HERBACEOUS STEMS I: TERMINAL MERISTEM GROWTH

I. INTRODUCTION TO GROWTH IN GENERAL

The body of the land plant is basically a column or cylinder. This is a good design. It has maximum structural strength but a minimum surface area. These qualities meet the plant's needs for supporting and displaying its leaves (to light) and for limiting water losses. The root too is cylindrical, growing like a needle threading through the soil.

When the plant grows, there are only two ways to increase the size of this cylinder, lengthen it or widen it. Increase in length is referred to as *primary growth*. Primary growth in the stem also includes the formation of new leaves and buds. Increase in width (girth) is referred to as *secondary growth*. Flowering plants with mostly primary growth are herbs. Many dicot herbs have some secondary growth and are hard-stemmed. Dicots with primary and extensive secondary growth are woody—the trees and shrubs.

In dicots, primary growth is localized at the tip of each stem and root (as you saw in Exercise 6). At these tips (apices) there is a zone, called the *apical* or *terminal meristem*. Here cells actively divide during the growing season. Here also are produced growth-promoting hormones known as auxins. In monocots, too, primary growth is localized in apical meristems, in tips of roots and stems. However, the stem tip often remains just below or at ground level, so it's not usually seen. Monocot stems also commonly have a second meristem located at the base of each internode. It is called an *intercalary meristem*.

Monocots are mostly herbs. A few, most notably palms and the Joshua tree *(Yucca brevifolia)*, are tree-like but do not have woody structure like dicots.

II. TERMINAL MERISTEM GROWTH OF THE DICOT HERB

A. GENERAL EXTERNAL APPEARANCE

▶ ACTIVITY 1 **Examine** a representative dicot herb, such as *Coleus blumei*. Others readily available, but possibly somewhat less simple to analyze, are geranium *(Pelargonium* sp.); or any young

plant of bean *(Phaseolus vulgaris)*, cucumber *(Cucumis sativus)* or squash *(Cucurbita* sp.). Gently squeeze and bend the stem. Is it hard and rigid or soft and flexible? _____.
Find the terminal (apical) buds. Where on the plant are these located? _____.
What other structures on the stem are comparable to the terminal bud? _____
What structures on the stem produce branches? _____

Look at the length of the internodes along the stem of *Coleus* (or those plants on hand). Where do you find the longest internodes, between the older (lower) leaves or between the younger (smaller) leaves? _____. Stem elongation is mostly due to elongation of the cells of the internode, particularly those at the base of the internode. Thus lengthening internodes lengthen the stem.

B. ABNORMAL GROWTH CAUSED BY AN HERBICIDE

▶ ACTIVITY 2   **Observe** young bean plants or other dicot plants that have been sprayed with the herbicide 2,4-D. Which part(s) of the plant look abnormal? _____.
Do the leaves look normal?   YES   NO.   The herbicide is a super auxin chemical. It overstimulates the meristem, and one of the results is malformed leaves. The bending and twisting of the stem and petiole (caused by uneven growth induced by the 2,4-D) is another characteristic symptom of 2,4-D injury and is known as EPINASTY.

C. MICROSCOPIC APPEARANCE OF PRIMARY GROWTH ZONE OF THE DICOT STEM

▶ ACTIVITY 3   **Study** a prepared slide with a longitudinal section of the stem tip of *Coleus blumei*. **Find** the APICAL MERISTEM which is the central dome or crest of cells near the top of the section. It is flanked by two partly developed outgrowths, the LEAF PRIMORDIA. A LEAF PRIMORDIUM (singular) is a leaf in its initial growth stage. It originates from dividing subsurface cells on each side of the dome-like meristem. These cells divide faster than those in the center of the dome, and consequently the young leaves grow and extend upward beyond the stem apex (meristem). Depending on where the technician's knife cut through the stem tip, you may or may not see the first (youngest) pair of leaves. Commonly the knife cut misses them, so you see only the next older pair of leaves. These will be very long, and extend far above the meristem.

Note the definite, orderly arranged layers of cells in the meristem. Certain herbicides kill plants by altering this orderly arrangement.

One meristem layer produces the EPIDERMIS. Another meristem layer, the PROCAMBIUM, is easy to identify because it appears as two narrow, darker tracts of longitudinally elongate cells. These tracts extend up into the embryonic leaves but are best seen in sections lower down the stem. The PROCAMBIUM gives rise to the PRIMARY XYLEM and PRIMARY PHLOEM. If enough of the stem has been mounted on the slide, you can see xylem cells in the more mature sections. However, this usually isn't the case.

The central region of the stem is known as the PITH. The region between the vascular tract and the epidermis is known as the CORTEX. Cortex and pith are derived from a tissue, called ground meristem.

Look for small, crest or dome-shaped bulges of tissue in the angle (axil) between a leaf and the stem. Each of these is a LATERAL or AXILLARY BUD PRIMORDIUM. When these bud primordia mature and grow, what do they produce? _____

Note the stump of tissue below and on the outside of the base of each leaf. This is the remnant of the severed base of the leaf at the next lower node. In a section this young, the nodes are close together.

Label the longitudinal section of the stem tip in Figure 8.1, using the capitalized terms defined in the preceding description.

III. PATTERN OF STEM GROWTH OF THE MONOCOT HERB

➤ ACTIVITY 4   **Examine** monocot plants, such as corn *(Zea mays)*; any turfgrass or grassy weed (if growing in season); or any of the common house plants such as dumbcane *(Dieffenbachia* sp.), spiderwort *(Tradescantia* sp.), wandering Jew *(Zebrina* sp.), *Rhoeo* sp., or *Philodendron* sp. Is there a terminal bud on any of these plants?   YES   NO. Are there any axillary buds?   YES   NO. Are there branches on any of these plants?   YES   NO.

   Carefully pull apart the leaves of a young corn plant having at least three leaves, or use the common house plants, such as purple heart *(Setcreasea pallida)*; wandering Jew *(Zebrina* sp.), or *Rhoeo* sp. Note the way the bases or sheaths of the leaves extend down over the internodes. After removing all the leaves, you will be left holding a stem with knobby ridges along it. What are these ridges? _____

   Each leaf seems to be wrapped around the next younger, inner leaf. The elongation of each leaf is due to the division and expansion of cells at the base of the leaf sheath. The stem lengthens because the internode lengthens due to expansion of new cells formed by the INTERCALARY MERISTEM at the base of the internode.

   The stem apical meristem of a grass, such as corn, can be visualized as a mound of cells sitting at the bottom of a well. The sides of the well are formed by the tube-like leaves growing upward from the flanks of the mounded meristem. Each new leaf that is formed by the meristem comes up in the center of the "tube" of the previously formed leaf. Now, if you have followed this description, and it is still a bit confusing, you can understand why we don't ask you to study a microscopic section of the growing tip of a monocot stem!

➤ ACTIVITY 5   **Complete** Table 8.1 showing the comparison of the stem of monocot and dicot herbs.

EXERCISE 8   **STUDENT NAME** _____

QUESTIONS

1. Practically all woody flowering plants are dicots or monocots? _____

2. Do monocots have axillary buds? _____

3. Is the formation of leaf primordia and stem tissues orderly or random? _____

4. How would you recognize 2,4-D injury to a plant? _____

_____

5. What is the mode of action of 2,4-D, and what tissue does it affect? _____

_____

6. Elongation of monocot stems is due mainly to elongation of cells where? _____

7. Primary growth causes increase in girth or length? _____

8. Give two examples of monocots that have a tree growth form. (a) _____;

    (b) _____

9. Why does your lawn continue to send up blades of grass even though you mow it all summer? _____

_____

10. If you ran the mower over your Dad's young tomato plants would they continue to grow? _____

    Explain: _____

11. Do monocot plants branch? _____

12. Do monocot plants have a vascular cambium? _____

13. Secondary growth causes increase in girth or length? _____

14. What is the basic shape of the body of the land plant? _____

15. Give two reasons as to why this shape is a good design. (a) _____;

    (b) _____

16. What are auxins? _____

17. What part of the plant produces auxins? _____

18. What is the basic growth difference between an herb and a woody plant? _____

_____

19. Are terminal meristems zones of year-round cell division? _____

20. Among the dicots, there are both herbaceous and woody species, but the monocot species are practically

all _____.

Table 8.1   Comparison of Stems of Herbaceous Monocots and Dicots

Feature	Dicot	Monocot
Branches (present or absent)		
Terminal buds (present or absent)		
Axillary buds (present or absent)		
Primary growth (present or absent)		
Secondary growth (present in many or absent in most)		
Location of the apical meristem		

STUDENT NAME _____

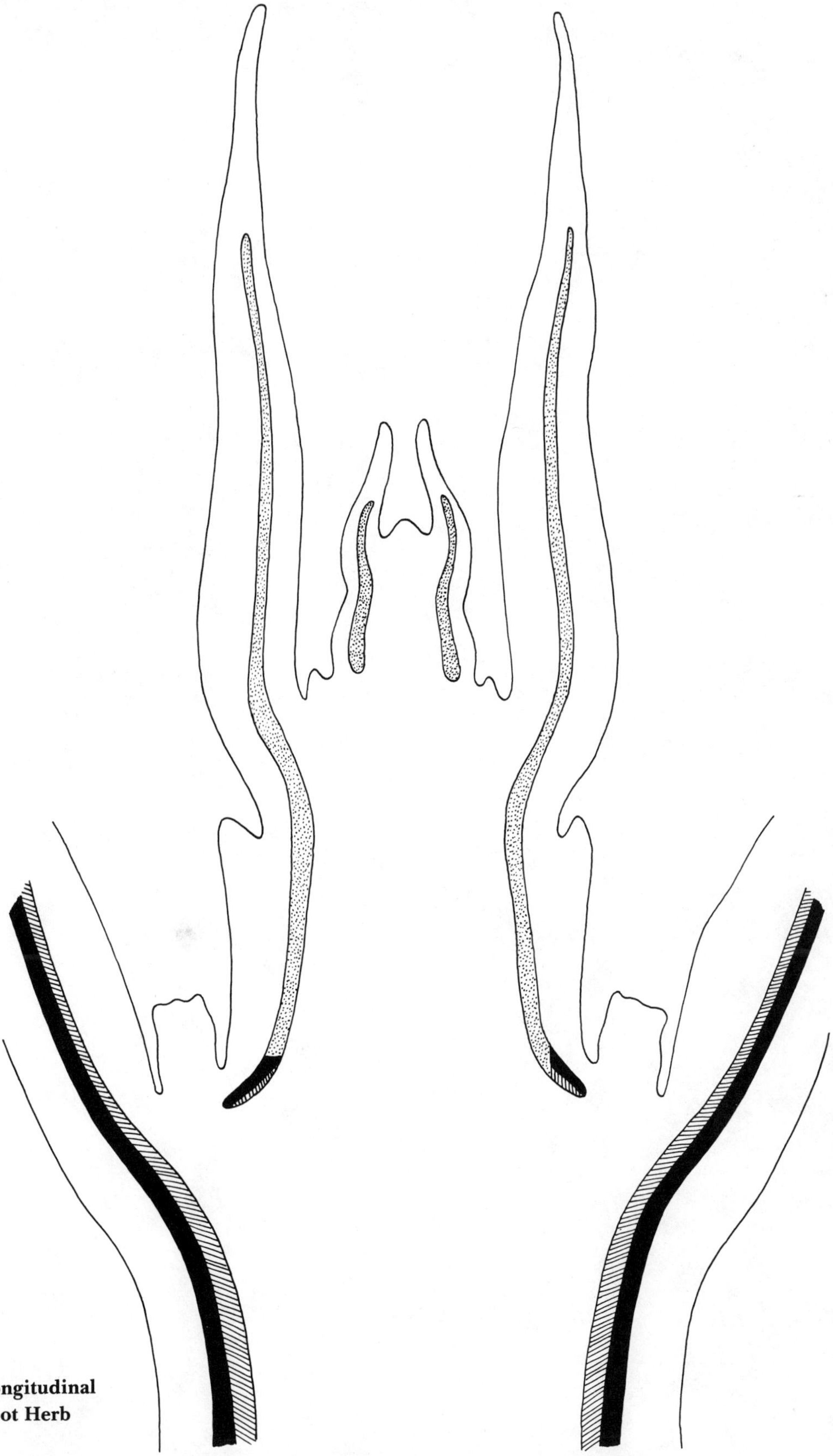

**Figure 8.1   Longitudinal
Section of a Dicot Herb**

EXERCISE 9

HERBACEOUS STEMS II: STEM ANATOMY OF HERBACEOUS DICOTS AND MONOCOTS

The term "herb" is popularly interpreted to mean a plant grown for some medicinal purpose or for its sweet scent or flavor. But botanically, an herb is any seed plant that does not develop woody, persistent tissue. In other words, it has mostly the soft *primary body tissues* formed by terminal meristems.

Woody plants use most of their photosynthate (the food synthesized) to build a permanent, woody framework supporting a canopy of leaves. Herbs use most of their photosynthate to make seeds, fruit, roots, tubers, and other storage organs. In a sense, the woody plant construction is designed mainly for perpetuating the individual plant, whereas the herbaceous plant construction is designed mainly for perpetuating the population.

We consider the herb design a more specialized way of life. Humans have benefited from the herbaceous design because it is chiefly from herbs that we derive our food.

Herbaceous dicots evolved from woody dicots. This came about by the evolutionary decline of the vascular cambium, the meristem that forms secondary tissue. In herbs, cambial growth is limited to one season or part of one season, or it is totally absent.

I. STEM OF THE HERBACEOUS DICOT

When you look at the microscopic sections of the stems of herbs, don't get "lost" in the details of the anatomy. The *main point* is for you to see that the arrangement of tissues is an efficient design which integrates five different functions: (1) *support*, (2) *conduction*, (3) *storage*, (4) *photosynthesis*, and (5) *growth*. You have already seen (in Exercise 8) how the stem is designed for growth. In this exercise you will look at the tissue patterns designed for the other four functions.

► ACTIVITY 1   **Examine** a slide showing the cross section of a young stem of a representative dicot herb such as sunflower (*Helianthus* sp.) or geranium (*Pelargonium* sp.). **Refer** to Figure 9.1 as a guide.

A. SUPPORTIVE FUNCTION OF THE STEM

In an herb support is achieved by (1) fibrous tissue (collenchyma cells and sclerenchyma cells), (2) the arrangement of fibrous tissue, and (3) turgor pressure. Fibrous tissue occurs near the outer region of the stem. This arrangement creates a cylinder—the strongest type of columnar support acknowledged by engineers.

Fibrous Tissue and Its Functional Arrangement

the collenchyma layer   **Find** the COLLENCHYMA LAYER, a band of thick-walled cells just within the EPIDERMIS, the surface layer of cells. COLLENCHYMA CELLS are thick-walled, elongate, and particularly suited to support the stem when it is young and growing. They give strong, stable support, yet are plastic (stretchable), adjusting to stem elongation. In mature stems, the collenchyma cell walls are hard. Sometimes collenchyma occurs in strands visible as ridges along stems or petioles. The "strings" you can pull from a celery stalk are strands of collenchyma. **Label** the collenchyma in Figure 9.1 (A).

sclerenchyma–bundle cap fibers   **Focus** on the FIBROVASCULAR BUNDLES. These are individual tracts of tissues running lengthwise in the stem and appearing in cross section as cell clusters, called BUNDLES. The bundles are arranged orderly in a ring close to the outer zone of the stem.

On its outer side, each bundle is capped by a mass of very thick-walled SCLERENCHYMA CELLS, known as the BUNDLE CAP FIBERS. The rest of the bundle consists of the vascular tissue (xylem and phloem). The arrangement of the bundles near the periphery of the stem forms a cylinder. The tough sclerenchyma cap reinforces each tract (bundle) as it extends the length of the stem. This pattern is comparable in principle to the construction found in reinforced concrete pillars where steel rods are embedded in the concrete to increase its strength.

SCLERENCHYMA CELLS (also known as FIBERS) are very long with thick, rigid, but highly elastic walls. They can take pulls and pushes, but will rebound to their original length. The fibers intermesh and form a cable-like system. This gives the stem strength, yet also permits it to yield flexibly to the stress of wind and the weight of leaves and bobbing fruit. In some plants sclerenchyma is valuable material. We use it to make rope, twine, and fabrics. The sclerenchyma of the marijuana plant is the hemp fiber of commerce used for cordage. The sclerenchyma of the *Corchorus* plant is the jute fiber used for making rough fabrics, such as burlap. The sclerenchyma of the *Linum* plant is the fine fiber of linen. **Label** the bundle cap with the sclerenchyma cells in Figure 9.1 (A).

Turgor Pressure

Turgor pressure is the pressure developed by the water present in plant cells. It keeps plant parts (leaves, flowers, and herbaceous stems) rigid or turgid.

Note the large thin-walled PARENCHYMA CELLS that fill in the center of the stem and also the area around each bundle. These large cells contain water which builds up a turgor pressure that keeps the stem firm. **Label** a paremchyma cell in Figure 9.1 (A).

B. STORAGE FUNCTION OF THE STEM

Look again at the thin-walled parenchyma cells that fill in all the areas not occupied by sclerenchyma or fibrovascular bundles. Besides their role in turgor pressure, parenchyma cells have two other functions, (1) they store materials, and (2) all together they serve as a lightweight filler material that keeps the fibrovascular bundles in place. Large, water-filled vacuoles make the cells look almost empty of contents. Starch grains (stained blue or purple) in many cells indicate their food-storage role.

Different "place names" have been given to the parenchyma tissue where it fills in the different parts of the stem. In the very center of the stem, within the circle of fibrovascular bundles, parenchyma is given the name PITH. Between the fibrovascular bundles, it is named PITH RAYS. Between the epidermis and the outer edge of the ring of fibrovascular bundles, it is named CORTEX. The cortex typically has supporting tissues, besides parenchyma cells. **Label** the different parenchymatous regions of the stem in Figure 9.1 (A), using the capitalized terms defined in the preceding description.

C. THE CONDUCTIVE FUNCTION OF THE STEM

Locate again the ring of fibrovascular bundles. The conducting tissues in the bundle are the XYLEM and PHLOEM. Between them lies the VASCULAR CAMBIUM.

The Phloem

Locate the zone of thin-walled cells that lies just inside the bundle cap. This is the PHLOEM. **Note** that there are two sizes of cells present, the larger ones are the SIEVE TUBE MEMBERS, and the smaller ones are either COMPANION CELLS or PHLOEM PARENCHYMA CELLS. Sieve tube members are long cells stacked one above the other, forming a pipe-like SIEVE TUBE. The sieve tube cells have perforated end walls, called SIEVE PLATES, through which cytoplasmic strands extend from one member to the next. Food and other materials are carried in these strands. Sieve plates are not commonly visible. COMPANION CELLS are specialized cells whose nuclei are thought to have some control over the functioning of the cytoplasm in the sieve tube member which has no nucleus. PHLOEM PARENCHYMA cells serve as filler tissue and for storage. **Label** all the phloem cell types in Figure 9.1 (A), using the capitalized terms defined in the preceding description.

The Xylem

The XYLEM makes up the innermost tissue of each fibrovascular bundle. Most xylem cells are angular in outline and have thick walls. These are the water-conducting cells, TRACHEIDS and VESSEL ELEMENTS. Vessel elements are larger in diameter than tracheids. Vessel elements are barrel-shaped cells stacked one above the other, forming a pipe-like VESSEL. Water moves up through the vessels and also through the more slender tracheids, but at a slower rate. Located here and there among the tracheids and vessel elements are smaller, thinner walled, rounded XYLEM PARENCHYMA cells. **Label** the xylem cells in Figure 9.1 (A), using the capitalized terms defined in the preceding description.

The Vascular Cambium

Find the VASCULAR CAMBIUM, a thin layer of four-sided, flattened looking cells located between the phloem and the xylem. **Note** that this band of flattened looking cells occurs not only within the fibrovascular bundle, but also in the spaces (pith rays) between the bundles. The cambium within the bundle is called FASCICULAR CAMBIUM; the cambium lying between the bundles is called the INTERFASCICULAR CAMBIUM. As the sunflower grows, the interfascicular cambium forms xylem and phloem in the pith ray regions, thus "filling in" the spaces between the bundles. Eventually, instead of having separate bundles of xylem and phloem, the plant has a single complete ring of xylem and phloem tissue in the stem. **Label** the two types of cambia in Figure 9.1 (A), using the capitalized terms defined in the preceding description.

D. PHOTOSYNTHETIC FUNCTION OF THE STEM

The stems of herbs are usually green due to the presence of chloroplasts. **Examine** the outermost cells of the stem, the EPIDERMIS, and also those cells forming the COLLENCHYMA BAND, and other cells in the vicinity. Any of these cells can have chloroplasts, but depending on how the tissue was stained and prepared for the slide section, you may or may not see the chloroplasts. The epidermis contains openings, STOMATES, flanked by two GUARD CELLS. Carbon dioxide and oxygen pass through the stomates. You may or may not be able to see stomates and guard cells.

Stem photosynthesis is mostly supplementary in mature plants, but it contributes significantly to the sustenance of seedlings and many desert plants. In Figure 9.1 (A) **indicate** which of the cells are photosynthetic cells.

► ACTIVITY 2   **Examine** twine, rope, bags, or cloth made from jute, hemp, or linen fibers.

► ACTIVITY 3   **Examine** and compare young seedlings of bean or corn that have been well supplied with water versus those that have not been watered.

Observe a demonstration showing the rapid wilting effect of a young, soft-stemmed plant exposed to a high amount of illumination versus one not so exposed.

Examine several kinds of vegetables that have been allowed to wilt versus those kept moist. Plants that readily show this wilting are celery, carrots, and lettuce.

As long as the cell membranes are still alive, a wilted plant can be made turgid again by providing it with water.

During the class period, see if you can "revive" any of the wilted specimens.

II. COMPARISON OF THE STEM OF A DICOT HERB AND A ONE-YEAR-OLD TWIG OF A WOODY DICOT

► ACTIVITY 4   **Compare** cross sections showing an older stem of a dicot herb such as geranium (*Pelargonium* sp.) or sunflower (*Helianthus* sp.) and a one-year-old portion of a twig of a dicot woody plant, such as species of *Tilia, Platanus,* or *Liriodendron.*

Based on the comparisons you've made, **circle** YES or NO for each of the following. Are fibrovascular bundles present in one-year-old twigs?   YES   NO. Are the bundles of the one-year-old twig as equal in size as in the herb?   YES   NO. Are the pith rays as wide in the one-year-old twig as in the herb?   YES   NO. Does the one-year-old twig show much interfascicular cambial growth?   YES   NO. Is it possible for the plant to develop one entire ring of vascular tissue subsequent to a stage in which there are separate bundles?   YES   NO. Is there a pith present in both stem types?   YES   NO. Herbs evolved from woody plants. Was this due to a decline in the activity of the interfascicular cambium?   YES   NO.

The similarity between these stems is due to the fact that both the herb and the twig in its first year bear leaves, and many of the separate vascular bundles are leaf traces. Leaf traces are portions of the vascular system of the stem that diverge out to the petioles of leaves. If the stem you are examining has been sectioned across a node, you can see some of these traces diverging off from the ring of vascular bundles. In most instances, however, the slides made for your study have been cut across an internode, so you cannot see the leaf trace diverging from a bundle.

Any portion of a woody twig that is older than one year does not bear leaves, hence has no leaf traces and does not resemble an herb stem. Instead of separate vascular bundles, it has a complete ring or rings of vascular tissue. Many herbs, too, in their mature growth stages of the late summer tend to form a complete ring of vascular tissue. Sunflower, as noted earlier, commonly does this.

III. THE STEM OF THE HERBACEOUS MONOCOT

Monocots have evolved from herbaceous dicots, and herbaceous dicots evolved from woody dicots. As you have seen, herbaceous dicots show partial or almost complete decline of the vascular cambium, and monocots show a total absence of this tissue. In the absence of cambium, most monocots are herbs.

► ACTIVITY 5   **Examine** a slide showing a cross section of the young stem of corn (*Zea mays*) or of lily (*Lilium* sp.). Using low power, locate the tissues and cells as described in the following outline of stem functions: *conduction, support, storage,* and *photosynthesis*. **Refer** to Figure 9.2 as a guide.

A. CONDUCTIVE FUNCTION OF THE STEM

Circle YES or NO. Are the vascular bundles arranged in a ring as in a dicot?   YES   NO. Are they in some regular pattern?   YES   NO. Is there a pith?   YES   NO. Is there a cortex? YES   NO. Are the vascular bundles all the same size?   YES   NO. Are the bundles in the center larger than those near the periphery?   YES   NO. Are the bundles more densely clustered at the periphery?   YES   NO.

Each of these vascular bundles is a leaf trace or several combined leaf traces. If the stem was cut across a node, a leaf trace may be seen diverging outward.

In monocots the network of vascular strands in the stem is very complex and has not yet been as adequately deciphered as in dicots. In the stems of some monocots, such as wheat (*Triticum* sp.) and rye (*Secale* sp.), the vascular strands are distributed in a ring as in dicots, but more frequently their complicated network shows up in cross section in a scattered pattern, as seen in corn.

Change to high power and **study** the details of a single vascular bundle. **Locate** the following tissues and cell types.

The Xylem

Xylem tissue consists of the thick-walled cells, the VESSEL MEMBERS, many of which in series form the tube-like VESSEL. Two or three of these vessel elements commonly occur in an arrangement that could imaginatively be compared to the "eyes and nose" of a "face" (if you imagine that the entire vascular bundle resembles a face). Between the larger vessel elements are smaller vessel elements and also thinner walled XYLEM PARENCHYMA. Typically there is a large round hole (forming the "mouth" of our imaginary "face"). This hole was caused by the collapse of cells stretched to the breaking point during the time the stem was elongating. **Label** the cell types in the xylem of the vascular bundle in Figure 9.2 (A), using the capitalized terms defined in the preceding description.

The Phloem

Phloem tissue is located on the outer side of the xylem in what would be the "forehead" of the "face". Two types of cells are clearly visible here, the larger SIEVE TUBE MEMBERS and the smaller COMPANION CELLS. **Label** the cell types in the phloem of the vascular bundle in Figure 9.2 (A), using the capitalized terms defined in the preceding description.

The Sclerenchyma Sheath

Thick-walled sclerenchyma cells (fibers) surround the xylem and phloem, forming a protective sheath which also serves to help support the stem.

B. SUPPORTIVE FUNCTION OF THE STEM

You have already noted that each vascular bundle is enclosed by a SCLERENCHYMA SHEATH which is a strong protective sleeve composed of long sclerenchyma fibers (cells). The scattered arrangement of the bundles, while basically due to the complex vascular network, effectively "installs" the fibrous support tissue (the sclerenchyma) in a manner somewhat comparable to that of embedded steel rods in steel reinforced concrete columns. These "rods" of support fibers are stabilized in place by the large rounded PARENCHYMA CELLS which all together serve as filler. Individually, each parenchyma cell stores food and other materials.

The sheathed vascular strands (bundles), like the ones seen here in corn, are removed from the leaves of the monocot *Agave* grown commercially in Mexico. The extracted strands are known as sisal fiber. Sisal is used to make twine, sacks, rugs, and fiberboard. Such bundles are also removed from the petioles of the monocot *Musa textilis* grown commercially in the Philippines. These fibers are known as Manila hemp, which is used in making very strong cordage, twine, brown wrapping paper and bags, and manila envelopes.

Examine the extreme outer edge of the stem. Here are many small SCLERENCHYMA CELLS. In older stems these cells are even more numerous. They, together with the densely crowded, fibrous-sheathed, small vascular bundles, form the hard RIND of the corn stalk. **Label** the sclerenchyma sheath of a vascular bundle and the peripheral sclerenchyma cells in Figure 9.2 (A).

C. STORAGE FUNCTION OF THE STEM

Examine the large, thin-walled PARENCHYMA CELLS that fill in all stem areas not occupied by the vascular bundles and sclerenchyma. These parenchyma cells are collectively referred to as the GROUND PARENCHYMA. The large clear area in each cell is the VACUOLE which stores water and other materials. Turgor pressure in these cells helps keep the stem tissues firm. Are there starch granules present?   YES   NO. In the sugar cane plant (*Saccharum officinale*), which is a grass-like corn, these parenchyma cells are filled with sucrose which is extracted by crushing the stem between fluted rollers. The young corn stem is similarly filled with sugary water, but as the plant matures it moves most of its sugars and other foods into the kernels of the ear. **Label** the ground parenchyma of Figure 9.2 (A).

D. PHOTOSYNTHETIC FUCTION OF THE STEM

As with herbaceous dicots, monocot herbs have green, photosynthetic stems. The outermost parenchyma cells contain chloroplasts, and the epidermis has stomates with guard cells. However, only an occasional cross section on a slide will show stomates or guard cells, so your chances of finding any are slim. Similarly, chloroplasts do not show up readily in these prepared sections. **Indicate** which parts of the stem in Figure 9.2 (A) are photosynthetic cells.

▶ ACTIVITY 6   **Examine** products made from sisal or Manila hemp.

▶ ACTIVITY 7   **Complete** Table 9.1 comparing the stems of dicot and monocot herbs.

Table 9.1   **Comparison of the Stems of Dicot and Monocot Herbs**

Feature	Dicot	Monocot
Vascular bundle arrangement		
Vascular cambium present or absent		
Specific supporting structures or tissues		
Filler tissue		
Storage tissue		

EXERCISE 9   **STUDENT NAME** _____

QUESTIONS

1. Why in general are herbaceous plants more valuable to man as a food source than are woody plants?

_____

2. In general why are herbaceous plants smaller than woody plants? _____

_____

3. What is the botanical definition of an herb? _____

_____

4. The tissue arrangement in an herbaceous stem is functional. In what way(s)? _____

_____

5. What are the three ways in which support is achieved in an herbaceous stem? (a) _____

_____; (b) _____; (c) _____

6. Explain the physical difference between collenchyma and sclerenchyma. _____

_____

7. Name three commercial fibers made from the sclerenchyma fibers of the stems of dicot herbs.

(a) _____; (b) _____; (c) _____

8. Name two commercial fibers that are obtained from the sheathed vascular strands of the leaves or

petioles of monocots. (a)_____

_____; (b) _____

9. Parenchyma tissue in herbaceous stems serves which two purposes? (a)_____

_____; (b) _____

10. A series of sieve tube members connected end to end form a _____

11. A series of vessel members connected end to end form a _____

12. The separate vascular bundles of a young woody twig gradually "merge" into one complete ring of vas-

cular tissue. This occurs because new xylem and phloem are laid down in the region of the _____

_____ by the active division of the _____

13. What is meant by the primary body of a plant? _____

_____

14. Do woody plants have a primary body? _____

15. One-year-old woody twigs and herbaceous dicots have almost identical internal stem anatomy because

_____

16. Herbaceous dicots have evolved from _____ and show a decline or complete

loss of the _____

17. Monocots have evolved from (a) woody monocots; (b) herbaceous dicots; or (c) woody dicots? Circle (a),

(b), or (c).

18. Do all monocot stems look like the stem of corn in cross section? _____

19. In construction engineering of monocot herbs, is support in the stem achieved by the same general

means as in herbaceous dicots? _____ Whether you answered YES or NO, specify what means are

used. _____

_____

20. Biologically, which is the more progressive way of life, the woody plant design or the herb design? _____

_____ Which is the better design for perpetuating the population? _____

21. What is the strongest type of columnar support acknowledged by engineers? _____

22. In what design of what tissue in the herbaceous dicot stem do you find this type of support? _____

_____

23. Define turgor pressure. _____

_____

24. Which tissue gives the plant most of its turgor? _____

25. Sometimes during the summer, your tomato plants and garden flowers wilt, but they usually regain
turgor if you water them, or if it rains. In the autumn, an early-season overnight frost will wilt your
tomato plants. Even though the next few days are warm and you give the plants adequate water, they

won't regain turgor. Why? _____

EXERCISE 9   **STUDENT NAME** _____

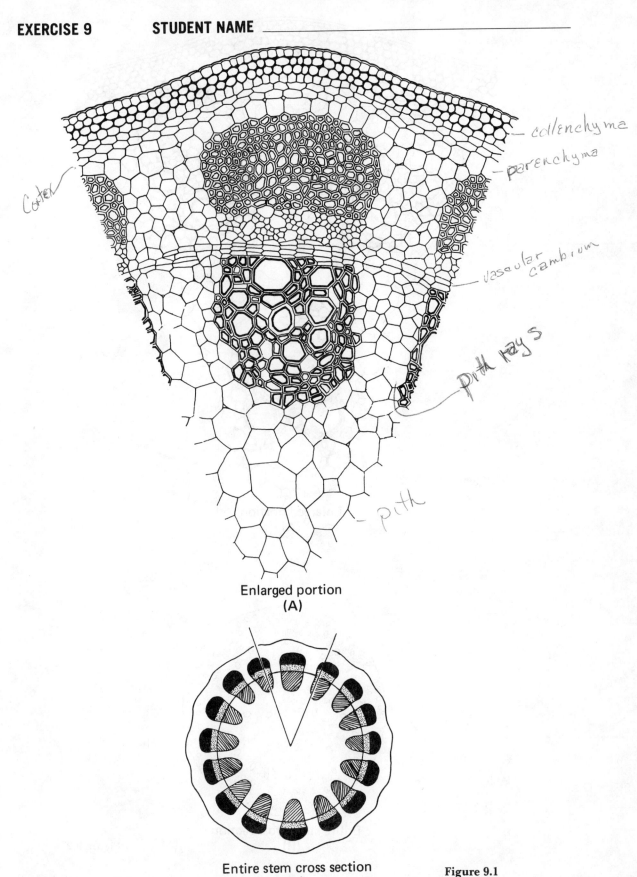

collenchyma

parenchyma

Cortex

vascular cambium

Pith rays

pith

Enlarged portion
(A)

Entire stem cross section
(B)

Figure 9.1

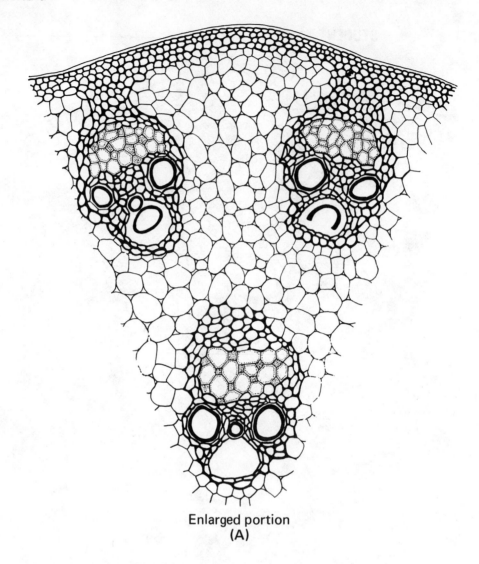

Enlarged portion
(A)

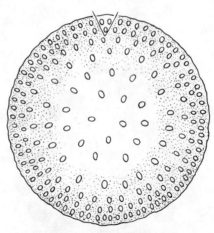

Entire stem cross section
(B)

Figure 9.2

EXERCISE 10

WOODY STEMS I:
EXTERNAL FEATURES
OF YOUNG TWIGS

The woody plant form has existed on earth for millions of years. It predominated at a time when world climate was universally mild. As the climate fluctuated, and droughty and cold spells became common, the tree's way of life was not as well suited for survival as that of the herb. Many were destroyed.

In today's climate, woody plants living in zones with drastic seasonal differences survive the cold or dry season by going dormant. However, their woody framework still stands vulnerable to wind, ice, snow, or drought. The herb escapes these forces by not making any permanent woody framework. Instead, it is short-lived and quickly produces the seeds of the next generation.

Herbs are fast plants. Trees, in contrast, are slow. Their growth rate and net photosynthetic rate are slower than that of herbs. They are slow to reproduce, typically requiring years of growth before they first flower and produce the seeds of another generation. Most of their photosynthate and energy are spent on building the woody framework and making a new set of leaves each year.

Although we consider the life style of an herb to be biologically a more successful way of life, this is not to say that the tree or shrub is a lost cause. In climates where there are no adverse seasons and there is little or no freezing weather, woody plants are overwhelmingly dominant. Also, there are many woody plants that are exceedingly well adapted to survival in severe climates.

I. THE YOUNG WOODY STEM—THE TWIG

Twigs are the youngest parts of the tree or shrub.

► ACTIVITY 1   **Examine** leafless twigs taken from dormant trees. Suitable species for study are horse chestnut or buckeye (*Aesculus* sp.); walnut (*Juglans* sp.); tree of heaven (*Ailanthus altissima*); ash (*Fraxinus* sp.); and tulip tree (*Liriodendron tulipifera*). **Identify** the following features.

A. MAIN TWIG FEATURES

Terminal Bud

A TERMINAL BUD occurs at the tip of a twig and is responsible for its growth in length. If your twig has branches, each branch will also have a terminal bud. **Note** the BUD SCALES. These are modified leaves. They enclose and protect the embryonic leaves and apical meristem within the bud. Depending on the bud type or the plant species, the bud may also contain embryonic flowers.

At the onset of the growing season, bud scales flare out and fall off as the young stem and leaves grow. The internodes between successive leaves elongate, thereby increasing the twig length. Some plants, like *Aesculus* sp., have a solitary terminal bud. Others, like oaks (*Quercus* sp.), ashes (*Fraxinus* sp.), and maples (*Acer* sp.), have two or more lateral buds clustered closely about the terminal bud, giving the appearance of several terminal buds. Still others, like sycamore (*Platanus* sp.), have no terminal buds; nearby lateral buds bring about twig elongation. **Label** the terminal bud(s) on Figure 10.1.

Terminal Bud Scale Scars

A TERMINAL BUD SCALE SCAR is a band of closely grouped scars encircling the twig like so many rings. Each such band of scars marks the position of the terminal bud of a previous year. The scars are those left after the bud scales fell off. **Label** the terminal bud scale scars on Figure 10.1 and indicate with a bracket ONE SEASON OF GROWTH.

Measure (in millimeters) the length of stem produced in each of the last two to three seasons of growth and record the figures.

Youngest (or most recent) amount of growth	_____	mm
Next older growth region—amount of growth	_____	mm
Next older growth region—amount of growth	_____	mm
Next older growth region—amount of growth	_____	mm

Are all regions equal in length?   Yes   No.

Leaf Scars

A LEAF SCAR is a scar left on the twig where the petiole or base of the leaf was once attached. The scar is a corky layer, sealing off the living tissues beneath. Leaf scars are of many different shapes and sizes, depending on the species, and are a key feature used to identify woody plants. Will new leaves arise from these leaf scars?   YES   NO. **Label** a leaf scar in Figure 10.1.

Vascular Bundle Scars

Continue to examine one of the leaf scars on your specimen. **Note** the pinhead-size, corky bumps within the leaf scar. These are the scars of the broken ends of the vascular strands (bundles) that passed between leaf and stem. The numbers and arrangement of these VASCULAR BUNDLE SCARS vary with different species. They, like leaf scars, are useful characters for identifying woody plants during their dormant season when leaves are absent. **Label** the vascular bundle scars in Figure 10.1.

Nodes and Internodes

The leaf scar occurs at a position on the stem called the NODE. The node is defined as that region on a twig where one or more leaves are borne. The portion of the twig between any two successive nodes is the INTERNODE.

Compare the lengths of the internodes of one growing season with those of another growing season. Are they equal in length each year?   YES   NO. **Label** a node and bracket an internode in Figure 10.1.

Lateral Buds

Note the bud on the upper margin of each leaf scar. Since this bud occurs on the side of the twig, it is called a LATERAL BUD. When the leaf is present, this bud is in the axil (upper angle) between the petiole and the stem, so it is also known as an AXILLARY BUD. Lateral (axillary) buds are made of the same parts as a terminal bud.

Lateral buds are formed at the same time as the leaves on a twig. But they do not grow when the leaves grow because they are inhibited by a hormone from the terminal bud. As the twig lengthens, the distance between lateral buds and the terminal bud increases. This results in less inhibition by the terminal bud, and lateral buds then may grow and produce branches.

Sometimes lateral buds fail to grow for an indefinite period and become buried in the bark as the tree ages. These are called LATENT BUDS. Under certain conditions they may start growth and form small leafy shoots on old large limbs.

Lenticel

Examine the surface of your·twig (or larger limbs, if available). Note that all over the surface there are rounded, pinhead-like or linear, slit-like markings. These are the LENTICELS. Lenticels are raised areas of loose cork. In trees with deeply furrowed barks they are not visible. They are avenues for entry of air into the woody parts just as stomates admit air into leaves. A heavy growth of moss, lichens, or vines on the trunk of a tree can interfere with its lenticels and in extreme cases the tree may die. What other agents might injure trees by entry through the lenticels? _____. **Label** the lenticels in Figure 10.1.

B. SOME COMMON ADDITIONAL TWIG FEATURES

Thorns, Spines, and Prickles

► ACTIVITY 2   **Examine** twigs of honey locust (*Gleditsia triacanthos*); hawthorn (*Crataequs* sp.); black locust (*Robinia pseudoacacia*); and crabapple (*Malus* sp.). Many woody species have thorns, spines, or prickles. Thorns are sharp-pointed modified twigs. Spines are either modified stipules or modified leaves, depending on the plant type. Prickles are just epidermal or cortical outgrowths, as in roses. **Examine** the twigs of a honey locust. Are the thorns branched or unbranched? _____. Is the position of the thorn the same as that of an axillary bud? _____. How can you tell? _____. On the basis of this observation, you can conclude that the thorn of honey locust is a modified _____. Is thorn the correct name for these? _____. Should they be called spines? _____.

Thorns are usually objectionable features for shade trees. Cultivated varieties which characteristically lack thorns are more usually grown and sold by nurseries. The "Moraine" locust is a patented, thornless, fruitless honey locust. **Examine** the twig of a black locust. Do these sharp-pointed processes occur singly or in pairs? _____. Where is the leaf scar in relation to these processes? _____. On the basis of their position, you can conclude that these sharp-pointed processes are modified _____ and therefore should be called _____ . **Examine** twigs of hawthorn. Do these sharp-pointed processes occur singly or in pairs? _____. How are they positioned in relation to the leaf scar? _____. What do you think these are? _____. **Examine** twigs of crabapple. Their sharp-pointed processes are modified twigs and are therefore properly called _____.

Spur Shoots

Examine a branch of a mature crabapple, apple, or pear and identify spur shoots. Spur shoots are short, usually stocky dwarf branches with crowded leaf scars. They grow very slowly compared to other twigs on the tree. Commonly, as in apple and pear, flower buds, and hence fruit, are produced on spur shoots whose short, stocky form is well suited for holding the weight of the fruit.

II. PROPAGATION BY WOODY TWIG CUTTINGS

► ACTIVITY 3   If specimens are available, **examine** twig cuttings of species of willow (*Salix*); poplar (*Populus*); forsythia (*Forsythia*); privet (*Ligustrum*); honeysuckle (*Lonicera*); or spirea (*Spiraea*) which have been rooted in sand. Most woody ornamental plants are propagated by twig cuttings (slips) rather than by seed. The rooted twig will grow, branch, and produce an entire plant. This is a faster and easier way to maintain and produce a given variety of plant, rather than by seed. Plants grown from cuttings will have the same features as the parent plant. The grower can guarantee that his rooted cuttings will produce plants with the characteristics he has advertised and promoted in his catalog.

III. ADDITIONAL EXAMPLES OF WOODY TWIGS

► ⎰ACTIVITY 4   **Examine** twig specimens of several different species of woody plants to see the great variety in

1. Leaf scars: Size and shape
2. Vascular bundle scars: Number and arrangement
3. Terminal buds: A single bud or otherwise; size and shape
4. Lateral buds: Size and shape
5. Lenticels: Shape.

IV. PRUNING TWIGS AND BRANCHES

► ACTIVITY 5   **Practice** pruning if time permits. It's worthwhile knowing how. This may be done on the standing tree or shrub or on limbs brought into the classroom. For a large branch, cut back to its juncture with another. Cut it flush off the other branch. Don't leave a stub. Small twigs can be pruned the same way or can be cut at a point just above any one of its lateral buds. **Make** all cuts slanting, never straight across the stem. (It heals better this way.)

Incorrect pruning methods stimulate the growth of latent (and also new) buds near the cut end of a limb. These buds will produce a porcupine-like cluster of twigs growing near the cut end of the limb. Such limbs have simply been shortened some place along their length rather than carefully cut off at the point of their juncture with another limb. This bushy, often ugly, porcupine-like growth, known as the POLLARD EFFECT, disfigures the tree, can reduce its vigor, and only creates more work for the next hapless pruner. **Examine** samples or photographs of limbs or branches that have been indiscriminately cut back in length and which have produced the pollard effect.

EXERCISE 10   **STUDENT NAME** _____

QUESTIONS

1. Which is considered to be biologically the more successful way of life, the tree or the herb? _____

2. Which has the slower growth rate, the tree or the herb? _____

3. Which is slower to reproduce itself, the tree or the herb? _____

4. The photosynthate of the tree is chiefly used for what purpose(s)? _____

   _____

5. On mountain tops and in the Arctic one finds mainly herbs and perhaps a few dwarf trees. List the

   climatic elements of these regions that prevent growth of trees. _____

   _____

6. Why are herbs better as food crops for humans than are trees? _____

   _____

7. How do trees contribute to:

   a. Human economy? _____

   b. The landscape? _____

   c. The environment? _____

8. Which grower must wait longer to get a return on his investment in planting: the Nebraska wheat

   farmer or the Michigan apple grower? _____

9. What features on the twig would you count in order to determine how many years of growth are present

   on the twig? _____

10. Suppose an ecologist wanted to determine the previous five years' climatic conditions of a watershed
    (that is, the forested upland head) of the Columbia river. What twig feature(s) could he rely on in mak-

    ing some of his determinations? (a) _____; (b) _____

11. List three twig features that are useful for identifying woody plants during their dormant season when

    no leaves or flowers are present. (a) _____; (b) _____;

    (c) _____

12. Another name for a lateral bud is _____

13. Define a thorn. _____

14. What is the function of a lenticel? _____

_____

15. Does a twig have only primary growth, only secondary growth, or both? _____

16. Which portion of the twig would have primary tissue only? _____

17. Which portion of the twig would have both primary and secondary tissues? _____

18. Besides the scales, what are the other parts of a bud? _____

_____

19. In winter or in the dry season (in the tropics and subtropics) how could you determine the leaf arrangement of a tree since at this time it would be leafless? _____

_____

20. What inhibits the growth of lateral buds? _____

21. Which buds on a twig are more inhibited from growing: those nearer to, or those farther from, the terminal bud? _____

22. When pruning a fairly large branch off a tree or shrub, where should you make the cut? _____

_____

23. How should cuts be made, straight across the stem or on a slant? _____

24. Explain what is meant by the pollard effect. _____

_____

25. Many cultivated varieties (cultivars) of woody ornamentals are valued because they are sterile and don't produce any fruit or seeds to litter the ground. How then can the nurseryman continue to grow and

offer such plants for sale? _____

26. Poorly pruned, pollarded trees are commonly seen. In city streets, where are such trees usually located?

_____

EXERCISE 10   STUDENT NAME _____

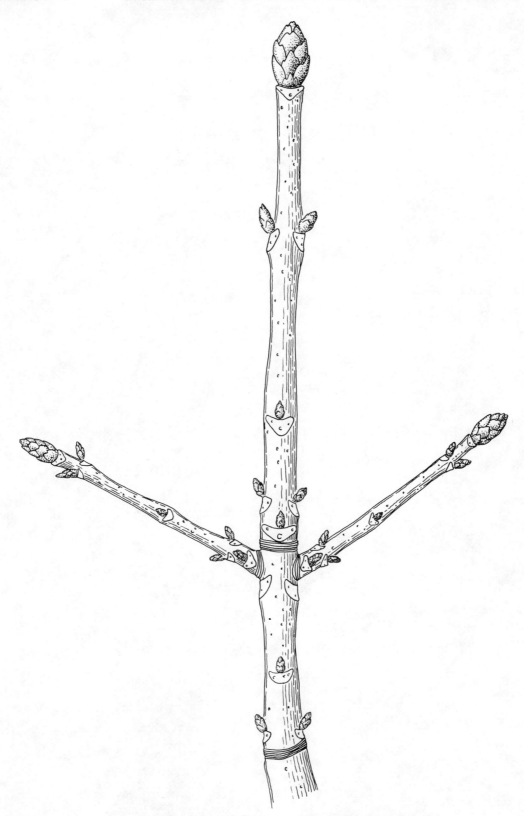

Figure 10.1   External Features of Woody Twig

EXERCISE 11

WOODY STEMS II: INTRODUCTION TO WOODY STEM ANATOMY AND THE GYMNOSPERM STEM

I. INTRODUCTION TO WOODY STEM ANATOMY

A woody plant is a permanent structure that yearly increases its leafy canopy through primary growth (which, as you learned in Exercise 10, is apical growth). Supporting this canopy requires a reinforcement of the woody framework. Secondary growth (growth that increases width) provides this reinforcement.

As the plant grows more massive through secondary growth, most of its cells become buried by the added layers and cannot be supplied with oxygen.

It would seem then that there might be a physiological limit to the size that a woody plant can grow. It turns out that the plant has no such limit. This dilemma has been "solved" by the way plants evolved. Only the few layers of cells nearer the outside are alive, while those nearer the center are in a nonactive state—dead. Most of these inactive cells remain structurally intact and furnish the additional support needed each year.

The *woody stem* has three basic parts, each composed as shown.

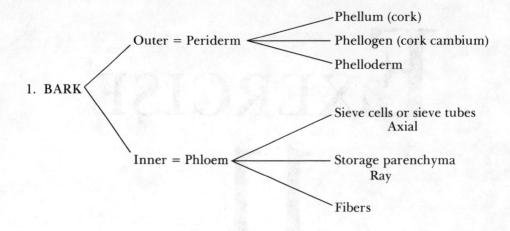

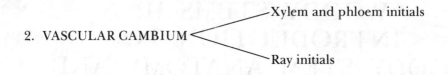

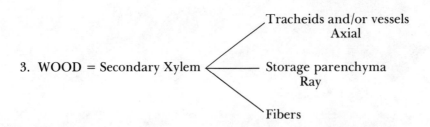

The main functions of these parts are *conduction*, *storage*, *support*, *growth*, and *protection*, each as indicated:

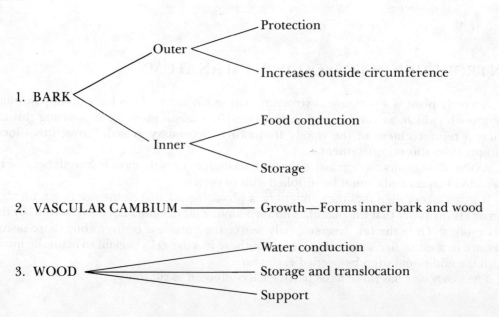

In the stem, most cells are elongate rather than cube-shaped and are oriented in one of two ways, transverse or longitudinal. Longitudinally oriented cells are those oriented parallel to the long axis of the stem. They make up the AXIAL SYSTEMS which form the bulk of the stem. Cells oriented crosswise to the long axis of the stem make up the RAY SYSTEM.

Longitudinal Axial System Cells	Transverse Ray System Cells
1. Tracheids 2. Vessel members (form tubes called vessels) 3. Sieve cells 4. Sieve members (form tubes called sieve tubes) 5. Fibers 6. Axial xylem parenchyma 7. Axial phloem parenchyma 8. Xylem and phloem initials of the vascular cambium 9. Phellogen 10. Phellum 11. Phelloderm	1. Xylem ray parenchyma 2. Phloem ray parenchyma 3. Ray initials of the vascular cambium

II. GYMNOSPERM STEM

Gymnosperms [pines (*Pinus* sp.); hemlocks (*Tsuga* sp.); spruce (*Picea* sp.); and the like] are ancient plants. Internally their stems are similar to those of the woody angiosperms (flowering plants), but they are generally much simpler and more homogeneous. They have no vessels nor sieve tubes, relatively little parenchyma, and many lack fibers. For this reason you will find it easier to begin your study of woody stems with a gymnosperm.

A. XYLEM CELL TYPES IN A GYMNOSPERM

► ACTIVITY 1   **Examine** a prepared slide of macerated wood of some gymnosperm such as pine (*Pinus* sp.). This will enable you to see the general shape of the water-conducting TRACHEIDS and their side walls; you will also see the RAY PARENCHYMA cells.
     Locate a TRACHEID. A tracheid is an elongate, slender, needle-like cell with long, tapering end walls. It is a dead cell, lacks a protoplast and has reinforced walls. **Note** the circular PITS on the (radial) side walls. Pits are permeable cavities in the wall. In the plant, tracheid cells overlap each other, with the tapered end walls of one cell touching the radial wall of another. Water passes from tracheid to tracheid through pores in the pits. This zigzag routing of water means that the water moves slowly, compared to the rates typical of angiosperms in which water moves through large tube-like vessels.
     Find a RAY PARENCHYMA CELL. This is a parenchyma cell with a living protoplast that stores foods. In the plant, the ray parenchyma cells border the tracheids at right angles and connect with them through pits. They are much shorter cells than tracheids, their length being only one to three times the width of a tracheid. Their end walls are not tapered.

B. CROSS SECTION OF THE WOODY STEM OF A GYMNOSPERM

► ACTIVITY 2   **Examine** a prepared slide showing a cross section of a three-year-old (or older) stem of pine (*Pinus* sp.). Don't get "lost" in the details of the anatomy. **Refer** to Figure 11.1 as a guide and **label** the parts as you proceed. The *main* point of this study is for you to see that the arrangement of the axial and radial systems is an efficient design. It integrates five different functions, *support*, *conduction*, *storage*, *protection*, and *growth*.

The Wood

In ordinary use the term "wood" means any hard part of a tree or shrub. But botanically, WOOD is SECONDARY XYLEM only.

the water-conducting function of the wood   Water is conducted in the axial tracheid system. **Focus** on the central portion of the stem where you can easily recognize concentric rings of very orderly arranged cells. These cells are the TRACHEIDS which are more rectangular than round in cross section. The rings are known as ANNUAL RINGS or GROWTH RINGS, and each represents one year's formation of xylem. Each year, the vascular cambium adds a new ring around that of the previous year. **Note** the gradation in size of the cells in any one ring. Tracheids at the inner margin of each ring are much larger than those at its outer edge. The larger tracheids were formed early in the growing season and are known as EARLYWOOD or SPRINGWOOD. The smaller tracheids, formed later in the growing season, are known as LATEWOOD or SUMMERWOOD. The sharp contrast between the last formed latewood cells of one growing season and the first formed earlywood cells of the following season delineates the boundary of a growth ring. How old is your section of wood? _____. How many growth rings are there? _____. Are all these rings secondary wood? _____. Explain: _____ _____. What is the functional advantage to the plant of the larger sized cells in the earlywood? _____ _____

Most of the water moves up the plant in the outermost ring, that is, the current season's tracheids. The more recent of the previous years' annual rings contain water, but the plant does not rely on them for its current needs. Do any of the tracheids have a protoplast?   YES   NO. **Label** a growth ring in Figure 11.1. **Label** detail diagram **C** of Figure 11.1, using the capitalized terms defined in the preceding description.

the support function of the wood   **Switch** to high power and examine the walls of the tracheids. **Note** that the walls are relatively thick. These are the SECONDARY WALLS which give the cells a rigid reinforcement. They are also encrusted with a material known as lignin which makes them hard and dense. Which has the thicker walls compared to the overall cell size, (a) latewood or (b) earlywood? (Circle one.) Which do you suppose has the greater strength, (a) latewood or (b) earlywood? (Circle one.) Tracheids thus perform two functions: (a) _____ and (b) _____. **Label** the secondary wall of a tracheid in detail diagram **B** or **C** of Figure 11.1.

the storage function of the wood   **Note** the single-file rows of thin-walled cells that radiate intermittently across the growth rings somewhat like the spokes of a wheel. These radial rows are called RAYS. Each ray is a sheet of parenchyma cells that extends vertically in the stem. Ray cells store sugars, fats, and other foods. In the spring these stored materials are transferred to the actively dividing cambium or to the apical meristems.

Note also that many of these rays extend into the phloem which lies outside the xylem. Rays in the phloem have the same structure and function as those in the xylem. As the tree enlarges, new rays are constantly formed by the cambium so that the density of rays is maintained approximately the same. **Label** the xylem ray in detail diagram **C** of Figure 11.1.

other features of the wood—pith and resin canals   **Focus** on the exact center of the stem within the innermost growth ring. This central core is the PITH. Originally it contains parenchyma cells, and these may or may not still be present in the stem section you are examining. The life span of the pith cells varies with the species. With age, pith cells commonly accumulate tannins and crystals (waste products). In many plants, the pith becomes hollow or chambered. **Label** the pith cells in detail diagram **D** of Figure 11.1.

Switch to low power and **observe** the large cavities scattered throughout the stem. These are the RESIN CANALS or RESIN DUCTS (not found in all gymnosperms). They extend lengthwise in the stem. **Switch** to high power and **note** that the cavities are lined with parenchyma cells. The parenchyma cells secrete a resinous substance which in pines is known as oleoresin. Its usefulness to the plant is unknown. It may possibly repel or attract insects. Turpentine and rosin are manufactured from oleoresin and are still used in some high-quality paints, though they have been largely replaced by synthetics. **Label** the resin duct in detail diagram **B** of Figure 11.1.

The Vascular Cambium—Growth Zone of the Stem

The woody part of a stem grows only in girth. The VASCULAR CAMBIUM increases the girth of the stem by forming layer after layer of xylem and phloem around the original primary cylinder. In the temperate zones of the world cambial growth is cyclic with periods of activity alternating with periods of relative rest. Cambial activity is under the control of hormones which reactivate it each spring. Some of these hormones are known to come from the developing buds and leaves.

Focus on the latewood of the outermost growth ring in the xylem. Immediately outside this latewood you will find a band of flattened looking cells. This is the VASCULAR CAMBIAL ZONE, a band of unexpanded meristematic cells. It is about four to eight cells wide. It consists of one VASCULAR CAMBIUM layer of self-perpetuating cells, called CAMBIAL INITIALS, flanked by layers of dividing daughter cells. Daughter cells on the outside of the cambial layer are called PHLOEM INITIALS and develop into phloem cells; those on the inside of the cambial layer are called XYLEM INITIALS and develop into xylem cells. **Label** the vascular cambial zone in detail diagram **A** of Figure 11.1.

The Bark

BARK is defined as all those tissues that lie outside of the vascular cambium. It consists of three tissues, phloem, cortex (which may or may not be present), and the periderm (present only in older stems). The phloem is considered to be the INNER BARK and the periderm the OUTER BARK.

the food-conducting function of the inner bark (the phloem)   **Focus** on the region just outside the vascular cambial zone. Here are angular, thin-walled cells in somewhat orderly rows. Scattered among them are larger, round cells filled with dark-colored material. Cells farther out are flattened and in a zigzag pattern. If you have located tissue fitting these descriptions you have found the phloem. **Locate** the following phloem cell types.

SIEVE CELLS are vertically elongate cells which in this cross section are the angular, thin-walled cells in orderly rows. Sieve cells conduct photosynthates to actively growing buds and young leaves or to roots or storage rays. Sieve cells are named for their numerous SIEVE AREAS, the porous zones in their walls. Strands of protoplasm extending through these pores connect functioning adjacent sieve cells and ray parenchyma. Food tranfport in sieve cells is thought to occur in the form of a flowing solution of foods within the protoplasm. The protoplasm itself doesn't flow. Exactly how this is done is still unresolved.

Observe the outer regions of the phloem. These outer, older layers may still be alive but are not all actively functioning. Many of the sieve cells are crushed and the rays are distorted into a wavy, accordian-pleated shape. Normally only the innermost (youngest) layer of phloem is functional. The outer, older layers are incorporated into the outer bark, which may eventually be sloughed off. **Label** the functional phloem in detail diagram **A** of Figure 11.1.

the storage function of the inner bark (axial and ray parenchyma)   **Locate** the round, thin-walled cells filled with dark-colored material. These are the AXIAL PHLOEM PARENCHYMA CELLS. When alive, these cells store foods transferred to them from adjacent or nearby sieve cells or other parenchyma. When foods are mobilized for growth, these foods are recirculated back through sieve cells or through ray parenchyma to the cambium and developing buds and leaves. **Label** an axial phloem parenchyma cell in detail diagram **A** of Figure 11.1.

Focus on the inner (youngest) region of phloem and find the large, radially elongated cells with prominent nuclei. The length of one of these cells spans about five of the adjacent sieve cells seen in cross section. These are the RAY PARENCHYMA CELLS. They are lined up in a transverse

row, called a RAY. Can you identify any ray in the outer crushed, nonfunctioning phloem?
YES   NO. Crushed rays give the old phloem a zigzag pattern. Phloem rays are similar in form
and function to xylem rays.

the cortex   The cortex is primary tissue, produced by the apical meristem. In a young, non-woody stem, one of the functions of the cortex is support, but as the plant ages, this function is accomplished by the wood produced by secondary growth. Certain of its cells may go through a dedifferentiation (lose their distinctive shape) and become meristematic again. These meristematic cells become organized into the CORK CAMBIUM (PHELLOGEN). As the tree gets even older, however, it is commonly old phloem cells that become cork cambium.

the growth and protective function of the outer bark—the periderm   The CORK CAMBIUM produces CORK CELLS (PHELLEM) to the outside of the stem and thick-walled parenchyma cells, PHELLODERM to the inside. These three tissues, cork cambium, cork, and phelloderm, comprise the OUTER BARK—the PERIDERM. You may not be able to identify the cork cambium as easily as the vascular system. Mature cork cells are dead and approximately square in outline with thick walls. They are usually arranged compactly with no intercellular spaces. What advantage is this? _____. Cork cell walls are impregnated with suberin, a fatty material that makes them impervious to water. Cork also has thermal insulating qualities. If your pine stem section came from a stem only three years old, the amount of periderm will be negligible.

EXERCISE 11   STUDENT NAME _____

QUESTIONS

1. Latewood is usually stronger than earlywood. What material is abundant in latewood that makes this

   wood so strong? _____

2. Leaves of gymnosperms are designed to limit or impede water loss. This leaf design more or less com-

   pensates for what internal feature of the wood? _____

3. Gymnosperms growing in exposed sites need some kind of protection during winter because a severe
   wind can kill many of their leaves that turn a brown color, described as winter burn. Explain why the

   leaves are killed by the wind. _____

4. Explain how the woody plant body design can allow for unlimited growth with no physiological limita-

   tions. _____

   _____

5. Do any animals have comparable unlimited growth? _____

6. The annual, new leafy framework of a woody plant is the result of primary or secondary growth?

   _____

7. Which tissue is vital for the survival of a woody plant? _____

8. Define bark. _____

9. Define a growth ring. _____

   _____

10. Why isn't there an accumulation of phloem growth rings as there is for xylem? _____

    _____

11. In gymnosperm wood, the main support comes from what cell type? _____

12. What is (are) the function(s) of the ray system of woody plants? _____

    _____

13. The diameter of the woody cylinder is continually increasing. How is it that there is always enough bark

    to cover the outside? _____

14. If you carved your initials in the bark of a tree when you were ten years old, would these initials be

higher up on the trunk when you reach 16 years? _____ Explain your answer.

_____   _____

15. Which system supplies nutrients to the actively growing cambium: the sieve tubes or the ray system?

_____

16. Is bud opening and renewed cambial activity correlated? _____ How? _____

_____

17. Name two tissues of primary growth that are expendable in the mature tree.

(a) _____; (b) _____

EXERCISE 11   **STUDENT NAME** _____

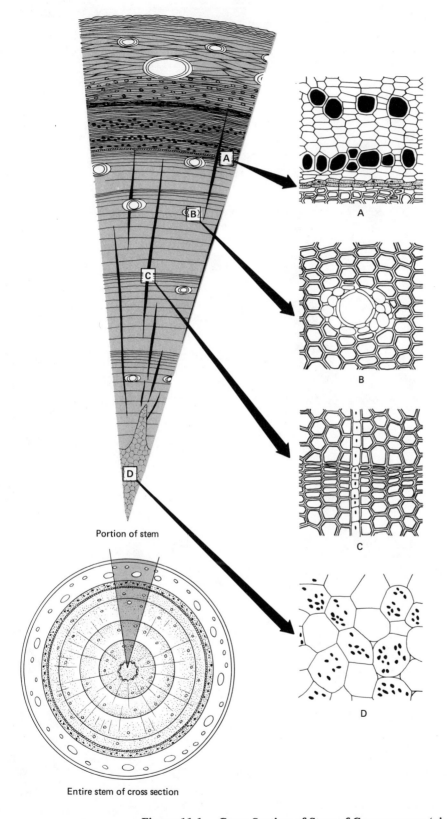

Portion of stem

Entire stem of cross section

Figure 11.1   Cross Section of Stem of Gymnosperm (pine)

Exercise
12

WOODY STEMS III:
ANATOMY OF THE ADVANCED
WOODY STEM—THE ANGIOSPERM STEM

The general organization of a woody angiosperm stem is essentially the same as that of a gymnosperm. The main difference between them is that angiosperms have many more specialized cells, particularly those cells concerned with water and food conduction and with support. In gymnosperm wood, tracheids provide much of the support, in addition to conducting water. In angiosperm wood, more fibers are present for support and, besides tracheids, there are vessel members (cells) that are specialized for rapid water flow. The sieve tube cells of angiosperms can move foods faster than the sieve cells of gymnosperms.

With these improved cell types, angiosperms (both woody and herbaceous) have become adapted to a wider variety of habitats than gymnosperms. They tend to grow faster than gymnosperms, and they can also afford to have broader leaves whose photosynthetic output exceeds that of the smaller, needle-like leaf of the gymnosperm. Even though this broader leaf is more prone to evaporative water loss, it can be quickly replenished by the rapid water delivery vessel system in the stem. In studying the stem of a woody angiosperm, therefore, we mainly want you to see the cell specializations which have helped to make the angiosperms such a successful group of plants.

I. XYLEM CELL TYPES

➤ ACTIVITY 1   **Examine** a prepared slide of macerated angiosperm wood so you can see and compare individual cells in their entirety rather than as segments in a cut section.
   Look for four cell types.

   1. WOOD FIBERS (sclerenchyma cells) are very long, slender cells with tapering ends. They are support cells. Their walls are thick with many pits, so small, however, that little water passes through them. Fibers are very long cells, and most are broken in parts here.
   2. VESSEL ELEMENTS are shorter, wider, barrel-like cells with thin side walls that have many pits. The end walls have one large circular pit or a few smaller, more rectangular

shaped pits. In intact wood, vessel elements are connected end to end, forming a long and wide tube, the VESSEL. A vessel is capable of rapid conduction of water.

3. TRACHEIDS are long, narrow cells with tapered ends. They resemble wood fibers somewhat but are not as long and, therefore, not as easily broken during the process of macerating the wood. If you find a cell with both its tapered ends intact, it is probably a tracheid and not a wood fiber. Tracheids conduct water, but not as rapidly as a vessel.

4. XYLEM RAY PARENCHYMA CELLS are short cells with straight-sided end walls. They have a granular appearance due to the presence of cytoplasm. **Note** the pits in their side walls. What is the function of the ray parenchyma? _____

II. WOOD ANATOMY

► ACTIVITY 2   **Examine** a slide of oak (*Quercus* sp.) wood showing three sections of cut: cross, radial, and tangential.

Focus on the CROSS SECTION which is recognizable because it consists of portions of growth rings. **Note** the extremely large, circular, or (in some cases) angular cells. These are the VESSEL MEMBERS, cells "streamlined" for water transport, with wide diameters for high volume flow and without cytoplasm or end walls for fast, unimpeded flow of water. Vessel members are stacked up one over the other in longitudinal series and joined end to end to form a water tube, the VESSEL. Transport velocities in vessels are rapid, several meters per hour—much faster than in a tracheid system in which water moves in a zigzag route through small pores of narrow tracheids that border each other but are not joined to form distinct tubular units.

Note the appearance of any one growth ring. As in pine, the inner part of each ring is the EARLYWOOD, the outer part is the LATEWOOD. In angiosperms, earlywood and latewood are not as easily distinguished as in most gymnosperms. In some species, vessels are mostly in the earlywood, and fibers make up most of the latewood. This arrangement is called RING POROUS WOOD. In other species, vessels occur throughout the entire width of the ring. This is called DIFFUSE POROUS WOOD.

Focus on the smaller cells between the larger vessels. These are the WOOD FIBERS or TRACHEIDS. Fibers have thick walls and are very small in diameter. Cells of the same small diameter but with thinner walls are tracheids. Fibers are largely responsible for the hardness and strength of angiosperm wood. It's these cells that give us the hardness needed for baseball bats, hockey sticks, skis, and handles for tools. There are angiosperms, however, with very few fibers in their wood which is, therefore, a soft wood. Balsa wood of the tropical *Ochroma* tree is a good example.

Switch to higher power (if necessary) and find the narrow rows of parenchyma cells that radiate across the growth rings. These are the XYLEM RAYS. A few cells of the XYLEM RAY PAREN-CHYMA system may also be visible in this section. All ray cells are parenchymatous and serve for storage.

Label the vessels, wood fibers, xylem rays, and tracheids in the cross section shown in Figure 12.1. Also **label** the figure for the type of section it represents.

Move the slide so that the RADIAL SECTION is in view. This section is represented in Figure 12.2. The stem has been cut lengthwise and along a radius. Here the VESSELS appear as wide, vertically elongate cavities. In the intact stem they are much longer, but in this section you see only portions of them. The very narrow, columnar cells with tapered ends are the WOOD FIB-ERS. **Note** how densely packed the fibers are and how they compose the bulk of the wood. Some angiosperms like basswood (*Tilia* sp.) and willow (*Salix* sp.) have a large number of vessels per unit volume of wood with fewer fibers than you see here in oak. Wood from such trees is much softer than oak. Although the U.S. Department of Agriculture classifies angiosperm wood as "hard woods" and gymnosperm wood as "softwoods," many angiosperm woods are actually as soft as, or softer than, gymnosperm wood.

Look for bands of cells running crosswise to the fibers and vessels. These are the XYLEM RAYS. **Label** the vessels, wood fibers, and xylem rays in the radial section shown in Figure 12.2. Also **label** the figure for the type of section it represents.

Move the slide so that the TANGENTIAL SECTION is in view. In this section the stem has been cut lengthwise in a plane perpendicular to its radius. VESSELS and WOOD FIBERS look the same here as they do in the radial section, but the XYLEM RAYS appear as lens-shaped clusters of cells. Some are wide massive clusters, some are narrow slender clusters only one or a few cells wide. Since rays run horizontally in the stem along the various radii, in this vertical tangential cut, you are seeing them cut across their long axis in cross-sectional view.

Label the vessels, wood fibers, and xylem rays in the tangential section shown in Figure 12.3. Also **label** the figure for the type of section it represents.

III. CROSS SECTION OF A WOODY ANGIOSPERM STEM

A. THE PHLOEM

➤ ACTIVITY 3   **Examine** a slide showing the cross section of some woody angiosperm stem such as basswood (*Tilia* sp.). *Tilia* is not representative of all woody angiosperms. **Find** the PHLOEM which is the darker stained region near the outer part of the stem. It is made up of triangular-shaped sections. Those triangular sections that "point" inward are the PHLOEM RAYS. Other RAYS, which are only a few cells wide, form narrow rows of parenchyma cells as they cross the phloem region in general. **Note** that all the rays in the phloem are continuous with rays in the xylem (which is the homogeneous-looking central core with the growth rings).

Wedged between the inward-pointing phloem rays are other triangular sections that "point" outward. In these sections are bands of heavily stained cells with very thick walls and pin-point-like lumens (the central part of the cell bounded by the walls). These heavily stained cells are the PHLOEM FIBERS. Alternating with these bands of phloem fibers are narrower bands of larger cells with thinner walls. These large, angular cells with very little cytoplasm and with thin walls are the SIEVE TUBE MEMBERS. They are wide, columnar cells connected end to end in a longitudinal series to form SIEVE TUBES. The end walls of sieve tube members are not completely open as in vessel members but have large porous end areas, called SIEVE PLATES. Sieve plates are not commonly seen on these slides. Any one of the several sieve tube members that make up a sieve tube can deliver foods at a rate of some five times its own volume per second. Hence the flow through the entire tube is very fast.

Find the COMPANION CELLS. These are smaller cells usually found at the corners of sieve tube members, hence the name "companion cell." They have thin walls, considerable cytoplasm, and a nucleus that occupies most of the cell. The nuclei of companion cells are thought to exert some control over the normal functioning of the sieve tube members which have lost their nuclei in the process of cell maturation. The absence of a nucleus in such a physiologically active cell as the sieve tube member is inexplicable.

Label the phloem tissue and all its specialized cells in detail diagram **B** of Figure 12.4.

B. THE XYLEM

The center of the cross section is occupied by a pith and three or more concentric layers of xylem. EARLYWOOD and LATEWOOD are easily recognized. A VASCULAR CAMBIAL ZONE separates the XYLEM and PHLOEM. Basswood is a softwooded angiosperm because its xylem has many more vessels than fibers. The structural differences between the various xylem cells are not readily apparent in *Tilia*. It is, therefore, not as good an example as oak for an introductory study of angiosperm wood (xylem tissue). Hence we do not ask that you study the basswood xylem in detail. Simply note that, as in all woody dicots and gymnosperm of the temperate zone, the xylem is in concentric growth rings and RAY PARENCHYMA radiate across them. **Label** detail diagrams **C**, **D**, **E**, and **F** of Figure 12.4, using the capitalized terms defined in the preceding description.

C. THE CORTEX AND PERIDERM OF THE WOODY ANGIOSPERM STEM

Examine the outermost portion of a three-year-old *Tilia* stem.

Cortex

The CORTEX is the zone of loosely arranged, relatively large parenchyma cells found just external to the outermost band of phloem fibers. Depending on how old the stem is, the cortex may have been replaced by the periderm. Is the cortex a primary or a secondary tissue? (Circle one).

Periderm

The PERIDERM is the dense band of cells just under the stem surface. All the periderm cells are rectangular, with their long axes paralleling the circumference of the stem. From the phloem outward, the PERIDERM consists of (a) four to six layers of PHELLODERM cells that are thick-walled and filled with cytoplasm; the phelloderm is a secondary cortex formed by the cork cambium; (b) one to two layers of thin-walled, nucleated cells filled with cytoplasm and making up the CORK CAMBIAL ZONE; and (c) three to four scale-like layers of dead CORK CELLS. **Label** detail diagram **A** of Figure 12.4, using the capitalized terms defined in the preceding description.

► ACTIVITY 4   **Complete** Table 12.1.

EXERCISE 12   STUDENT NAME _____

QUESTIONS

1. In angiosperm wood, the main support comes from what cell type? _____

2. In gymnosperm wood, the main support comes from what cell type? _____

3. What are the water tubes of angiosperms known as? _____

4. What is the name of the cell type that makes up this water tube? _____

5. What is the name given to the food-conducting tubes of angiosperms? _____

6. What is the name of the cell type that makes up this food-conducting tube? _____

7. It is conceded that the water- and food-conducting system of angiosperms is more efficient than that of

   gymnosperms. What advantages does this give the angiosperm over a gymnosperm? _____

   _____

8. What are the two structural features of the vessel member that make it a superior water-conducting cell to

   the tracheid? (a) _____; (b) _____

9. Angiosperms have relatively thin flat leaves. Gymnosperm leaves have thick cuticles and are prism-

   shaped so as to reduce water losses. How do you account for this difference? Explain in terms of their

   wood. _____

   _____

Table 12.1   Comparison of the Woody Stem of Angiosperms and Gymnosperms

Feature	Gymnosperm Wood	Angiosperm Wood
Comparative makeup of the wood (simple or complex)		
Chief water-conducting cell		
Comparative diameters of the water-conducting cells		
Comparative transport velocities of these water-conducting cells		
Number of different functions performed by the water-conducting cell		
Chief food-conducting cell		
Comparative diameters of these food-conducting cells		
Cell type for support	tracheids only	tracheids & vessels
Comparative growth rates		

EXERCISE 12   **STUDENT NAME** _____

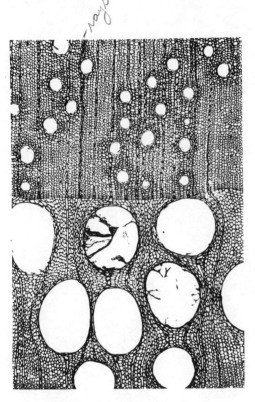

Figure 12.1   Cross Section

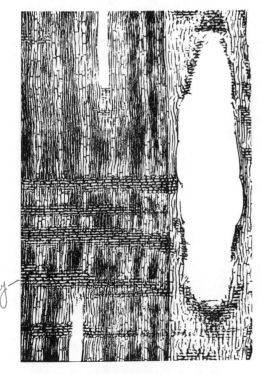

Figure 12.2   Radial Section

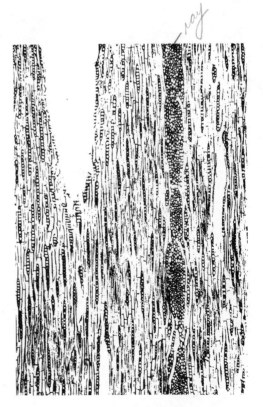

Figure 12.3   Tangential Section

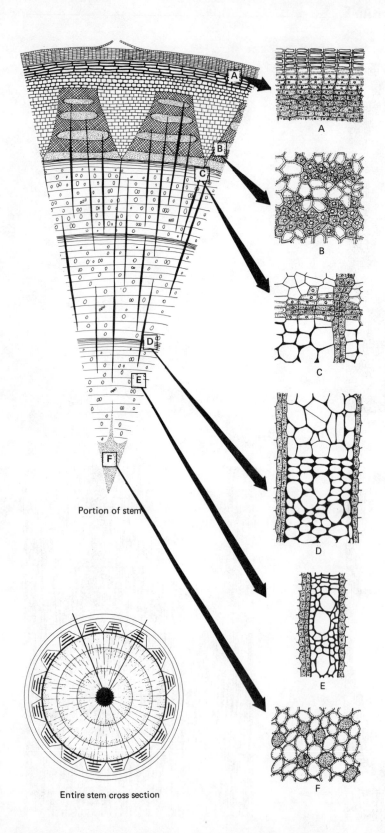

Portion of stem

Entire stem cross section

Figure 12.4   Cross Section of Stem of Woody Angiosperm (*Tilia* **sp.**)

EXERCISE 13

WOODY STEMS IV:
WOOD GRAIN AND TREE-RING DATING

In preceding exercises you have studied the cellular detail of wood. The present exercise should be more fun. You'll look at wood more in the way you might do on a day-to-day basis. When you choose your plywood paneling, when you buy a high-quality wood table, when you select anything made of wood, you pay a pretty stiff price. What you are mainly paying for is the particular grain of the wood, the species, and, of course, the labor costs.

Knowledge of wood grain is useful information to you. The physical visible properties of wood have always been admired by people. And since the 1950s scientists have been using the growth rings of trees to discover the weather of the past. This is known as the science of dendrochronology, or tree-ring dating.

I. GROWTH RINGS AND VASCULAR RAYS

➤ ACTIVITY 1   **Examine** blocks of wood or twigs or limbs cut across the long axis of the stem. Each light ring you see is earlywood whose cells are large; each dark ring is latewood whose cells are small. One pair of dark- and light-colored rings represents one annual ring (or growth ring). In any one growth ring, which is outermost, earlywood or latewood? (Circle one.) Was the amount of growth the same each year? YES NO. What is the indicator of the amount of growth? _____. What is the tissue that makes up all these rings? _____. Is there any phloem in these rings? _____. What are the lines of cells running across the growth rings like the spokes of a wheel? _____

Label a growth ring in Figure 13.1 on each of the three planes of cut shown (transverse, radial, and tangential).

II. HEARTWOOD AND SAPWOOD

At the very outer edge of your wood section is a dark brown, rather crusty zone, the bark, which you'll study later. We simply point it out here to help direct you where to look. **Concentrate** on looking at the whole general area of growth rings inside the bark. Can you see any color differences in this zone?   YES   NO. If you have a good sample, the more central of the growth rings will be a darker color. This dark-colored central region is the HEARTWOOD. Heartwood no longer contains water or living cells. Various compounds and coloring materials have infiltrated it and protect it from internal decay. As long as no air gets to it, it is practically indestructible. Although dead, it is structurally sound and the backbone of the tree. When used for lumber, it is more durable than SAPWOOD.

SAPWOOD is the lighter zone of growth rings. It lies between the heartwood and the bark. The relative widths of sapwood and heartwood vary in different plants. Sapwood is not necessarily a narrower zone than heartwood. In sapwood, some of the storage cells are still alive and some of the vessels and tracheids still contain water. But most of the water moving up the tree is conducted only in the outermost growth ring of the sapwood.

In the standing tree, sapwood is more resistant to decay than heartwood. When trees are pruned, the heartwood is exposed to air and is the first to succumb to fungus or insect invasion. This is the reason why you should paint cut surfaces of large limbs whenever you prune a tree. Any asphalt-base paint can be used. Don't use creosote; it kills tissues. **Label** the heartwood and sapwood in Figure 13.1.

III. THE BARK

BARK is the brown outermost zone of your specimen. **Look** carefully. Can you see a crack separating the bark from the sapwood?   YES   NO. This crack marks the position of the VASCULAR CAMBIUM. All the soft, delicate vascular cambial cells have died and disappeared since this wood was cut.

BARK is defined as all the tissues that lie outside the vascular cambium. It is made up of an INNER BARK and an OUTER BARK. The OUTER BARK is the outermost protective corky material.

INNER BARK is all the phloem which the plant has made all its life. Without a microscope you can't tell the difference between the inner and outer bark. Let's say that about half the thickness of the bark represents phloem. **Look** how thin the phloem is. It is negligible compared to the amount of xylem. Some of the difference in thickness between xylem and phloem exists because the xylem is hard and uncompressed, whereas the phloem is soft and gets squashed by the expanding xylem.

There's definite engineering "sense" to this. What is the function of all the old xylem, (a) support; (b) water conduction? (Circle one.) Xylem cell walls have lots of cellulose and lignin. Phloem cell walls do not. Is there any point in saving phloem tissue for future structural use?   YES   NO. **Label** the bark in Figure 13.1.

IV. THE FIGURE OF WOOD—THE "GRAIN"

The figure (grain) of many kinds of wood is economically valuable because of its beauty. The particular grain is determined by the plane of cut of the tree trunk. In different planes of cut, the xylem rays, the fibers, and the growth rings present very different and identifiable patterns.

▶ ACTIVITY 2   **Refer** to Figures 13.2, 13.3, and 13.4 as a guide and **examine** different cuts of wood. Wood sawed with a tangential cut is one in which the plane of the cut is at right angles to the radius of the log (see Figure 13.2). The plane of this cut intersects the growth rings in such a way that they appear in patterns that look like parabolas or portions of parabolas. Wood sawed with a radial cut is one in which the plane of the cut is along a radius of the log (see Figure 13.3). Obviously you can't make too many radial cuts from a log. So, to get the most lumber out of a log, sawmills mostly cut logs tangentially. It is never cut transversely as shown in Figure 13.4.

Label the growth rings and the vascular rays in Figures 13.2–13.4. Also **label** each figure with the name of the plane of cut that it represents.

➤ ACTIVITY 3   **Examine** "Hough" slides, which are very thinly cut sections of wood, and identify the planes of cut.

➤ ACTIVITY 4   **Examine** pieces of plywood. Plywood is made from several layers of thinly sliced wood (veneer) which are glued together. The grain of the wood in alternate veneer layers runs at right angles. This gives the plywood much greater strength than a single piece of wood of the same thickness. Because of the way it is cut, common fir or pine plywood always shows a tangential cut grain pattern. More expensive paneling may have the surface layer made from a hardwood such as birch or pecan. The pattern of this layer may be either tangential or radial. Cutting timber into plywood is a much more efficient use of the total log.

V. DATING THE PAST WITH TREE RINGS

Tree-ring dating is based on the general principle that a living tree forms a new layer or ring of wood each year. Rings made in wet years are wider than average; those made in dry years are narrower than average. The year-by-year sequence of varying ring widths forms a unique nonrepeating pattern. These rings are visible on pencil-thin cores of wood bored out of a tree trunk by means of a special instrument.

The same ring pattern occurs in trees growing within a few hundred kilometers of each other. Scientists have learned how to use ring patterns to tell climate as far back as 8400 years ago. Basically, the way it is done is as follows. First the scientist determines the dates of the rings on a wood core taken from a relatively young tree as, for example, **A** in Figure 13.5. Some of its earliest formed growth rings will match some of the later formed rings of an older tree, such as **B** in Figure 13.5. In turn, the earliest rings of **B** can be matched with the later rings of the older wood of sample **C**, and so on. Thus it becomes possible to put dates on older and older rings of older wood. The oldest woods that have thus been dated give us direct information on the past climate of a given area. We can see cycles of wet and dry years. By knowing the history of past cycles of precipitation, we can better analyze trends of change in our current climate. This is expecially important for agriculture and watershed management.

Tree-ring dating or dendrochronology has proved to be the only tool that gives precise year-to-year history of past climate. It is more accurate than radiocarbon dating.

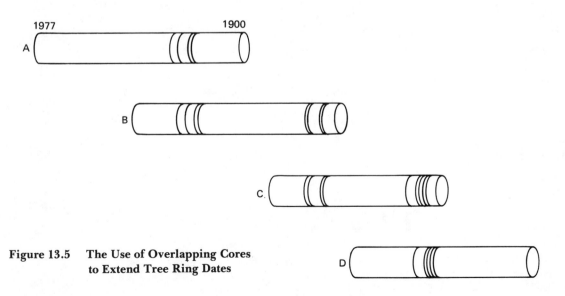

Figure 13.5   The Use of Overlapping Cores to Extend Tree Ring Dates

➤ ACTIVITY 5   **Try** your hand at tree-ring dating. **Examine** cores which have been taken from trees of the same species growing in similar environments and preferably as near each other as possible. You will be given a pair of cores. One of the pairs will have a date marked on the mount. This represents the year that the tree was cored or cut down to obtain the sample. **Hold** the sample with the dated end to your right. Which end of the core is the inside of the tree? The left end or the right end? (Circle one.) Which is the outside of the tree, the right or the left end? (Circle one.) Write down the year of the oldest (innermost) growth ring.

Remember that this oldest ring that you can see on the core may not represent the first year in which the tree was alive. Why not? Look at Figure 13.6. This shows that cores taken at levels **A**, **B**, and **C** will not give the true age of the tree, and that if 1956 is the first year of growth, you need to take a core from level **D** (near the ground) to get all the annual rings made by the tree.

Examine the second core of the pair. It will have a code number or other identification to show that it can be matched with the dated sample. Which end, right or left, of this core was nearer the outside of the tree? (Circle one). Which end represents the oldest wood, right or left? (Circle one.) Count and write down the number of years' growth represented by this second core. _____. The second core will probably overlap the first one, that is, it will share some years in common with the first one, but it will also extend further back (or forward) in time.

Can you find the years that are shared in common? The simplest way to do this is to set down the cores alongside each other (being sure you have both cores with their most recent years kept oriented at the right). Slide one to the left or right until some of the ring series seem to match. Can you find such a match this way? With some types of wood this is possible, and is just as easy as indicated. If you have made a match, can you now date the youngest (that is, latest formed) ring of the second sample? What year was it formed? _____. What is the date of the oldest ring of the second sample? _____. Does this core have at either of its two ends rings

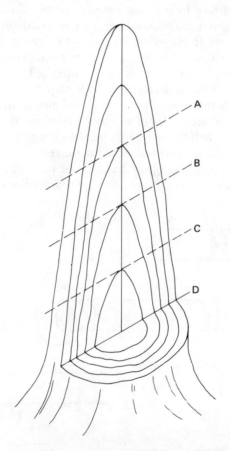

Figure 13.6

that are older than the dated core?   YES   NO. Are there any rings that are younger than those in the dated core?   YES   NO.

If you were unable to get a match in this simple way, you were probably in the majority. Even professionals don't use this method. For really old samples there are just too many rings to try to count. Dendrochronologists use techniques somewhat like the following procedure.

➤ ACTIVITY 6   **Measure** the overall length of the dated core in millimeters _____. **Divide** this figure by the number of rings you have counted. The number is _____. The average ring width, in millimeters, is _____. **Enter** this average as the basic value on the upper graph in Figure 13.7. **Mark** the graph above and below this basic value so that each ring is represented as a dot above or below the average. **Plot** out the value (above or below average) for each year of the dated core. **Draw** a line connecting the dots of each year. **Repeat** this for the undated core of the pair on the lower graph, using a different numerical scale if necessary. Remember to *adjust* values on this second sample so that *its* average is at the midline, and *its* larger and smaller rings are graphed above or below the midline.

Now, **cut** the second graph off the bottom of the page along the marked line. **Slide** this graph, with heavy lines or dots for each year, along the upper graph. The patterns of larger and smaller rings should match up, approximately, at some point. You have then reached the same point as when the cores, themselves, matched. Is the second sample, overall, older or younger than the first? (Circle one.) What years are represented by the second sample? _____ to _____.

It is by techniques similar to these that our knowledge of many past events can be dated precisely. Computers are commonly used for the scanning and matching. Suppose an archeologist or historian wishes to date the year that a village was destroyed by an earthquake. The wood of the beams in the houses, furniture, tool handles, and even firewood may be dated to within one or two years in many cases. The accuracy is far greater than that possible with Carbon-14 dating, to the point that dendrochronology is often used to check the accuracy of refined C^{14} techniques during their development.

EXERCISE 13   STUDENT NAME _____

QUESTIONS

1. Is heartwood physiologically active? _____

2. Boards cut from which type of wood (heartwood or sapwood) need a longer period of drying?

   _____ Explain why: _____

   _____

3. In general which kind of tissue is more resistant to decay, living tissue or dead tissue? _____

4. Which part of a living tree therefore is a more resistant to decay, (a) heartwood or (b) sapwood? Circle

   (a) or (b).

5. Why is lumber cut from heartwood more resistant to decay than lumber cut from sapwood? _____

   _____

6. Define bark. _____

   _____

7. Which is outermost, the earlywood of 1977 or the latewood of 1976? _____

8. In tree ring dating methods, basically one tries to get a match in ring series of the younger part of a tree

   of known age with the older part of a tree of older and unknown dates. Circle one: (a) true; (b) false.

9. Figure 13.8 represents the six faces of a block of wood. Only one transverse (cross) section has had the
   ring and ray pattern drawn in. Cut out the figure along the dotted lines and fold the faces to make a
   six-sided block. Draw the ring and ray patterns as they would appear on each face. Glue or tape the faces
   together along the tabs.

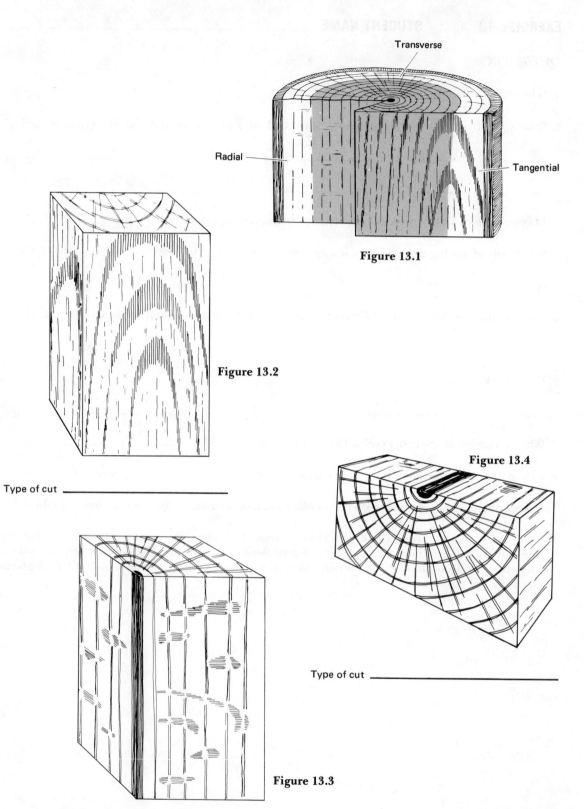

Transverse

Radial

Tangential

Figure 13.1

Figure 13.2

Type of cut _____

Figure 13.4

Type of cut _____

Figure 13.3

Type of cut _____

EXERCISE 13   STUDENT NAME _____

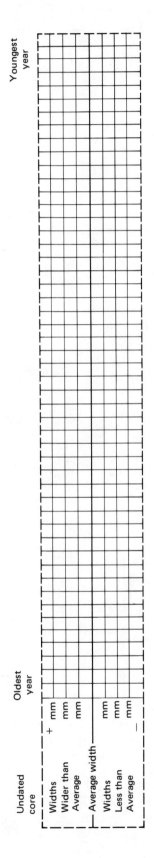

Figure 13.7

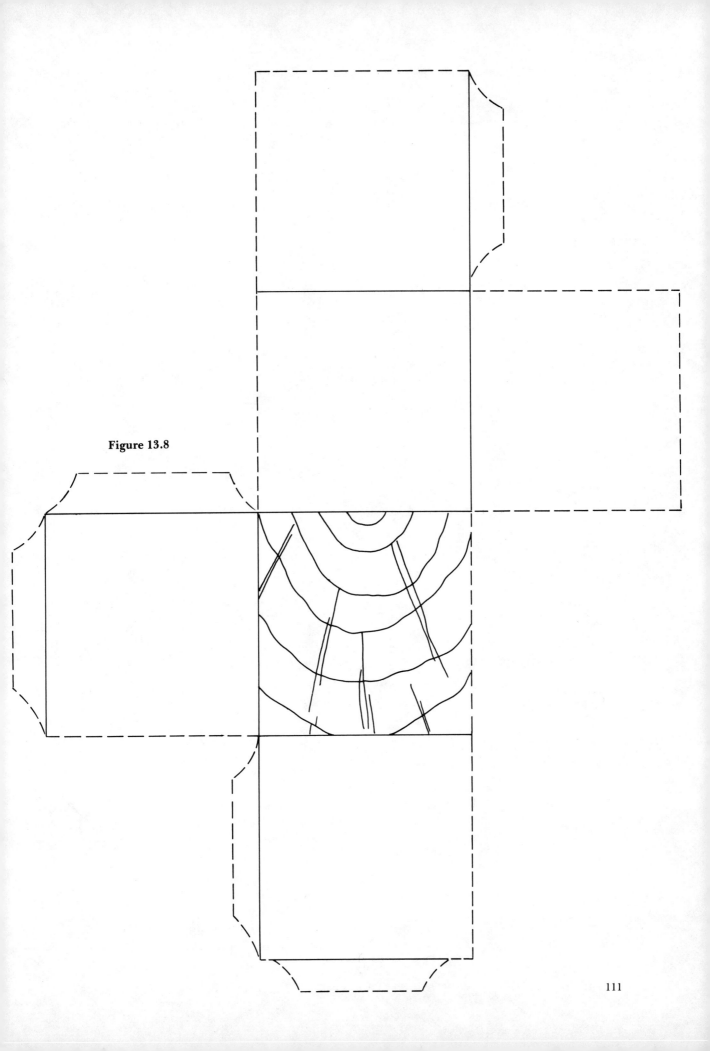

Figure 13.8

Exercise 14

LEAVES I:
EXTERNAL FEATURES

The size and shape of a leaf is not completely arbitrary. In many cases (though not all) it is an adaptation to habitat. Just as many desert animals have big ears which dissipate heat, whereas their arctic cousins have small ones, so with many plants differences in leaf size and/or shape are adaptations to temperature and moisture stresses. This is best seen in plants living in hot dry, cold dry, or warm wet climates. In temperate moist regions there seems to be little correlation with climate, and no one leaf type is typical.

Plants are immobile and cannot avoid drastic rise and fall in air temperature and soil moisture. Plant parts most vulnerable to such changes are the leaves, flowers, and buds. In regions or in periods of intense sun heat, leaf temperature could soar drastically. But it turns out that for many plants the size and shape of their leaves help prevent this temperature jump. The leaf can also prevent a rapid heat loss in cold periods.

One way that leaf temperature is lowered is by evaporative cooling, that is, the loss of water from the leaf, called transpiration. But in dry habitats, cooling by transpiration would waste the plant's water. Hence in such places, leaf temperature is lowered in another way—by losing heat directly to the air. The leaf can also gain heat from the air.

This exchange of heat with the air is by a process called convection. It works like this. When a leaf and the air are both warmed by the sun, the air always heats up more slowly and is thus always a few degrees cooler than the leaf. Excessive rise in leaf temperature is reduced by losing heat to the air. Conversely, when a leaf and air both lose heat (as at night), the air cools down more slowly and is thus always a few degrees warmer than the leaf. An excessive drop in leaf temperature can be prevented by gaining heat from the air.

Convection flows are effective only near the edge (margin) of the leaf. Therefore, the greater the extent of the margin, as compared to the leaf area, the greater the effect of convection heat exchange. For this reason, the more irregular the margin, or the smaller and narrower the leaf, the more easily such heat exchange occurs. Thus plants living in deserts or droughty regions commonly have leaves that are narrow, small, or highly lobed or divided into smaller parts or leaflets.

The bigger and broader the leaf, the less readily does convection heat exchange occur. A big broad leaf is more suited to evaporative cooling. Thus in the shady understory of wet tropical forests we find plants with the biggest leaves in the world. Typically they are not lobed or divided, and their margins are even.

A third way to control leaf temperature is seen in cacti and succulents. They store large amounts of water. The water absorbs most of the sun's heat instead of the plant tissues.

People use leaf size and shape as a means of identifying many different plant species. Leaves, along with flowers and buds, are referred to as key characters. Books called "keys" use these key characters arranged in a certain step-by-step sequence which leads a person to the identification of a particular plant.

I. LEAVES OF PLANTS NATIVE TO HOT DRY AND/OR COLD DRY REGIONS

➤ ACTIVITY 1   **Examine** in the laboratory or in the field the leaves of plants native to hot dry or cold dry regions. **Record** as many of the features as are applicable in Table 14.1. Examples to study are cone-bearing plants (conifers), such as pine (*Pinus* sp.); spruce (*Picea* sp.); fir (*Abies* sp.); or juniper (*Juniperus* sp.).

If available, **examine** sagebrush (*Artemisia tridentata*) and creosote bush (*Larrea divariata*). Note that in each case the leaf is small, with very little surface area, but by comparison has a considerable amount of edge (margin). The conifer leaf is needle-like [Figure 14.5 (B)], scale-like [Figure 14.5 (F)], or awl-like [Figure 14.5 (C)]. These leaf types show an efficient design for which method of temperature regulation, (a) convection or (b) transpiration? (Circle one.) What other features of conifer leaves indicate their adaptation to dry regions? _____.

➤ ACTIVITY 2   **Examine** cactus plants and the cactus-like spurges; also succulents, such as species of *Sedum*, *Haworthia*, *Crassula*, *Kalanchoe*, or *Aloe*. In cactus type plants all photosynthesis is carried on by the green, fleshy, watery stem. The leaves are reduced to spines that discourage most animals from eating the watery stem. Succulents have thick, fleshy, watery leaves or stems. How does this type of plant operate to keep the internal temperature from rising to lethal levels? _____

How do you explain the almost identical appearance of the American cacti and the African spurges, two groups that are unrelated and native to different continents? _____ _____.

Why do you think that many terraria planted with a variety of small-leaved plants are unsuccessful? _____ _____.

II. LEAVES OF PLANTS NATIVE TO WARM WET REGIONS

➤ ACTIVITY 3   **Examine** plants native to the wet subtropics and tropics. **Record** your observations in Table 14.1. These plants are commonly grown as house plants in homes in the temperate zone, but the original habitat of most is the moist, shady floor of a tropical forest. Examples of such plants are rubber plant (*Ficus elastica*), fiddle-leaf fig (*Ficus lyrata*); dumbcane (*Dieffenbachia* sp.); caladium (*Caladium* sp.); rex begonia (*Begonia rex*); banana (*Musa* sp.); staghorn fern (*Platycerium* sp.); and others. This type of leaf is an efficient design for which method of temperature regulation, (a) convection or (b) transpiration? (Circle one.) Write a sentence or phrase defining this method. _____ Give one reason why these plants are particularly suited for house plants? _____ _____ Most of these leaves have what kind of a margin (edge), (a) even or (b) uneven? (Circle one.) Is this the type of margin characteristic of large-leaved plants of the wet tropics?   YES   NO.

III. LEAVES OF PLANTS NATIVE TO THE MOIST TEMPERATE ZONE

A. LEAVES OF DICOT PLANTS

► ACTIVITY 4   **Read** the following sections and **refer** to Figures 14.1–14.9 as a guide. **Examine** the plants that have been set out for you and **record** the leaf characters in Table 14.1.

Leaf Parts

The leaf consists of a flattened laminar portion, called the BLADE, and a stalk, called the PETIOLE, which attaches the blade to the stem [Figure 14.1 (B)]. If the blade is attached directly to the stem, the leaf is described as SESSILE. If the leaf you are examining is still attached to its stem, look for the AXILLARY BUD [Figure 14.1 (B)], a small bud situated in the angle (axil) between the base of the petiole and the stem. When this bud matures, it will produce a branch.

In many plants there are two small leaflike appendages at the junction of the petiole and the stem. These are STIPULES [Figure 14.1 (C)]. Their function varies and is not always clear cut. They may serve to protect the leaf when it's young; they are usually photosynthetic, but sometimes they are modified into spines.

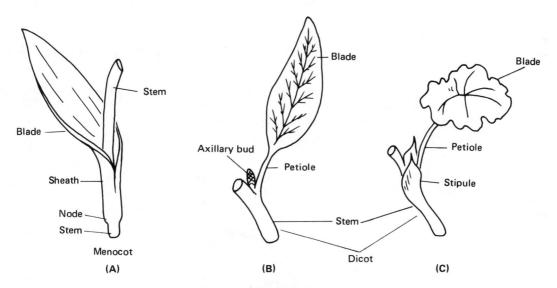

Figure 14.1   Simple Leaves

Leaf Types

Leaves are either SIMPLE or COMPOUND. A simple leaf has a blade consisting of one piece (Figure 14.1). A compound leaf has a blade that is composed of a number of segments, called LEAFLETS (Figure 14.2). There are two basic kinds of compound leaves, PINNATELY COMPOUND and PALMATELY COMPOUND. In a pinnately compound leaf the leaflets occur in a linear sequence lined up along both sides of a central axis, called the RACHIS. If the rachis is unbranched, the leaf is described as PINNATE [Figure 14.2 (A)]. If the rachis is branched one or more times, the leaf is described as DOUBLY PINNATE [Figure 14.2 (B)] or TRIPLE PINNATE. The honey locust tree (*Gleditsia triacanthos*) commonly has both single and doubly pinnate leaves.

A compound leaf that is PALMATE [Figure 14.2 (C)] is one in which three, five, seven, or more leaflets are all attached at one point near the tip of the petiole, and they radiate out from this tip. A palmate leaf with three leaflets is commonly referred to as TRIFOLIOLATE. The botanical name (genus) for the common field clover (*Trifolium*) really describes the trifoliolate

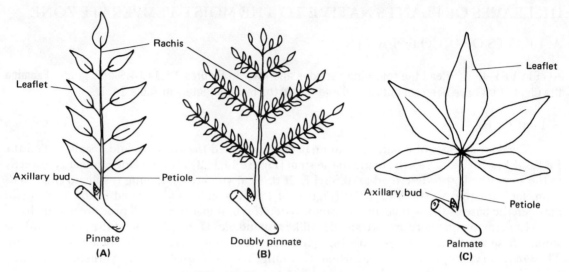

Figure 14.2   Compound Leaves

leaf of the plant. When you find the proverbial "lucky four-leaf clover," you really have a four-leaflet clover. Both simple and compound leaves may occasionally occur on the same plant, as in Boston ivy (*Parthenocissus tricuspidata*).

some distinctions between simple and compound leaves and between branches and pinnately compound leaves   It is sometimes difficult to distinguish between a branch and a compound leaf. Figure 14.3 and the following explanations will help in this matter.

1. Buds occur in the axils of leaves but not in the axils of leaflets.
2. All the leaflets of a compound leaf occur in the same plane, whereas the blades of adjacent simple leaves are not usually all oriented in the same plane.
3. Very large pinnately compound leaves sometimes resemble an entire branch. The branch will have a terminal bud, the leaf won't.

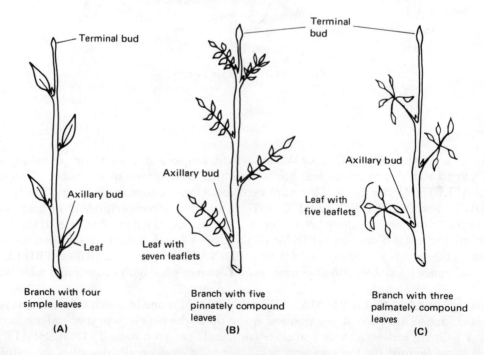

Figure 14.3   Branch Arrangement

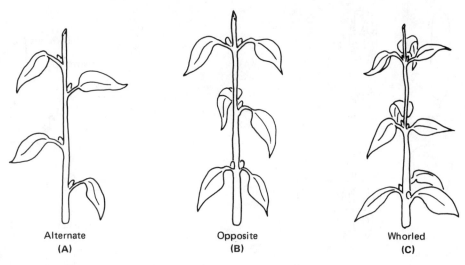

Figure 14.4   Leave Arrangement

Leaf Arrangement

Leaves are attached to the stem in different arrangements (Figure 14.4). In the ALTER-NATE (or SPIRAL) arrangement, one leaf occurs at each node [Figure 14.4 (B)]. This is the most common arrangement. The OPPOSITE arrangement with two leaves at a node [Figure 14.4 (A)] is somewhat less common, and the WHORLED arrangement [Figure 14.4 (C)] with three or more leaves at one node is the least common, particularly in woody plants. It is, however, not uncommon in herbaceous plants.

Leaf Margin

The leaf margin is the leaf edge. There is a great variation in margins. Some are shown in Figure 14.6. An ENTIRE margin is an even or regular edge. A LOBED margin is a deeply inde-nted edge, making the blade lobed. Indentations between lobes are called SINUSES. All other margin types are different degrees of dissection, ranging between entire and lobed.

Leaf Venation

Venation refers to the arrangement of veins in the leaf (Figure 14.7). In the NET VENA-TION characteristic of dicots, there are one or more large, main veins from which smaller veins branch and interconnect forming a mesh-like network. The PINNATE VENATION pattern [Fi-gure 14.7 (A)] has one main vein extending from the base of the leaf to its tip and bearing many lateral parallel branch veins. The PALMATE VENATION pattern [Figure 14.7 (B)] has several large main veins that radiate out from a common point at the base of the leaf. PARALLEL vena-tion is typical of monocots and is discussed in Section III. B.

Leaf Shapes, Tips, and Bases

The shape, the tip or apex, and the base of the leaf or leaflet are extremely varied but fairly constant for any given species. Some different types are shown in Figures 14.5, 14.8, and 14.9.

B. LEAVES OF MONOCOT PLANTS

The leaves of monocots are usually distinguished by the following features.

1. They usually consist of a narrow BLADE whose base is a SHEATH that wholly or partly encloses the stem [Figure 14.1 (A)]. Very few monocots have leaves with a distinct petiole.

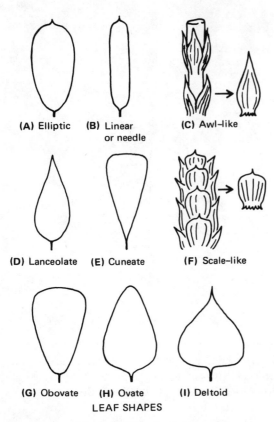

LEAF SHAPES

(A) Elliptic   (B) Linear or needle   (C) Awl–like

(D) Lanceolate   (E) Cuneate   (F) Scale–like

(G) Obovate   (H) Ovate   (I) Deltoid

Figure 14.5   Leaf Shapes

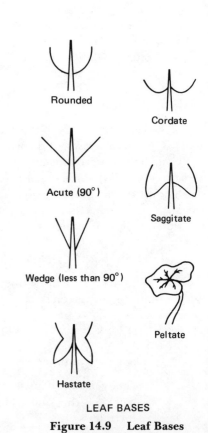

LEAF MARGINS

Figure 14.6   Leaf Margins

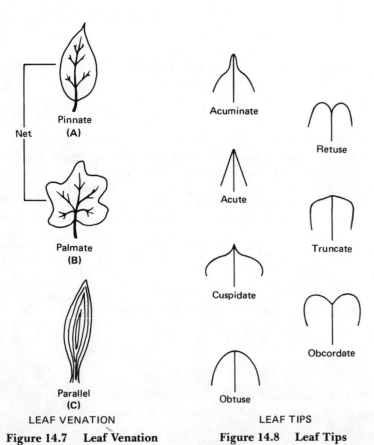

LEAF VENATION

Net

Pinnate (A)

Palmate (B)

Parallel (C)

Figure 14.7   Leaf Venation

LEAF TIPS

Acuminate

Acute

Cuspidate

Obtuse

Retuse

Truncate

Obcordate

Figure 14.8   Leaf Tips

LEAF BASES

Rounded

Acute (90°)

Wedge (less than 90°)

Hastate

Cordate

Saggitate

Peltate

Figure 14.9   Leaf Bases

2. They have parallel veins extending the length of the leaf [Figure 14.7(C)]. There are exceptions to this, however; many monocots have net venation and some dicots have parallel venation.
3. The blade and sheath are considered to be a highly modified, flattened petiole because they are parallel-sided, have parallel veins, and the blade is rarely lobed.
4. In many, especially the grasses, there is a persistent meristem at the base of the leaf which allows it to grow indefinitely.
5. They rarely show true opposite arrangement and are normally alternate or scattered.

➤ ACTIVITY 5   **Examine** the leaves of representative monocot plants and **record** the leaf charac- ters in Table 14.1. Suggested plants are any grass, sedge (*Carex* sp.), or rush; any orchid (*Orchis* sp.); gladiolus (*Gladiolus* sp.); lily (*Lilium* sp.); or tulip (*Tulipa* sp.). Common house plant sugges- tions for study are dumbcane (*Dieffenbachia* sp.); spiderwort (*Tradescantia* sp.): wandering Jew (*Zebrina* sp.); waterweed (*Anacharis canadensis*); *Sansevieria* sp.; *Dracaena* sp.; *Rhoeo discolor*; and pineapple (*Ananas comosus*). Most of the grasses of the temperate zone are basically adapted to open areas of full sunlight and occasionally to regular periods of drought. What structural de- sign (shape, dimensions, etc.) does the grass leaf show as an adaptation to hot, drought condi- tions? _____
_____. Think about the way a grass leaf is oriented on the stem. Is it (a) vertical or (b) horizontal? (Circle one.) In what way does this control the temperature of the leaf? _____
_____.

EXERCISE 14   **STUDENT NAME** _____

QUESTIONS

1. The leaves (needles) of conifers are adapted to which type(s) of climate, hot and dry; hot and wet; cold

   and dry; or cold and wet? _____

2. In regions with dry climates, which method of temperature regulation is predominant in leaves of

   plants, convection or transpiration? _____

3. List the structural features of a leaf adapted to dry climate _____

   _____

4. Plants with very large leaves and even margins are common to what type(s) of climate, hot and dry; hot

   and wet; cold and dry; or cold and wet? _____

5. Such leaves are adapted to which method of temperature regulation? _____

6. Can a leaf be both cooled and warmed by convection? _____

7. When a leaf is warmed by convection, does the heat come from the sun or the air? _____

8. Does there seem to be any correlation between climate and leaf type among plants native to the moist

   temperate zones? _____

9. What leaf shape or type predominates in the moist temperate zone? _____

10. Describe the attachment of the typical monocot blade to the stem. _____

    _____

11. What is a "key"? _____

    _____

12. What is one way to distinguish between a large pinnately compound leaf and a branch? _____

    _____

13. How does a simple leaf differ from a compound leaf? _____

    _____

14. Does a four-leaf clover really have four leaves? _____ What does it have? _____

15. The cells of cacti have special proteins that are resistant to high temperature. The cells also store abun-

    dant water. What function does the water serve? _____

    _____

Table 14.1 **External Features of Leaves**

Name of Plant	Type of Climatic Correlation (if any)	If Compound, Pinnate or Palmate	If Simple, Shape of Blade or Needle	Arrangement	Margin of Leaf or Leaflet	Apex of Leaf or Leaflet	Base of Leaf or Leaflet	Venation

EXERCISE 15

LEAVES II: INTERNAL STRUCTURE

Even though land plants are well adapted to life out of water, photosynthetic reactions demand aqueous conditions. This is a holdover from ancestral green algae in which photosynthetic cells were in direct contact with water. The terrestrial plant is not surrounded by water, but its photosynthesizing cells still require an aquatic medium. This is because the CO_2 necessary for photosynthesis is found in the cell not as a free gas, but dissolved in water. Thus you will see that the leaf of the land plant is really a watery "culture" of chlorenchyma (cells with chloroplasts) sandwiched between two protective layers, the upper and the lower epidermis.

The epidermis is porous to allow passage of carbon dioxide and oxygen. Water is unavoidably lost through these pores, and the potential exists, therefore, for the "culture" to dry out. That this does not happen as readily as you might expect is due to many factors which will be considered in this and later exercises.

You have seen that the external features of the leaf can, in certain cases, help to abate temperature stress. In this exercise you will see how the internal features and epidermis of the leaf help to abate moisture stress. You will study the photosynthetic tissues and the ways they are modified to receive light and maintain a saturated internal atmosphere.

I. THE MESOMORPHIC LEAF

The majority of flowering plants are mesophytes, that is, plants requiring moderate amounts of water regularly throughout the growing season. Their leaves are described as mesomorphic.

A. THE DICOT LEAF

➤ ACTIVITY 1   **Examine** a prepared slide of the cross section of a leaf of a representative dicot such as lilac (*Syringa* sp.) or privet (*Ligustrum* sp.). Using the lowest magnification on the microscope, orient the slide so that the section appears as a flat, ribbon-like mass extending horizontally across your field of view. If the midrib or main vein "bulges" downward, then you have the leaf correctly oriented. **Refer** to Figure 15.1 as a guide.

In this section you are seeing all the cell layers that occur from the top to the bottom of the leaf. The outermost layers at the top and at the bottom are the UPPER EPIDERMIS and the LOWER EPIDERMIS. All the cells between them comprise the middle of the leaf, the MESOPHYLL, that is, the wet culture of chlorenchyma and veins. **Change** to higher magnification and **examine** these in detail.

Upper Epidermis

The UPPER EPIDERMIS wraps around the leaf and is continuous with the lower epidermis, but the two differ in certain ways. The upper epidermis is a layer of closely packed cells. This compact fit helps prevent excessive water loss and provides mechanical support. The outer walls of the cells are covered by a waxy, noncellular layer, the CUTICLE. This further retards water loss. Do you see any openings or pores in the upper epidermis?   YES   NO. Beneath the upper epidermis are the cells of the mesophyll containing abundant small bodies, the chloroplasts. Do the epidermal cells also contain chloroplasts?   YES   NO. **Label** the upper epidermis and cuticle in Figure 15.1.

Mesophyll

MESOPHYLL is all the portion of leaf between the upper and lower epidermis. It is the part that can be likened to a water culture, since it is usually saturated with water and water vapor. It is composed of chlorenchyma, veins, a certain amount of fibrous tissue, and a labyrinth of free spaces. Chlorenchyma are thin-walled parenchyma cells containing abundant chloroplasts. The colorless walls and glassclear protoplasm of these cells reflect and refract the incoming light rays, scattering them in zigzag paths. This ricocheting of each ray lengthens its path and multiplies its chances of striking a chloroplast before it passes out of the leaf. The chlorenchyma in the upper part of the mesophyll is known as PALISADE TISSUE; in the lower part it is known as SPONGY TISSUE. **Bracket** the mesophyll of Figure 15.1.

palisade tissue   This is a layer of evenly spaced, cylindrical cells whose long axes are at right angle to the upper epidermis. This arrangement puts the cells at an acute angle to the incoming sun rays which are then not so intense. Full sunlight often inhibits photosynthesis. So, light intensity that is less than maximum results in more efficient rates of photosynthesis.

Most photosynthesis of the leaf occurs in these palisade cells which have more chloroplasts than spongy tissue cells. Although they appear to be close packed, palisade cells are slightly set apart so that the surface of each is exposed to the wet atmosphere of the mesophyll. In what practical way does this contribute to the reaction of photosynthesis? _____

Usually only one layer of palisade cells is present in the mesomorphic leaf. However, leaves daily exposed to direct sunlight (as opposed to those partially or periodically shaded) tend to have two or more layers of palisade. This makes the leaf thicker. **Label** the palisade tissue in Figure 15.1.

spongy tissue   This is the lower portion of the mesophyll. It is a more open zone of irregularly shaped cells and large spaces. Bumps or lumpy extensions of each cell touch those of other cells. The limited surface contact between cells leaves much of the cell surface free or exposed. What is the advantage of this? _____
_____. In growing conditions where the leaf is shaded this random arrangement of cells and their more or less cube shape favor maximum light absorption. The large air spaces in the SPONGY TISSUE allow for rapid, easy flow of carbon dioxide, oxygen, and water vapor. Specialized pores, called STOMATES or STOMATA (singular; stomate or stoma), occur all over the lower epidermis and control the passage of these gases. **Label** the spongy tissue in Figure 15.1.

veins   VEINS are cylindrical strands of vascular tissue and occur mainly in the mesophyll. They run in all directions, so you will see some in cross section (circular in outline) and some in profile view (as a more or less complete or incomplete horizontal band). **Label** the small vein in Figure 15.1.

Examine the large MIDVEIN in the center of the leaf.

The XYLEM CELLS in the vein are easily recognized by their thick, angular, red-stained walls and absence of cell contents. Xylem brings water into the mesophyll. PHLOEM CELLS are much less distinct; they are thin-walled, usually of of smaller diameter than xylem cells, and located below the xylem, that is, on the side between the xylem and the lower epidermis.

Study the ring of large, thin-walled parenchyma cells that encircle and may radiate out from the veins. These cells are the BUNDLE SHEATH CELLS. **Note** that they are also in close contact with both the palisade and spongy tissue. Bundle sheath cells are important because they are thought to serve as flow channels between veins and chlorenchyma. This increases conduction in the leaf. Sheath cells envelop the vein over most of its length and close over and around its tip. In many veins these bundle sheaths extend to the upper and lower epidermis and help provide mechanical support for the blade. In large veins sclerenchyma fibers are part of the sheath. If you look at the midrib you will see above and below it a mass of thick-walled SCLERENCHYMA cells. Such a "reinforced" vein serves as a flexible, yet strong, girder-like support for the leaf. **Label** the midvein and its parts in Figure 15.1, using the capitalized terms defined in the preceding description.

lower epidermis   The LOWER EPIDERMIS is similar to the upper epidermis, except that it has many small pores or STOMATES which control the passage of gases in and out of the leaf and the escape of water vapor. Each stomate is flanked by two sausage-shaped GUARD CELLS. Guard cells are smaller than the adjacent epidermal cells and have a few chloroplasts and, most noticeably, a very prominent nucleus. **Find** the guard cells and stomates. Guard cells and stomates are best seen in an epidermal peel which you will make in Activity 2. Changes in the shape of the guard cells result in an opening and closing of the lens-shaped pore between them. Such shape changes in the guard cell are due to changes in its water content. Stomates are opened as a result of the guard cell's response to light. They are usually closed at night or if it is windy, due to a loss of water pressure in the guard cells. When light and heat intensities are high, the leaf can stabilize its temperature somewhat by evaporating water from its mesophyll out through the stomates. **Label** the lower epidermis in Figure 15.1, using the capitalized terms defined in the preceding description.

stomates

▶ ACTIVITY 2   **Using** a razor blade, peel off a strip of the lower epidermis from a leaf of wandering Jew (*Zebrina* sp.) or any species of *Sedum*; German ivy (*Senecio mikanioides*), corn (*Zea mays*); or other suitable plants and make a wet mount of the tissue. Use the technique demonstrated by your instructor. Make sure that the outer surface of the epidermis is up when you place the tissue on the slide. Note the large epidermal cells which will have different shapes and patterns or arrangement depending on the species you have. In *Sedum* the cells have "wavy" contoured walls and are arranged like the pieces of a jigsaw puzzle; in corn the cells are rectangular and arranged in parallel rows. Do the epidermal cells have chloroplasts?   YES   NO. **Find** the guard cell pairs and the pore between them. Do the guard cells have chloroplasts?   YES   NO.

estimating the number of stomates on a leaf   Different species of plants have different numbers of stomates per unit area of leaf surface and on different parts of the same leaf.

▶ ACTIVITY 3   **Using** any number of different plant species, **count** the stomates seen in an epidermal peel. **Compute** the average number of stomates per square centimeter of the leaf. **Divide** the counting among the members of the class and **record** the averages in Table 15.1. Proceed as follows. Count the number of stomates visible in the field of view using the 40× objective and the 10× ocular on the microscope.

The diameter of this field is   0.5 mm

The radius of this field is   0.25 mm

The area of this field is $\pi r^2 = 3.1 \times (0.25)^2$

$= 0.194 \text{ mm}^2$

Field of view

The formula to be used is

$$\frac{\text{number of stomates seen}}{0.194 \text{ mm}^2} = \frac{X \text{ number of stomates}}{1.0 \text{ mm}^2}$$

Example: Suppose you counted 7 stomates in the field of view:

$$\frac{7}{0.194 \text{ mm}^2} = \frac{X}{1.0 \text{ mm}^2}$$

$$0.194 X = 7$$

$$X = 36 \text{ stomates per square millimeter or}$$

3600 stomates per cm²

(since 100 mm² = 1 cm².)

Table 15.1   Examples of Average Numbers of Stomates on Mesomorphic Leaves

Name of Plant	Average Number of Stomates in 400× Field	Average Number of Stomates per cm² of Leaf Surface	Heat Loss (cal/min)

Under conditions of high temperature stress, the leaf can keep its temperature lowered by evaporative cooling. A typical water loss rate from a leaf is 0.0005 gram of water lost per square centimeter per minute. This is an equivalent loss of about 0.3 calorie, or 2 percent of 1 watt heat, lost each minute. For fun, you might want to compute the figures and fill in the last column in Table 15.1.

B. SUN AND SHADE LEAVES OF MESOPHYTIC PLANTS

The leaves of a plant, particularly a tree or shrub, are not all equally exposed to the sun. In a tree, for instance, the leaves in the outer part of the crown (sun leaves) receive much more sunlight than those in the lower or deeper recesses of the crown (shade leaves). More water is lost by sun leaves than by shade leaves because of the exposure to the sun's heat. The quantity of light influences the development of the palisade tissue, and the quantity of water influences the growth of all the cells of the leaf: the more water available, the larger are the cells, and consequently the larger (in breadth and length) is the overall size of the leaf.

► ACTIVITY 4   **Examine** the leaves of *Zebrina* sp. which has been grown in bright sunlight and compare with those grown in shade. Which leaves are larger? _____. Which are thicker? _____. **Examine** a prepared slide showing both sun and shade leaves of any species in cross section. **Compare** the two leaf types and **complete** Table 15.2. From information given in this exercise and your own observations, you should be able to distinguish between the sun and shade leaf on the slide.

Tobacco plants whose leaves are to be used as cigar wrappers are grown under a cloth canopy supported over the crop by an elaborate wooden framework. What characteristics does this give to the leaf that makes it worthwhile to go to all this expense of building the shade frame?

_____

_____

C. THE HIGH-EFFICIENCY LEAF

The monocots most valuable to man are those known as the grasses. Chief among these are sugar cane and the cereals—corn, wheat, rice, sorghum, etc. It is now known that many grasses,

Table 15.2   **Comparison of Sun and Shade Leaves**

Feature	Sun Leaf	Shade Leaf
Comparative thickness of each (comparative width of each)		
Number of layers of palisade parenchyma		
Number of veins (abundant or few)		
Size of veins (large or small)		
Cuticle (thick or thin)		

notably sugar cane, corn, and sorghum, have evolved mord efficient photosynthesis reactions, particularly those trjpping carbon dioxide. This exvlains their extreme value as a food source. Other plants with these high rates include some major weeds notably pigweed (*Amaranthus* sp.); lambs-quarters (*Chenopodium* sp.); tumbleweed (*Amaranthus albus*); and purslane (*Portulaca oleracea*). The more COMPACT MESOPHYLL of the leaves of these grass crops and weeds maximizes conduction between the cells and also minimizes any possibilities of drying. Numerous veins provide each cell with more direct access to water. Big SHEATH CELLS around veins actively accumulate starch in SPECIALIZED CHLOROPLASTS, and their close contact with veins aids in rapid removal of the photosynthate. All this adds up to higher rates of photosynthesis which results in greater yield and/or more vigorous growth. The corn leaf is a good example of a high-efficiency leaf.

➤ ACTIVITY 5   **Study** a prepared slide of a cross section of the leaf of corn (*Zea mays*). Is the mesophyll differentiated into palisade and spongy tissue?   YES   NO. The vertical position of the corn leaf reduces the need for palisade cells. Explain how this is so. _____
_____ Are there as many large air spaces in the mesophyll as in the dicot leaf previously studied?   YES   NO. Are the veins (a) more numerous or (b) less numerous than in the dicot leaf previously studied? (Circle one.) Would you say that the mesophyll is more compact?   YES   NO. Are there bundle sheaths around each vein?   YES   NO. Are these sheaths (a) large and prominent or (b) small and inconspicuous? (Circle one.) Are the chloroplasts of the sheath cells (a) larger or (b) smaller than those of other mesophyll cells? (Circle one.) These chloroplasts are specialized types, very active in accumulating starch.

Another feature worth noting in the corn leaf are the large, bubble-like BULLIFORM CELLS. These have thin walls and occur in the epidermis in rows parallel to the veins. During the hottest part of the afternoon, or in droughts, when water losses from the leaf are high, bulliform cells lose water and turgor more rapidly than other epidermal cells. As a result, they shrivel causing the leaf to roll or fold up. This reduces evaporation from those stomates on the leaf surface now enclosed in the roll. Bulliform cells occur in all grasses and most monocots.

Locate bulliform cells on the corn leaf and compare their appearance in an expanded leaf versus a rolled or folded leaf.

Label the corn leaf cross section in Figure 15.2, using the capitalized terms in the preceding description of features of a high-efficiency leaf.

II. THE XEROMORPHIC LEAF

The xeromorphic leaf is one that is designed so that its water losses are minimized. Plants that grow in dry, droughty places or in the actual desert (including the arctic cold desert) have xeromorphic leaves. Other plants that live in the wet tropics also have xeromorphic leaves. Surprising? No. In this case it is an adaptation for survival during the annual dry season.

The special tissue modifications for conserving water in the xeromorphic leaf are

1. Thick cuticle.
2. Epidermal cells with thick walls.
3. Several layers of sclerenchyma below the epidermis and in other regions.
4. Sunken stomates, that is, stomates inside grooves or cavities recessed down from the leaf surface.
5. Stomate cavities are often lined with hairs.
6. In some plants hairs on the leaf surface.
7. A more compact and uniform mesophyll with very few air spaces.
8. In some plants, palisade tissue on both sides of the leaf.

A. THE XEROMORPHIC LEAF OF PINE

Pines are adapted to both hot dry and cold dry habitats. Besides having most of the above listed modifications, the leaf of pine is shaped like a prism. This makes for a low ratio of surface to volume, which automatically reduces water loss. (Think about it. Which will dry out less quickly, (a) a rolled up wet towel or (b) a wet towel laid out flat? (Circle one.) Besides water regulation, why (or how) can temperature be regulated by this narrow prism shape? _____

► ACTIVITY 6   **Examine** a prepared slide of the leaf (needle) of pine. **Refer** to Figure 15.3 as a guide. Most pines bear their needles in clusters (bundles) of two, three, or five. Depending on the species, therefore, your slide will show the needles in groups of two, three, or five. One or two VEINS occupy the center of each needle. The veins are embedded in a mass of thin-walled cells thought to conduct water from the vein to the surrounding MESOPHYLL. **Note** how the mesophyll is compact. The cells are filled with resins and their walls have many indentations, which give them a wavy outline. What effect does this have on the rigidity of the leaf? _____ _____ RESIN CANALS occur here and there in the mesophyll. **Find** the SUNKEN STOMATES. **Review** the list of xeromorphic features, then reexamine the pine needle cross section and list those xeromorphic features that you can recognize.

_____   _____
_____   _____
_____   _____
_____   _____

Label Figure 15.3, using the capitalized terms defined in the preceding description.

B. THE XEROMORPHIC LEAF OF THE UNDERSTORY PLANTS OF THE WET TROPICS

The large-leaved plants growing in the moist shade of the tall trees of the wet tropics make up much of the understory. During the dry season many of the tall trees are able to conserve water by temporarily shedding their leaves. This opens up the canopy and exposes the understory plants to the sun's heat. Their leaves, therefore, have many xeromorphic features which reduce their water losses during this time.

► ACTIVITY 7   **Examine** living plants native to the tropics, such as rubber plant (*Ficus elastica*); fiddle-leaf fig (*Ficus lyrata*); and the large-leaved *Philodendron* sp. What tissue is responsible for making these leaves rather stiff? _____. Without this tissue, what would happen to the leaf when its moisture content was low? _____ _____. In temperate climates, these plants are successfully grown as house plants. How do you explain this? _____If prepared slides are available, **examine** the tissues of the xeromorphic leaves of these tropical plants.

III. THE HYDROMORPHIC LEAF

The hydromorphic leaf is found in plants that grow either submerged in water or with their roots continually in wet soil and their leaves floating on the surface of the water.

► ACTIVITY 8   **Review** the list of modifications given for the xeromorphic leaf. Think about the different conditions impinging on the leaf surrounded by, or floating on, water, and write a one-word or one-phrase description for each of the following features that you would expect to find in a hydromorphic leaf.

Now **examine** a prepared slide showing the cross section of the leaf of a hydrophytic plant, such as water lily (*Nymphaea* sp.) or water poppy (*Castalia* sp.). Do your observations agree with your descriptions? YES NO.

Modifications Expected to Occur in a Hydromorphic Leaf
1. Cuticle
2. Walls of the epidermal cells
3. Position of stomates
4. Amount of air spaces in the mesophyll
5. Amount of sclerenchyma
6. Number or veins (few or many)

IV. COMPARISON OF THE MESOMORPHIC, XEROMORPHIC, AND HYDROMORPHIC LEAVES

► ACTIVITY 9   **Fill** in Table 15.3 to show the major distinctions between leaves of these three types.

EXERCISE 15   STUDENT NAME _____

QUESTIONS

1. Why must photosynthetic cells (chlorenchyma) be bathed in a watery film? _____

_____

2. The leaf is primarily involved in photosynthesis. What other major role does the internal structure of

the leaf play? _____

3. How is the path of a light ray inside the leaf increased? _____

_____

4. What is the advantage of lengthening the path of light? _____

_____

5. What internal design helps to reduce the intensity of light rays entering the leaf? _____

_____

6. What is the value in reducing the intensity of light? _____

_____

7. In which part of the leaf do air and gases move freely about and enter and exit from the leaf? _____

_____

8. What is the role of sheath cells in a mesomorphic leaf? _____

_____

9. Give three features of the "high-efficiency" leaf that allow for higher rates of photosynthesis.

(a) _____; (b) _____; (c) _____.

10. What two general kinds of plants have been identified as having high-efficiency leaves?

(a) _____; (b) _____

11. Why are sun leaves thicker than shade leaves? _____

_____

12. Why are shade leaves broader and longer than sun leaves? _____

_____

13. How does a grass leaf cut down on its water losses on a hot day? _____

_____

14. Why do understory plants of wet tropical forests have xeromorphic leaves? _____

_____

_____

15. Why are understory plants of the wet tropical forests so successfully grown as house plants in the tem-

perate parts of the world? _____

_____

Table 15.3   Comparison of Mesomorphic, Xeromorphic, and Hydromorphic Leaves

Feature	Mesomorphic Leaf	Xeromorphic Leaf	Hydromorphic Leaf
Cuticle (thick or thin)			
Walls of epidermal cells (thick or thin)			
Amount of sclerenchyma (fibrous tissue) (little or much)			
Amount of air spaces in mesophyll (little or much)			
Nature of the mesophyll (thick or thin)			
Position of stomates (sunken or at surface)			
Overall size of xeromorphic versus mesomorphic leaf (large or small)			

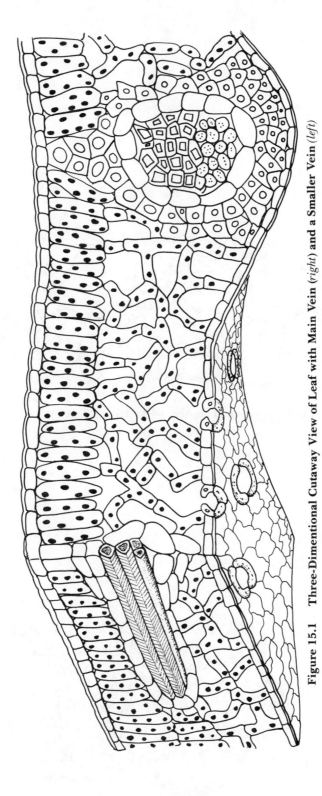

Figure 15.1   Three-Dimentional Cutaway View of Leaf with Main Vein (*right*) and a Smaller Vein (*left*)

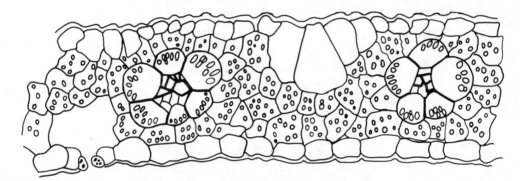

Figure 15.2   Cross Section of an Area of the High-Efficiency Leaf of Corn *(Zea mays)*

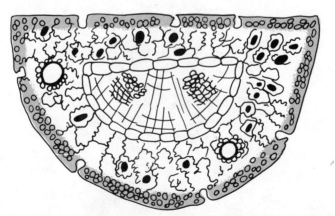

Figure 15.3   Cross Section of a Leaf (Needle) of Pine *(Pinus sp.)*. **Showing Its Special Cells and Features for Conserving Water**

Exercise
16

FLOWERS

The flower is a specialized short shoot. Its petals and other parts are modified leaves. But, it's not just a collection of parts. It is the most successful plant mechanism for sexual reproduction.

It is a harmonious unit that performs two functions, (1) precise pollen transfer and (2) production of fruit and seeds. With precise pollen transfer, there is greater success in cross fertilization. This leads to more hybridizing among plants and consequently a larger variety of types.

Parts of the flower are later converted into the fruit, an organ unique in design for seed dispersal. Efficient seed dispersal has given the angiosperms more opportunities for spreading their populations. Genetic diversity of these populations allows for adaptation to a wide variety of habitats.

Pollen transfer is achieved by transfer agents. These are insects, birds or certain other flying animals, wind, or water. Specific flowers have evolved that either attract and reward visiting insects or other animals, or that utilize wind or water currents. By comparison, the more primitive seed plants (pines, spruce, hemlocks, cycads, and so on) have haphazard, inefficient pollen transfer.

Fertilization is the union of the sperm with the egg. Transfer agents efficiently bring the male unit, the pollen (which contains the sperm), to the female unit, the egg, which is enclosed in the pistil. This is called pollination. Portions of the pistil are made so as to maximize the chances for catching or receiving the pollen and holding it.

Success and dominance of the approximately 350,000 known species of flowering plants is in large part due to the flower. Between 1000 and 2000 angiosperms fit into the human economy. About 10 percent of these are important in world trade, and 15 species provide the bulk of the world's food crops. Aside from the biological role of the flower, its aesthetic role in the beauty and pleasure of human life is immeasurable.

I. SIMPLE, SINGLE, COMPLETE, AND PERFECT FLOWER WITH RADIAL SYMMETRY

The flowers of most garden and house plants are complicated and not easy for the beginner (and even sometimes the expert) to understand. So, today you won't be studying the African violet (*Saintpaulia ionantha*); poinsettia (*Euphorbia pulcherrima*); zinnia (*Zinnia elegans*); rose (*Rosa* sp.); or geranium (*Pelargonium hortorum*).

▶ ACTIVITY 1   **Examine** some simple, single, regular flowers that are commercially available or in season such as petunia (*Petunia hybrida*); tulip (*Tulipa gesneriana*); gladiola (*Gladiolus* sp.); lily (*Lilium* sp.); primrose (*Primula* sp.); poppy (*Papaver* sp. or *Eschscholtzia* sp.); *Browallia* sp.; *Amaryllis* sp.; *Hoya* sp.; *Clivia* sp.; *Azalea* sp.; or *Magnolia* sp. **Refer** to Figure 16.1 as a guide.

The stalk that bears the flower is called the PEDICEL, and the top of the pedicel from which the floral parts arise is called the RECEPTACLE. **Label** the pedicel and receptacle in Figure 16.1.

Look at the flower's form and shape, that is, its SYMMETRY. Hold the flower so that you are looking directly down at the top of it. Are the parts circularly arranged around a central point or axis? If so, this arrangement is called REGULAR (or RADIAL or ACTINOMORPHIC) SYMMETRY.

Record the name of your plant and the symmetry of its flower in Table 16.1. **Examine** the flower parts.

Any one of the flowers suggested for you to study is a COMPLETE FLOWER because it has four whorls (circlets) of floral organs. These four organs are (1) SEPALS, (2) PETALS, (3) STAMENS, and (4) PISTIL(S). It is also a PERFECT FLOWER because it has both stamens and pistil(s).

The most conspicuous parts are the petals. Petals are thin-textured and white or colored. All taken together they are called the COROLLA. SEPALS are usually green, but not always so. They occur below the petals, that is, they are the lowermost whorl (circle) of floral organs. All taken together they are called the CALYX. Calyx and corolla together are called the PERIANTH.

If you don't see any green sepals, then your specimen is one whose sepals are the same color as the petals, and the only way to identify them is to remember that sepals are the most basal whorl of the flower.

Count the number of petals and sepals and record this in Table 16.1. If sepals and petals each total four or five or some multiple of these, then your flower is that of a dicot plant. If the petals and sepals each total three or some multiple of three, then your flower is that of a monocot plant. **Label** a petal and a sepal in Figure 16.1. **Record** whether your plant is a monocot or a dicot in Table 16.1. Sepals and petals are called ACCESSORY PARTS because they are nonessential for the actual reproductive process. What role do they play then? _____.

Smell the flower. If it is aromatic, it is due to volatile oils in the petals. These lure the insect or other pollinator to the flower. The aroma may also come from a gland or glands, called a NECTARY, which secretes the NECTAR, a sugary fluid that birds and insects like to eat.

Look for colored lines or spots on the petals. If these are present, do they occur in any particular pattern?   YES   NO. Do the lines or rows seem to converge down in the central depths of the flower?   YES   NO. What do you suppose is the function of these? _____ _____.

Carefully remove the sepals and petals and **find** the stamens encircling a pistil or pistils. These organs are concerned with reproduction. Since this flower has both stamens and pistils, it is called a perfect flower. A STAMEN usually consists of a slender stalk, the FILAMENT, which has at its top a POLLEN SAC or ANTHER. Depending on the maturity of your flower, the pollen sacs may or may not be shedding powdery or sticky yellow POLLEN GRAINS. **Label** a filament, anther, and pollen grain in Figure 16.1 and indicate the stamen by a bracket.

Count the stamens and record in Table 16.1. How does the number of stamens compare to the number of sepals? _____; to the petals? _____.

In the center of the flower **find** the PISTIL or PISTILS. A pistil is a pear-shaped or bowling-pin-shaped body, usually white or green. Its fat basal part is the OVARY. Inside the ovary (and not visible unless you dissect it) are one or more small cavities or partial cavities, called LOCULES. Locules contain one or more small white bodies, the OVULES. Each ovule contains the female embryo sac which in turn contains an EGG cell and other related cells. A pistil is made up of one or more units, called CARPELS. In this regard, it is either a SIMPLE or a COM-POUND PISTIL. **Dissect** the ovary if your flower specimen is suitable.

Extending from the ovary is a slender, columnar part, the STYLE. The tip of the style is called the STIGMA. It has a sticky surface which holds the pollen. After falling on the stigma, the pollen grain produces a long tube (pollen tube) that tunnels through the style by means of diges-tive enzyme action. It eventually reaches the ovules where it deposits the sperm nuclei that have been carried along in the tube.

In some flowers the stigma and (less commonly) the style is two or three parted. This is often an indication of the number of carpels composing the pistil. The carpel is the basic leaf that is modified to form the pistil. Some pistils are composed of more than one carpel. **Label** the ovary, style, stigma, ovule, embryo sac, and egg in Figure 16.1. **Indicate** the pistil by a bracket.

Note the position of the ovary in relation to the other parts. **Refer** to Figure 16.2 as a guide in this part. If the stamens, petals, and sepals arise from a level below the ovary, then the flower has a SUPERIOR OVARY and is described as a HYPOGYNOUS FLOWER [Figure 16.2 (A)] or a PERIGYNOUS FLOWER [Figure 16.2 (B)]. If they arise at a level that is above the base of the ovary, the flower has an INFERIOR OVARY and is described as an EPIGYNOUS FLOWER [Figure 16.2 (C)]. **Record** in Table 16.1 the type of flower you have in this respect.

➤ ACTIVITY 2   **Label** each flower in Figure 16.2, using the capitalized terms defined in the pre-ceding description.

II. SIMPLE, SINGLE, IMPERFECT FLOWER
WITH RADIAL SYMMETRY

Many plants have IMPERFECT FLOWERS, that is, flowers that do not contain both sta-mens and pistils. STAMINATE flowers have stamens only; PISTILLATE flowers have pistils only. The wax begonia (*Begonia semperflorens*) is a good example. It is a nice house and garden plant and easily available.

➤ ACTIVITY 3   **Obtain** some begonia flowers and **look** at them in side view. If the flower is pistil-late, it will have a large, three winged ovary which is below the five-parted perianth. The styles are bright yellow, wiggly-looking things above the perianth. If the flower is staminate, it has no structure below its perianth and the perianth is more "skimpy" than the female. It has only four parts, two large sepals and two small petals. The stamens are the numerous small yellow things in the center of the flower.

III. SIMPLE, SINGLE FLOWER WITH BILATERAL SYMMETRY

Flowers built on a right and left plan have BILATERAL SYMMETRY. The petals are un-equal in size and often fused together. Their arrangement gives the flower right and left mirror-image parts.

➤ ACTIVITY 4   **Examine** flowers of snapdragon (*Antirrhinum* sp.); the garden pea (*Pisum sativum*); or the sweet pea (*Lathyrus* sp.). **Note** in the snapdragon that two petals form an upper "lip" and three petals form a lower, spreading "lip". If you pull the lower lip down, you will expose the stamens and pistil. How do you suppose these articulating "lips" accommodate visiting insects?

Do you see any guide-line markings for the insect on the petals?   YES   NO. The petals of the snapdragon show partial fusion with each other. Fusion of petals is common among many flower types and is considered to be a more advanced stage of floral evolution than separate, unfused petals. **Complete** Table 16.1 with the data observed for the snapdragon.

IV. THE FLOWERS OF COMPOSITES

Daisies, sunflowers, asters, and dandelions are members of the composite family of plants. What you see is not really a single flower with numerous petals, but many tiny separate flowers clustered together on a common receptacle. This composite of many miniature flowers is called a HEAD. It superficially resembles a single flower.

The miniature flowers of composite plants are of two kinds, (1) the RAY FLOWER (Figure 16.6) in which the corolla looks like one petal, tubular at its base but mostly long and strap-shaped; and (2) the DISK FLOWER (Figure 16.4) in which the corolla forms a tube of fused petals. In some species, such as the sunflower (*Helianthus* sp.), daisy (*Chrysanthemum leucanthemum*), and aster (*Aster* sp.), the head is composed of both ray and disk flowers. In these plants the ray flowers occur at the periphery (perimeter) of the head. In daisies (Figure 16.3), for instance, the ray flowers form the white "petals." Disk flowers occupy the central portion of the head. In other species, such as dandelion (*Taraxacum officinale*) (Figure 16.5), the head consists only of ray flowers.

► ACTIVITY 5   **Examine** the head of a dandelion flower. Using tweezers, **pull off** some of the tiny flowers. This is best done by thrusting the tweezers deep down around the base of the small flowers and pulling out several at once. **Use** Figure 16.6 as a guide and **study** one flower. **Use** your hand lens. Note the strap-shaped corolla. The single strap is composed of several fused petals. It is tubular at its base. **Find** the pistil, stamens, and ovary. Are these ray flowers (a) perfect or (b) imperfect? (Circle one).

Examine the head of a sunflower, daisy, aster, *Cineraria* sp., or other plant that has both ray and disk flowers in the head. Use tweezers to extract some of the small flowers from the head, like you did with the dandelion. **Refer** to Figure 16.4 as a guide. **Examine** a disk flower. It has a slender tubular corolla of fused petals surrounding the pistil and stamens. **Slit** open the corolla with a razor blade and expose the pistil and stamens. **Note** that the anthers enclose the style and that the stigma is bifurcate (two-parted). Is the disk flower (a) perfect or (b) imperfect? (Circle one.) **Examine** a ray flower from this head. Does it generally resemble the ray flowers of the dandelion?   YES   NO. Are anthers present?   YES   NO. Is this ray flower (a)pistillate, (b) staminate, or (c) perfect? (Circle one.)

There are several advantages to the way the composite type flower is built: (1) the head is usually more conspicuous to pollinators than is a single flower; (2) it is probably less dependent on one particular kind of pollinator; (3) pollen is readily exposed to all comers; (4) insects need no special body or head parts to get at the nectar in the flowers; and (5) the construction of the head allows several flowers to be pollinated by a single insect during one visit. This head type, a composite of several miniature flowers, is unique to the composite family (hence its name). It, along with other characteristics of the group, makes the family the most highly evolved dicots among flowering plants.

V. THE GRASS FLOWER

Species in the grass family (Poaceae) have flowers that are very modified in comparison with the ones you have examined so far. The individual grass flower, the FLORET (Figures 16.7 and 16.8), is enveloped by a pair of modified leaves, the BRACTS. The outer and larger bract is termed the LEMMA and the inner and smaller one is termed the PALEA. The palea encloses the stamens and pistil.

A group of florets is termed a SPIKELET (Figure 16.9). At the base of each spikelet is a pair of modified leaves, called GLUMES. The number of florets in a spikelet varies with the species. A cluster of spikelets makes up an INFLORESCENCE (Figure 16.10). The arrangement of spikelets in the inflorescence varies with the species.

► ACTIVITY 6   **Examine** a grass inflorescence, preferably oats (*Avena sativa*). With the aid of the drawings provided (Figures 16.7–16.10) and the use of a hand lens, dissecting needle, and forceps, **identify** a spikelet and a floret. **Completely open** the floret and determine the answers to the following:

1. Are petals and sepals present in the floret?   YES   NO.
2. How many stamens are present in the floret? _____.
3. How many florets are there per spikelet? _____.
4. Describe the stigma part of the pistil. _____.
5. Is the flower (floret) colorful and aromatic?   YES   NO.
6. On the basis of this flower's morphology, identify the pollinating agent for grass.

_____

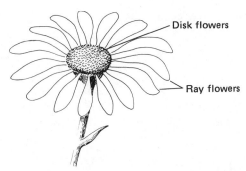

Figure 16.3   Daisy Head Composed of Both Ray and Disk Flowers

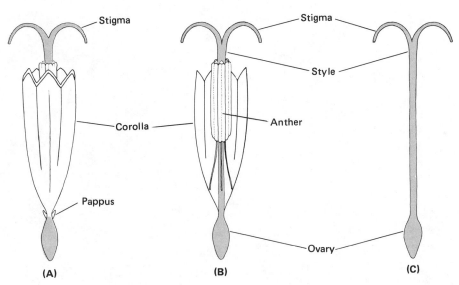

Figure 16.4   Disk Flower in Three Stages of Dissection

Figure 16.5   Dandelion Head Composed of Ray Flowers only

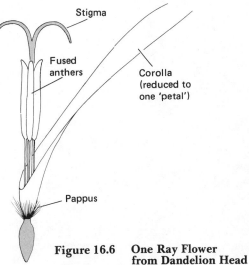

Figure 16.6   One Ray Flower from Dandelion Head

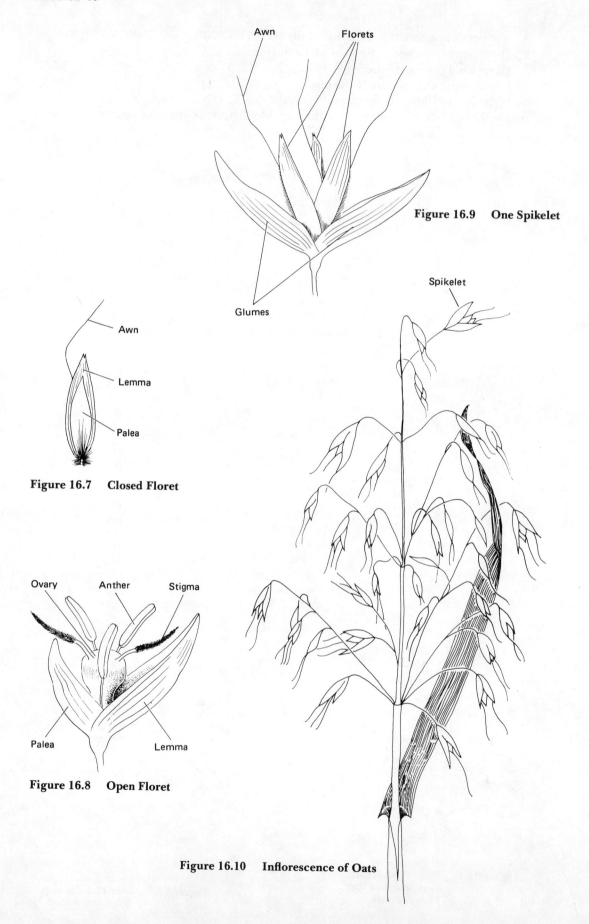

Figure 16.9   One Spikelet

Figure 16.7   Closed Floret

Figure 16.8   Open Floret

Figure 16.10   Inflorescence of Oats

EXERCISE 16   STUDENT NAME _____

QUESTIONS

1. The design of the flower is particularly suited for what two functions? (a) _____;

    (b) _____

2. What is meant by an imperfect flower? _____

    _____

3. What are the two broad categories of pollen transfer agents? (a) _____;

    (b) _____

4. What are the advantages of precise pollen transfer? _____

    _____

5. What organ is specifically suited to maximize the dispersal of seeds? _____

6. How does pollination in primitive seed plants, such as pines, compare to that of the angiosperms?

    _____

7. What part of the pistil is structurally modified to catch and hold the pollen? _____

8. How many species of angiosperms provide the bulk of the world's food crops? _____

9. Do any species of the more primitive seed plants serve as an important food source for humans?

    _____

10. What is meant by radial symmetry in a flower? _____

    _____

11. Define a hypogynous flower. _____

    _____

12. What is the collective term used for all the petals taken together? _____

13. What is the collective term for the calyx and corolla? _____

14. A flower is a specialized short shoot. Circle one: (a) true; (b) false.

15. If the petals and sepals of a given specimen are each three in number, the flower is that of a dicot or a

    monocot? _____

16. How does the pollen tube "force" its way down through the style? _____

    _____

Don't do

17. Is the pollen grain another name for the sperm? _____

18. What is the function of pigmented spots and/or streaks on the petals? _____

    _____

19. What comprises the head of a composite flower? _____

20. The construction of a composite flower has many advantageous features. List at least three.

    (a) _____ ; (b) _____

    _____ ; (c) _____ _____

21. Grass flowers are pollinated by what agent? _____

22. What is an individual grass flower called? _____

23. Grass flowers are grouped into spikelets which form clusters, called _____

24. Where is the embryo sac located? _____

25. What is a carpel? _____

    _____

EXERCISE 16

STUDENT NAME _____

Table 16.1   Variation in Floral Parts

Morphology	Lily	Yucca	Tulip					
Single flower or part of an inflorescence								
Flower type (perfect or imperfect)								
Number of sepals per flower								
Number of petals per flower								
Number of stamens per flower								
Ovary position (superior or inferior)								
Symmetry of individual flower								
Floral parts that show fusion								
Monocotyledon or dicotyledon								

Name of Plant

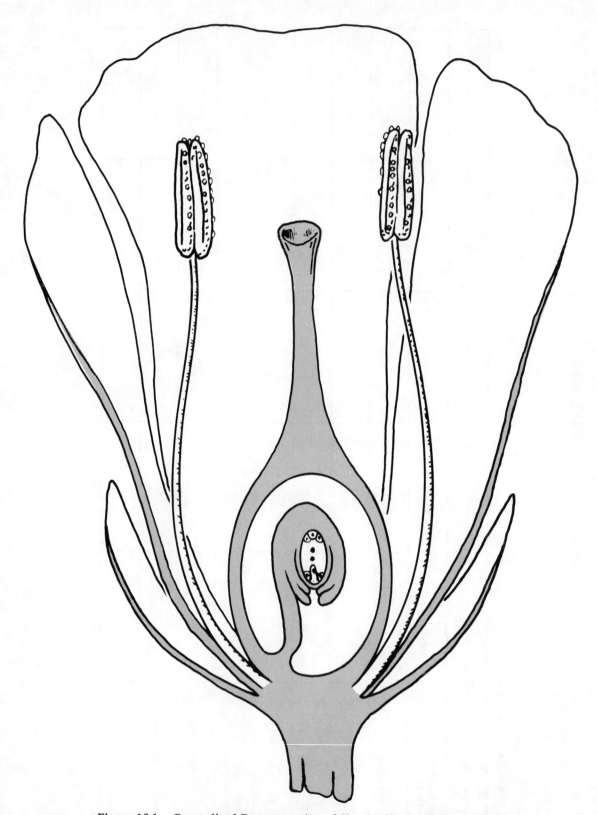

Figure 16.1   Generalized Representation of Simple, Single, Regular Flower

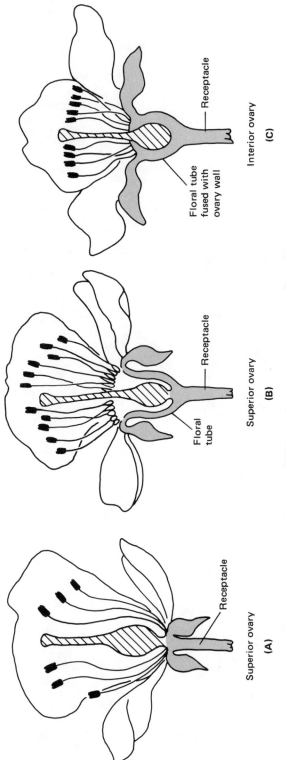

Figure 16.2   Different Positions of the Ovary in the Flower

EXERCISE 17

FRUITS

The fruit develops from the flower. It serves to protect and disperse seeds—the next generation of plants. Structurally, a fruit is the matured ovary or cluster of ovaries of the flower. In some plants, other floral parts are fused with the ovary and form part or most of the fruit. The seeds inside the fruit developed from the ovules of the flower. A seed is an embryonic plant with a supply of nutrients.

I. CLASSIFICATION OF FRUITS

The simplest classification of fruits is based on the number of ovaries and the number of flowers involved in the formation of the fruit. **Note** the outline given in Table 17.1.

Table 17.1   Classification of Fruits

Fruit Derived from	Fruit Type	
	if derived only from ovary (or ovaries)	*if derived mostly from other floral parts fused to ovary (or ovaries)*
One ovary of one flower	Simple	Simple accessory
Many ovaries of one flower	Aggregate	Aggregate accessory
Many ovaries of many flowers	Multiple	Multiple accessory

II. FRUIT DEVELOPMENT

When the fruit develops, the ovary wall, the PERICARP, may thicken and become differentiated into three layers. These layers may or may not be easy to distinguish, depending on the species. These three layers are the EXOCARP (outer epidermis), the MESOCARP (middle layer), and the ENDOCARP (inner layer). The peach is a good example of a fruit with three distinct layers. The peach skin is the exocarp, the fleshy part is the mesocarp, and the stony pit is the endocarp. The seed is inside the pit.

Examine different preserved or fresh specimens illustrating fruit development from the flower. These show the gradual stages of transition of the floral parts into the fruit. Some suggested examples are chilli pepper (*Capsicum frutescens* v. *longum*); horse chestnut (*Aesculus hippocastanum*); apple (*Malus* sp.); strawberry (*Fragaria* sp.); bean (*Phaseolus vulgaris*); maple (*Acer* sp.); parsnip (*Pastinaca sativa*); or others that have been assembled.

III. THE STRUCTURE OF SOME COMMON FRUITS

► ACTIVITY 1   **Dissect** and **examine** samples of five common fruit types: legumes, grain, berry, drupe, pome.

1. LEGUME   A simple dry fruit composed of one ovary and (usually but not always) readily splitting open along two sutures. *Examples*: pea (*Pisum sativum*); common bean (*Phaseolus vulgaris*); soybean (*Glycine max*); black locust (*Robinia pseudoacacia*); redbud (*Cercis canadensis*); wisteria (*Wisteria* sp.); peanut (*Arachis hypogaea*); and honey locust (*Gleditsia triacanthos*).
2. GRAIN (or CARYOPSIS)   A simple one-seeded dry fruit in which the pericarp is completely fused to the seed coat. *Examples*: corn (*Zea mays*); wheat (*Triticum aestivum*); and oats (*Avena sativa*).
3. BERRY   A simple fleshy fruit with a pericarp that is more or less fleshy throughout. *Examples*: grape (*Vitis* sp.); tomato (*Lycopersicon esculentum*); cucumber (*Cucumis sativus*); and banana (*Musa paradisiaca* v. *sapientum*).
4. DRUPE   A simple fleshy fruit with a stony endocarp. *Examples*: cherry (*Prunus avium*); apricot (*Prunus armeniaca*); and plum (*Prunus domestica*).
5. POME   A simple fleshy fruit derived mostly from floral parts other than the ovary. *Examples:* apple (*Malus* sp.); pear (*Pyrus communis*); hawthorn (*Crataegus* sp.); and rose hip (*Rosa* sp.).

IV. KEY TO SOME SIMPLE FRUITS

► ACTIVITY 2   **Examine** demonstration material of various fruit types. If time permits, **key** out some of the fruits that have been dissected for you. **Record** your observations in Table 17.2.

V. SEED AND FRUIT DISPERSAL

► ACTIVITY 3   **Observe** demonstration specimens of seeds and fruits that exhibit specialized features which aid in dispersal. One of the chief reasons why the flowering plants are the dominant, most diverse group of land plants is their capacity for wide dispersal. The ability to exploit new habitats plays a major role in the origin of species. Fruits, and even many seeds, exhibit an immense variety of mechanisms for dispersal by wind, water, birds, and mammals. Examples suggested for study are

1. WINGED FRUITS   Elm (*Ulmus* sp.); maple (*Acer* sp.); ash (*Fraxinus* sp.); dock (*Rumex* sp.); birch (*Betula* sp.); tulip tree (*Liriodendron tulipifera*); etc.

2. WINGED SEEDS   Trumpet vine (*Campsis radicans*) and catalpa (*Catalpa* sp.)
3. STICKY FRUITS   Mistletoe (*Phoradendron* sp.). Bits of fruit with seeds stick to the bird's beak. Seeds are transported quite some way by the bird and lodge in bark crevices when the bird cleans its beak.
4. PLUMED FRUITS   Dandelion (*Taraxacum officinale*) and goatsbeard (*Tragopogon* sp.).
5. PLUMED SEEDS   Milkweed (*Asclepias* sp.) and cottonwood (*Populus* sp.).
6. SPINY FRUITS   Cocklebur (*Xanthium* sp.); burdock (*Arctium* sp.); beggar tick (*Bidens* sp.); and sandbur (*Cenchrus* sp.).

VI. A KEY TO FRUIT TYPES[1]

1.	Fruit produced from several flowers crowded on the same inflorescence..**Multiple fruit** (*examples:* pineapple, mulberry, Osage orange)	
1′.	Fruit produced from a single flower 2	
2(1′).	Fruit derived from more than one pistil...**Aggregate fruit** (*examples:* buttercup, tulip tree)	
2′.	Fruit derived from a single pistil, **simple fruits** 3	
3(2′).	Fruit fleshy, usually indehiscent...4	
3′.	Fruit dry at maturity, dehiscent or indehiscent..7	
4(3).	Flesh of fruit consisting of accessory tissue that surrounds the papery carpels, giving the appearance of an inferior ovary ....**Pome** (*examples:* apple, pear, hawthorn)	
4′.	Flesh of fruit derived from ovary wall...5	
5(4′).	Pericarp with a fleshy mesocarp and a stony endocarp ....**Drupe** (*examples:* cherry, plum, coconut)	
5′.	Pericarp without a stony endocarp, more or less fleshy throughout...................**Berry** (*examples:* tomato, banana) See also 6 for modifications of the berry type:	
6(5′).	Septa (partitions) evident in cross section; the thick leathery exocarp separable from the inner layers of the pericarp..**Hesperidium** (*examples:* orange, lemon)	
6′(5′).	Septa not evident; exocarp leathery to hard or woody, but not separable from the inner layers of the pericarp..**Pepo** (*examples:* cucumber, gourd)	
7(3′).	Fruit indehiscent 8	
7′.	Fruit dehiscent 13	
8(7).	With one or more wings..**Samara** (*examples:* elm, ash, maple)	
8′.	Without wings 9	
9(8′).	Pericarp clearly differentiated into exocarp, mesocarp, and stony endocarp ....**"Drupaceous" nuts** (*example:* hickory)	
9′.	Pericarp not clearly differentiated into three layers, but fruit sometimes partly or completely enclosed by a papery or leathery husk or cup as in the acorn 10	

[1]*Directions for Using Key:* This key consists of pairs or couplets of choices, 1–1′, 2–2′, and so on. Compare your plant specimen with the descriptions given in couplet 1–1′. Choose whichever statement (1 or 1′) is more applicable to your specimen and go on to the next statement as indicated by the number at the extreme right. Continue through the key, each time choosing between a pair of choices until you terminate at the name of a plant. In case you get lost and forget which choices you've previously made, the numbers written in parentheses indicate your previous position in the key.

10(9').   Fruit two-carpellate, the indehiscent carpels separating at maturity but remaining at-tached to a common axis ....**Schizocarp**
(examples: carrot, dill)

10'.   Fruit one-carpellate, or of two or more united carpels that do not separate at maturity...11

11(10').   Fruits relatively large; pericarp thick and stony, seed at maturity separated from ovary wall ....**Nut**
(*examples*: hazel, acorn, chestnut)

11'.   Fruits relatively small, often very minute; pericarp thin, not stony, seed not com-pletely separate from ovary wall..12

12(11').   Pericarp completely fused to seed coat ....**Grain or Caryopsis**
(*example*: all members of the grass family)

12'.   Pericarp and seed coat united at only one point..**Achene**
(*examples*: dandelion, beggar-ticks, sedges, sunflower)

13(7').   Fruit derived from a simple pistil; one carpel 14

13'.   Fruit derived from a compound pistil; carpels two or more, united...........................15

14(13).   Dehiscent along two sutures (seams)..**Legume**
(*examples*: pea, sweet pea; some legumes are not readily dehiscent such as peanut and honey locust)

14'.   Dehiscent along one suture ....**Follicle**
(*examples*: milkweed, larkspur)

15(13').   Fruit two-loculed (with two seed cavities), the two valves splitting away from a persis-tent (remaining attached to the stalk) central septum (partition).......................**Silique**
(*examples*: shepherd's purse, mustard)

15'.   Fruit one- or several-loculed, dehiscing in various ways but without persistent central septum if the fruit is two-loculed ....**Capsule**
(*examples*: poppy, cotton, tulip, lily)

EXERCISE 17   **STUDENT NAME** _____

QUESTIONS

A. Check (✔) each statement as true or false.   True | False

1. A peanut is a nut. _____ | ✔
2. All legumes readily dehisce. ✔ | _____
3. A fruit that is a true nut has a thick, stony pericarp. ✔ | _____
4. The inner core of the apple is the ovary wall. ✔ | _____
5. Animal-dispersed fruits are necessarily colorful and attractive. ✔ | _____
6. Many seeds have structural adaptations for dispersal. ✔ | _____
7. A grain and an achene are approximately the same in their makeup. ✔ | _____
8. The fleshy part of the apple is the mesocarp. _____ | ✔
9. A cherry is a berry. _____ | ✔
10. All seeds are dispersed by fruits. _____ | ✔

B. Complete the following by choosing the correct word or phrase. Check (✔) either column a or b.   a | b

1. Peaches and almonds are both drupes. Is the hard shell enclosing the edible almond seed (a) an endocarp? (b) a stony pericarp? ✔ | _____
2. If each of the tiny black, gritty things on the surface of a strawberry is an achene (see the key), then the fruit type of the strawberry is (a) aggregate; (b) accessory aggregate. _____ | ✔
3. The seed is the matured (a) ovary; (b) ovule. _____ | ✔
4. The pit of a drupe is (a) the hard seed coat; (b) the endocarp. _____ | ✔
5. The coconut that is usually sold in the produce department in a grocery store is (a) the whole fruit; (b) just the endocarp and seed. _____ | ✔
6. Which term more accurately completes the statement: A simple fruit is derived from (a) one ovary; (b) one flower. ✔ | _____
7. A fruit that is formed from other floral parts besides the ovary is classified as (a) an accessory fruit; (b) a multiple fruit. ✔ | _____
8. The apple is (a) an accessory fruit; (b) a berry. ✔ | _____
9. Is a dehiscent fruit adapted to dispersal by (a) wind? (b) animals? ✔ | _____
10. Spiny fruits show particular adaptation for dispersal by (a) mammals; (b) birds. ✔ | _____

Don't do

Table 17.2   Characteristics of Some Common Fruits

Name of Plant	Name of Fruit	Dry or Fleshy	Dehiscent or Indehiscent	Structures(s) Other Than Ovary Involved in Fruit Formation	Fruit Type

EXERCISE 18

WATER RELATIONS: OSMOSIS AND DIFFUSION

Water relations experiments are interesting, easy to do, and of great practical value. But they are time-consuming.

This exercise may be completed in one or two three-hour class periods.

If done in one class period,

1. OSMOSIS, Sections II–IV, should be started early in the period.
2. DIFFUSION, Sections V–VIII, can be done while the osmosis experiments are running.

If done in two class periods;

1. DIFFUSION, Sections V–VIII, could be done in the first period (with reference to the terms given in Section I, Introduction).
2. OSMOSIS, Sections II–IV, could be done in the second period.

I. INTRODUCTION

In one sense, plants need water more than animals. Animals grow mainly by the synthesis of protein-building units. Plants grow by cell expansion from water pressure inside the cell. This occurs when the cell is young. Continued growth depends on the formation of additional young cells through the process of cell division. For this, water is also needed. Foods, hormones, nutrients, fertilizers, herbicides—everything that moves through the plant moves in water.

Water moves into, through, and out of the plant in two ways, by diffusion and by osmosis (which is a special type of diffusion). Today's exercise consists of a few experiments and demonstrations that illustrate how plants relate to water, in both their structure and function.

A. SOME TERMS AND DEFINITIONS

1. DIFFUSION   The net movement of a substance from a region where it is more highly concentrated to a region where it has lower or zero concentration. This movement is influenced by temperature and a number of other conditions.
2. OSMOSIS   The diffusion of a solvent (usually water) through a differentially permeable membrane.
3. DIFFERENTIALLY PERMEABLE MEMBRANE   A membrane that allows some substances to pass through them more readily than others.
4. OSMOMETER   An instrument or apparatus that is used to demonstrate and measure osmosis.
5. OSMOTIC PRESSURE   The pressure that theoretically could develop in a cell if it were under the perfect conditions of an unlimited supply of pure water available to flow into the cell.
6. TURGOR PRESSURE   Pressure that develops in a cell as a result of the osmotic uptake of water. This pressure is exerted against the cell wall and is equal to, but in opposite direction of, the wall pressure exerted against the cell contents.
7. TURGOR   The normal swollen condition of a living cell caused by internal water pressure. Tissues and organs develop rigidity or turgor because of the turgor pressure within the cells pushing outward against the cell wall.
8. FLACCID   The limp, flabby condition of cells, tissues, and organs due to the loss of water.
9. PLASMOLYSIS   The shrinkage of the protoplast of a living cell away from the cell wall as a result of diffusion of water out of the cell. Because the water diffuses through the cell's membrane, the passage is by osmosis.
10. TRANSPIRATION   The diffusion of water from plant tissues in the form of water vapor.
11. POTOMETER   An apparatus that is used to demonstrate and measure transpiration.
12. IMBIBITION   The inward diffusion of water into colloidal material, followed by swelling of the material.
13. GUTTATION   The diffusion of liquid water out of plant leaves.

Note: Because the experiments involving osmosis require time for completion, they will be started before the more simple diffusion experiments. Section V.A should also be done early in the class period. Other tests and demonstrations can be carried out in the interim between the start and finish of the osmosis experiment.

II. OSMOSIS

A. OSMOSIS IN A NONLIVING SYSTEM

The instructor will have ready for the class some "sausage" osmometers or students can make them. These are tubes made of cellulose acetate film. This film is readily permeable to water but less so to other substances. In this way it is analagous to a living cell membrane. What phrase is used to describe this type of membrane? _____

All the sausages contain a 10 percent sugar solution plus a dye for color. Those labeled as set (A) have for some time before class been immersed in a 20 percent sugar solution. Those labeled as set (B) have been immersed for the same length of time in pure water.

► ACTIVITY 1   **Work** in teams of two or more and **proceed** as follows.

1. Each student or group of students will work with one sausage from set (A) and one from set (B).

2. Remove sausage (A) and sausage (B) from the pan or beaker in which they have been soaking.
3. Wash off with water the one that was in the sugar solution.
4. Note each sausage as to whether it is flaccid or turgid. Record this in Table 18.1.
5. Dry and weigh each sausage.
6. Compile the weights for the class on the chalkboard and record in Table 18.1.
7. Immerse sausage (A) into a beaker or pan containing pure (tap) water.
8. Immerse sausage (B) into a beaker or pan containing a 20 percent sugar solution.
9. After one hour, remove the sausages, note their condition, dry, and weigh them.
10. Compile the class data on the chalkboard.
11. Record these data in Table 18.1.
12. Calculate the average net loss or gain in weight and record in Table 18.1.

Table 18.1   Changes in Weight and Turgor Pressure in Sausage Osmometers

	Set (A)		Set (B)	
	when taken from sugar solution	*after 1-hour immersion in water*	*when taken from water*	*after 1-hour immersion in sugar solution*
Condition of sausages (flaccid or turgid)				
	weight (grams)	*weight (grams)*	*weight (grams)*	*weight (grams)*
Individual sausages				
	Average weight change (from sugar solution into tap water) _____ grams (a) gain; (b) loss (circle one)		Average weight change (from tap water into sugar solution) _____ grams (a) gain; (b) loss (circle one)	

OBSERVATIONS

1. Which set of sausages initially contained the greater concentration of water? Circle one: (A) (B).
2. After immersion of set (B) sausages in the sugar solution, (a) water moved out of the sausage; (b) sugar moved out of the sausage. Circle (a) or (b).
3. When taken from the sugar solution, set (A) sausages had a (a) low; (b) high turgor pressure. Circle (a) or (b).
4. Weight changes in both sets (A) and (B) are due primarily to gains and losses of (a) sugar; (b) water. Circle (a) or (b).

B. OSMOSIS IN A LIVING SYSTEM

ACTIVITY 2   **Work** in teams of four students and **proceed** as follows.

1. Obtain a cork borer about 7 to 10 mm in diameter (or use a "french-fry" cutter) and cut four plugs of tissue about 3 to 5 cm long from potatoes, turnips, or rutabagas.
2. Make certain that all four plugs are from the same potato, turnip, etc., and are the same in length and diameter.
3. Measure the length and diameter of each plug to the nearest millimeter.
4. Weigh each plug to the nearest 0.1 gram.

5. Squeeze and bend (gently) each plug to note if it is turgid or flaccid.
6. Record these data in Table 18.2.
7. Place two plugs in tap water and the other two in a strong salt solution.
8. After 60 to 90 minutes remove the plugs from the liquids and note their degree of turgor.
9. Measure and weigh each plug.
10. Record all data for your group in Table 18.2.

Complete Table 18.2 with the comparison of the average changes in tap water versus the changes in the salt solution.

Table 18.2   Weight, Volume, and Tissue Condition Changes in Plant Tissue Due to Osmosis

	Time	Length (mm) plug 1	plug 2	Diameter (mm) plug 1	plug 2	Weight (grams) plug 1	plug 2	Tissue Condition (flaccid or turgid)
Osmotic changes in tap water	Start of experiment							
	End of experiment							
	Change: loss or gain							
Osmotic changes in a strong solute solution	Start of experiment							
	End of experiment							
	Change: loss or gain							
		Length (mm)		Diameter (mm)		Weight (grams)		
Comparison of average osmotic changes in tap water and strong solute solution	In tap water							
	In strong solute solution							

OBSERVATIONS

1. Are the results of this experiment comparable to the results obtained with the sausage osmometers? YES   NO.
2. If you wish to retain more of the water-soluble vitamins when you boil potatoes, how should you cook them? (a) in salted water; (b) in unsalted water? Circle (a) or (b).

III. DIFFERENTIALLY PERMEABLE MEMBRANES AND OSMOTIC PRESSURE

Now we will examine the role of the cell membrane during osmosis. We will also see how the cell's starch content affects the osmotic pressure differently from that of sugar.

Although water may freely diffuse across cell membranes, substances dissolved or suspended in the water may or may not do so. The membrane is selective: some substances pass through, others do not. The film, cellulose acetate, can be used to simulate the living cell membrane.

Damo

➤ ACTIVITY 3   **Observe** the two "sausage" osmometers that were set up sometime before or during the class period. Diagrams of the setups are shown in Figure 18.1.

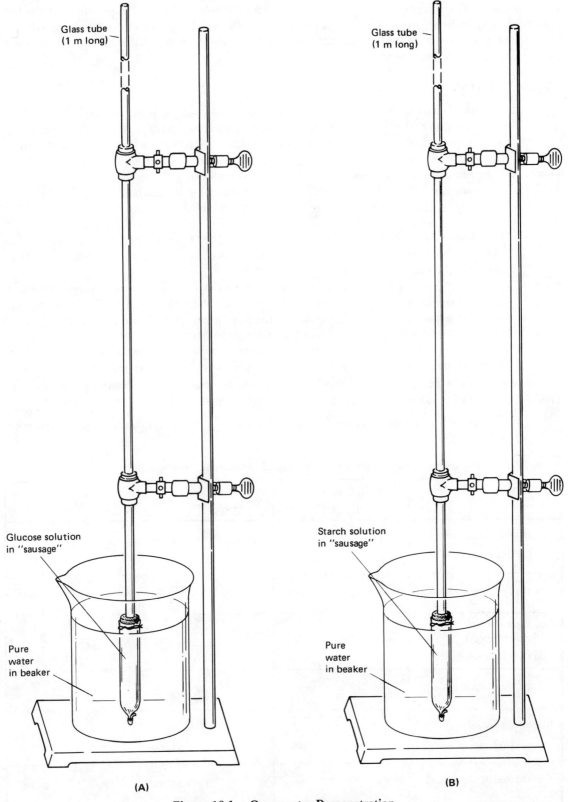

Glass tube
(1 m long)

Glass tube
(1 m long)

Glucose solution
in "sausage"

Starch solution
in "sausage"

Pure
water
in beaker

Pure
water
in beaker

(A)

(B)

Figure 18.1   Osmometer Demonstration.

The sausages are made of differentially permeable cellulose acetate. Sausage (A) contains a glucose solution; sausage (B) contains an equal volume of starch suspension. Each was weighed at the start of the experiment. A glass tube about 1 meter long is inserted in each sausage and securely tied to it. The beakers were filled with pure water at the start of the experiment. If, at the start of the experiment, any liquid moved up the glass tube, the height was noted and recorded.

Observe the instructor perform certain tests. We will now see what substances have, or have not, passed through these membranes.

OBSERVATIONS

1. Has starch passed through the membrane into the beaker of water in (B)?

    Test for starch: A sample (about 15 ml) of liquid is removed from the beaker and put in a test tube. A few drops of iodine–potassium iodide (IKI) are added. A blue color indicates the presence of starch (a blue-colored starch–iodine complex is formed).

    Result: (a) Starch present; (b) no starch. Circle (a) or (b).

    Conclusion: The membrane (a) is; (b) is not permeable to starch. Circle (a) or (b).

2. Has glucose passed through the membrane into the beaker of water in (A)?

    Tests for glucose:

    a. A strip of glucose indicator paper (trade name, "Tes-tape") is inserted into the liquid in the beaker. The new color in the paper as compared to the sample color chart indicates the presence of glucose.

    b. An alternative test: add a few drops of Benedict's solution to a test tube containing some of the liquid from the other beaker. The tube is heated in a beaker of boiling water. If sugar is present, a yellowish to reddish brown precipitate will form.

    Result: (a) Glucose present; (b) no glucose. Circle (a) or (b).

    Conclusion: The membrane (a) is; (b) is not permeable to glucose. Circle (a) or (b).

3. Does (a) starch or (b) sugar build up osmotic pressures? Circle (a) or (b).

Examine the osmometers one or more times during the class period. **Make** notations of any changes on Figure 18.1. Fill in Table 18.3 with the information as indicated.

4. Has water moved into the sugar solution in (A)? _____; into the starch suspension in (B)? _____

Table 18.3   Comparison of Sugar and Starch in Regard to Osmotic Pressure Buildup

	At Start of Experiment	After _____ Minutes	After _____ Minutes
Height of liquid in tube in (A) (sugar solution)	mm	mm	mm
in (B) (starch suspension)	mm	mm	mm
Condition of sausage (flaccid or turgid) in (A) (sugar solution)			
in (B) (starch suspension)			
Weight of sausage in (A) (sugar solution)	grams	grams	grams
in (B) (starch suspension)	grams	grams	grams

5. Both membranes (that of (A) and of (B)) are equally permeable to water. Do you agree? YES   NO.

6. It is known that starch does not readily dissolve in water. A starch solution, therefore, contains far more water molecules than dissolved starch molecules. This forms a dilute (that is, very watery) solution. If such a dilute (watery) solution is separated from pure water by a semipermeable membrane (as in the (B) osmometer setup), is there much difference in water concentration on either side of the membrane of the sausage? YES   NO.

7. In this situation, is there much tendency for the water to move into the sausage osmometer? YES   NO.

8. Can the sausage in (B), therefore, ever build up much osmotic pressure? YES   NO; become turgid? YES   NO.

9. In order for the plant to keep osmotic pressures low and within a manageable range for long-term storage situations, which, therefore, is the preferable storage form of carbohydrate, starch or sugar? _____

IV. PLASMOLYSIS AND WILTING

A. CELL CHANGES IN PLASMOLYSIS

► ACTIVITY 4   **Make** a wet mount of elodea (*Anacharis canadensis*), or species of *Rhoeo* or *Zebrina*. If using *Rhoeo* or *Zebrina*, strip off the epidermis from over the midrib or from the lower surface of the leaf and mount it. If using elodea, mount an entire leaf.

Examine the cells with the microscope. **Note** that in *Rhoeo* and *Zebrina* colored cell sap appears to fill the cells. A cell in this condition with the CYTOPLASM and CELL MEMBRANE pressed against the WALLS is described as a TURGID CELL. One large VACUOLE or several smaller vacuoles (not visible to you) occupy the cell and most of the cell sap is contained within them. A turgid cell has no EMPTY SPACE. In elodea the vacuole is usually obscured by the numerous CHLOROPLASTS in the CYTOPLASM.

Without moving the slide, **draw** off the water with a piece of paper towel and replace the water with a 5 percent salt solution. **Note** the change in any one cell.

OBSERVATIONS

1. Is water entering or leaving the vacuole? _____

2. Immediately after you added the salt solution, which had the greater concentration of water, the cell sap or the salt solution? _____

3. Is the protoplasm still pressed against the cell wall?   YES   NO.

4. The change that you see in the cell was caused by (a) loss of water; (b) influx of toxic salt. Circle (a) or (b).

5. A cell in this condition is described as _____

Fill in the contents of the cells in Figure 18.2. Let (A) represent a turgid cell and (B) a plasmolyzed cell, using the capitalized terms defined in the preceding description.

B. PLANT CHANGES (WILTING) RESULTING FROM CELL PLASMOLYSIS

► ACTIVITY 5   **Observe** herbaceous plants (or cuttings) such as geranium (*Pelargonium hortorum*); tomato (*Lycopersicon* sp.); coleus (*Coleus blumei*); etc., that have been placed in solutions of varying salt concentrations (0.0; 0.5; 3.0 percent). Alternatively, or in addition, **observe** plants that have received different concentrations of fertilizer. **Record** your observations in Table 18.4.

Table 18.4   Plant Changes Due to Cell Plasmolysis

	Salt Concentration			Fertilizer Concentration			
	0.0%	0.5%	3.0%	___ %	___ %	___ %	Plant Name
Degree of wilting (none = ○; moderate = +; or severe = ++)							

OBSERVATIONS

1. The cells of the wilted plants have (a) lost water; (b) lost turgor pressure; (c) gained salt; (d) been plasmolyzed. Circle as many as are correct.
2. The stems and leaves of the wilted plants have (a) become flaccid; (b) lost turgor pressure; (c) taken up salts. Circle as many as are correct.
3. If you were to return these plants to containers with pure water, many of the plants would regain their turgor because (a) water would move into the cells by osmosis; (b) salt would wash out of the cells. Circle (a) or (b).
4. This movement of water occurs because (a) the plant cells have increased in salt content; (b) the concentration of water in the plasmolyzed cells is less than that in the container. Circle (a) or (b).

V. DIFFUSION OF SOLIDS

A. DIFFUSION OF SOLIDS IN WATER AND ITS APPLICATION TO FERTILIZERS

ACTIVITY 6   **Work** alone or in groups, and **proceed** as follows.

1. Using forceps, place one crystal of potassium permanganate ($KMnO_4$) on the moist surface of agar in a petri dish (near the center of the dish). If agar plates are unavailable, use dishes with tap water instead.
2. Do not joggle or otherwise disturb the plate.
3. After 30 to 60 minutes, note the distance of diffusion of the permanganate.

OBSERVATIONS

1. What is the radius of the area of diffusion? _____ mm
2. How much time elapsed since the $KMnO_4$ was set on the agar (water)? _____ min
3. The rate of diffusion of the $KMnO_4$ through the agar (or water) was _____ mm/h
4. Would you say then that $KMnO_4$ is water soluble?   YES   NO (agar gel is a colloid containing water)

5. Is there still a fairly high concentration of $KMnO_4$ in the center of the dish?   YES   NO

6. If instead of $KMnO_4$, we were considering a soil fertilizer, the rate of diffusion (that is, its water-soluble properties) would become very important. There are a number of fertilizers, called "slow-release fertilizers." In these, the fertilizer is either coated with, or combined with, slowly soluble materials such as resins, plastic, glass, etc. The main effect of these on the fertilizer is to limit what? _____

7. Plant roots will suffer "fertilizer burn" when too much fertilizer is applied at any one time. How do you explain "fertilizer burn"? Is it a matter of (a) osmosis; (b) concentration; (c) both (a) and (b)? Circle (a); (b); or (c).

B. DIFFUSION OF MATERIALS IN STEMS

Water moves up stems mostly due to the evaporative pull of transpiration. Any material in the water is carried up the stem. The water itself must remain as an unbroken fine column. Xylem elements provide the conduit for this column.

► ACTIVITY 7   **Observe** stalks (petioles) of celery (*Apium graveolens*); the flowering stems of white varieties of carnation (*Dianthus* sp.); or *Chrysanthemum* species that have been set upright in beakers containing a water-soluble dye.

OBSERVATIONS

1. Describe the leaves and the petals. _____
   _____

2. Through what tissue of the stem does the dye move? _____

3. This movement up stems into the flowers is put to good practical use by florists preparing for certain holiday sales. Can you think of an example, and for what holiday(s)?
   _____

If time permits, **examine** thin cross sections cut from the midpoint of the celery petiole (several centimeters above the liquid level).

4. How would you describe the distribution of the dye, (a) spread about or (b) localized? Circle (a) or (b).

5. Name the structure in which the dye is visible. _____

6. Name the tissue (within the petiole) through which the dye is moving. _____
   _____

VI. DIFFUSION OF GASES

Oxygen and carbon dioxide are gases that DIFFUSE in and out of plants. Their concentration in the air around the leaves influences the rates of respiration and photosynthesis.

Commercial greenhouse operators in climates where the greenhouses remain closed much of the winter may experience reduced crop growth due to low levels of carbon dioxide. Addition of CO_2 to the greenhouse atmosphere is good practice and has become common among growers of roses, carnations, mums, foliage plants, etc. Large CO_2 generators produce the CO_2 which is then metered and allowed to diffuse through the house. Crop response is so good that many growers now consider adding CO_2 in winter just as essential as fertilizing or watering.

A. DIFFUSION OF GASES

► ACTIVITY 8   It is not as easy to demonstrate diffusion of CO_2 or O_2 as certain other gases, so the instructor will demonstrate *diffusion* of *hydrochloric acid gas* and *ammonia gas*. **Observe** the instructor's demonstration.

1. A large glass tube (50 to 70 cm long and 5 cm in diameter) is plugged at both ends with rubber or cork stoppers. To each stopper is attached a wad of cotton.
2. A few drops of one normal (1N) ammonium hydroxide (NH_4OH) are added to one wad of cotton and the same amount of 1N hydrochloric acid (HC1) to the other wad.
3. The ends of the tubes are marked for reference.
4. The stoppers are placed one in each end of the glass cylinder and the cylinder is held horizontal.
5. **Note** the time on your watch to the second when the tube is plugged at both ends.
6. **Observe** that a white band or cloud soon forms in the tube.
7. **Note** the time (to the second) that the white band first formed.
8. **Note** where the band appeared: in the center, or closer to one end, and which end.

OBSERVATIONS

1. The elapsed time for the diffusion was _____ seconds
2. The band appeared where? _____
3. The speed of diffusion of the two gasses is calculated by

$$\frac{\text{length of tube (cm)}}{\text{time (seconds)}} = \underline{\hspace{4cm}} = \underline{\hspace{2cm}} \text{ cm/sec (speed)}$$

4. How can you tell which gas moved faster? _____
_____
5. The HCl solution is HCl dissolved in water. NH_4OH is NH_3 (ammonia) dissolved in water. When vaporizing out of their water solutions, they diffuse as HCl and NH_3.
   a. What is the molecular weight of HCl? _____; of NH_3? _____
   b. Which is lighter in weight? _____
   c. Which, therefore, should move faster? _____
6. Does this correspond to your answer to question 4? _____
7. The white band is a precipitate formed between the two gases. What is it? _____
_____
8. Which would diffuse faster through a greenhouse? CO_2 or water vapor? _____
_____ Why? _____

B. TRANSPIRATION AND COMPARISON WITH EVAPORATION

▶ ACTIVITY 9   **Observe** the demonstration of transpiration and evaporation. Three bell jars are on glass plates. When originally set up, they were dry. Under one jar is a plant rooted in a sealed pot or peat cube; under the second, a wet sponge. The third jar is empty and serves as a control. All three jars have been under lights for at least 12 hours. **Compare** the relative amounts of condensed water on the inside of the bell jars.

OBSERVATIONS

1. The bell jar with the most condensed water is the one covering the _____
2. Is there water condensed under the jar covering the wet sponge?   YES   NO; under the control?   YES   NO
3. Water moved from the plant and the sponge as a gas, called _____
4. This loss of water from the sponge is called _____
5. The loss of water from the plant is called _____
6. Is the loss in both instances basically the same phenomenon? _____
7. Define transpiration and use the term you supplied for question 4. _____
_____

8. The sponge has many big pores; the leaf has many tiny pores (stomates). In this demonstration, which pore size is more effective in diffusing moisture? _____

9. Do you think a larger sponge would have lost more water than the leaves of the plant? YES NO

C. DEMONSTRATING AND MEASURING THE RATE OF TRANSPIRATION

► ACTIVITY 10   **Work** in teams of two and set up a potometer as shown in Figure 18.3. This is a simple (and even portable) device by which you can see and measure the uptake of water due to the pulling force developed by transpiration. **Proceed** as follows.

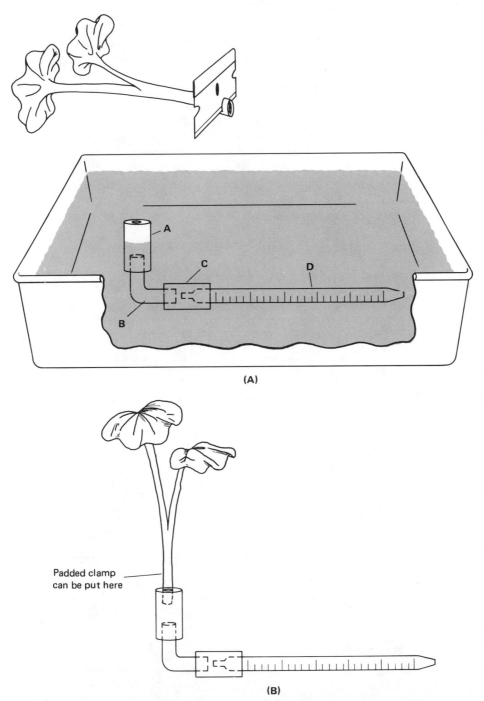

(A)

(B)

Figure 18.3   (A) Assembling the Potometer, (B) Potometer Assembled

1. **Assemble** all the parts of the potometer and put them together in a pan as shown in Figure 18.3 (A) under 5 cm of water. Figure 18.3 (A) shows a micropipette or capillary tube D connected to a glass or plastic elbow bend B by a piece of tubing C. The elbow is also connected to a 5 cm long piece of heavy gum rubber tubing A. The inside diameter of the tube A should be the same as the diameter of the stem of the plant that will be used.
2. **Force** all air out of the system by squeezing the tubing A and holding all the parts under water.
3. **Obtain** a leafy stem of some plant, such as geranium (*Pelargonium hortorum*); tomato (*Lycopersicon* sp.); or potato (*Solanum tuberosum*). Quickly put the cut end below the surface of the water in the pan and with a razor blade remove about 1 cm from the plant's stem [Figure 18.3 (A)].
4. Fit the stem into the tubing A. Keep everything under water while you do this.
5. If any air bubbles appear inside the apparatus, expel them by *gently* squeezing on the rubber tube A.
6. Lift the assembly (potometer) out of the pan. It should look like Figure 18.3 (B).
7. The potometer can be carried around the laboratory and exposed to different conditions of light and air movement.
8. If desired, a padded burette clamp may be used at the point shown to hold the potometer upright to a ring stand after removing it from the pan of water.
9. After removing the potometer from the water, **observe** the water being taken up by the plant. You will see this by looking carefully at the micropipette or capillary tube. As water is taken up by the plant, air will move into the tip of the pipette.
10. **Note** the time required for the air column to move toward the plant. (This may be expressed as mm/min if you use a capillary tube and ruler, or 0.01 ml/min if you are using a micropipette).
11. **Record** at least three successive minutes to see if the rate changes under normal classroom conditions.
12. Repeat for bright light or sunlight; repeat for dark; repeat for moderate air movement (not violent motion).
13. **Record** results in Table 18.5.

Table 18.5   Transpiration Rate under Various Conditions of Light and Air Movement

Air	Light	Transpiration Rate
Still	Classroom	
	Bright light	
	Dark	
Moving	Classroom	
	Bright light	
	Dark	

OBSERVATIONS

1. What conditions produced the fastest rate of transpiration? _____
2. What conditions produced the slowest rate? _____

3. Does gentle moving air speed up or slow down the rate? _____

4. What could excessive wind do to the water content of a plant? _____

5. Which type of plant is more prone to death by dessication on windy winter days, deciduous plants or evergreens? _____

6. What is the name of the pores in the leaf through which water is transpired? _____

7. During transpiration, does water exit from the plant as a liquid or a vapor? _____

VII. GUTTATION

A. LOSS OF WATER FROM LEAVES

▶ ACTIVITY 11   **Examine** seedlings of barley (*Hordeum vulgare*); wheat (*Triticum* sp.); or tomato (*Lycopersicon* sp.) set under a bell jar. The soil or growing medium was well watered when the demonstration was first set up several hours ago.

OBSERVATIONS

1. Localized at the _____ of the leaves are a few _____

2. These have been forced out of the leaves through pores, called hydathodes. Is this then a type of transpiration? _____

3. This phenomenon is guttation. It occurs when the rate at which water enters the plant is more rapid than the rate of water loss.
   a. What soil conditions would induce this? _____
   b. What atmospheric conditions? _____

4. The water in the xylem elements of roots is under osmotic pressure, called root pressure. Root pressure can often be observed when the rate of water loss is slow or nil. With the information provided in these questions and statements, list two environmental conditions and a plant factor that are responsible for guttation:
   (a) _____; (b) _____; (c) _____

5. In plants, is guttation just another name for dew? _____

6. Root pressure is very low and practically nonexistent in some plants. It is of minor importance in moving water up the plant. In temperate regions of the world, less than 1 percent of water loss is due to guttation. In what part of the world would you expect guttation to be very common? _____

VIII. IMBIBITION

A. DIFFUSION OF WATER INTO COLLOIDAL SUBSTANCES
FOLLOWED BY THEIR SWELLING

▶ ACTIVITY 12   **Examine** samples of material in the dry state and again after they have imbibed water. Suggested material is seeds, old mature pine cones, and wood veneer, etc. (Pine cones will float and need to be held under water with a rock, for example.)

OBSERVATIONS

1. Cellulose is a colloid. What part of the plant cell contains cellulose? _____

2. Are living cells necessary in order for imbibition to take place?   YES   NO.

3. Are there any living cells in the veneer?   YES   NO

4. In seeds are living cells involved in imbibition?   YES   NO; are dead cells involved? YES   NO

5. In what way(s) is the swelling effect of imbibition important to seed germination?

_____

6. Pine cones are a good illustration of a plant structure that can imbibe water and lose it at different stages of its growth. These gains and losses serve particular functions. Cones produce seeds and disperse them. Mature seeds can be found in the woody segments. Match the two cone conditions with the two functions.

Function	Cone Condition	Matching Pairs
a. Seed development	1. Dry with segments spread apart	
b. Seed dispersal	2. Swollen with imbibed water, and segments closed in on cone	

EXERCISE 18   **STUDENT NAME** _____

QUESTIONS

1. When you plant a young (for example, five-year-old) tree in the spring, which is the greater threat to its survival? (a) dessication; or (b) insufficient photosynthesis? Circle (a) or (b).

2. One fine summer day you go to a nursery and select a tree for purchase, but the nurseryman prefers to wait until late October to dig it up and then plant it in your yard. Give at least two advantages to fall

    planting: (a) _____; (b) _____

    _____

3. Certain fertilizers with controlled rates of diffusion are called _____

4. Many herbicides can kill the entire plant because they move through the vascular tissue. How do they

    move? By (a) osmosis; (b) diffusion? Circle (a) or (b).

5. When trees are transplanted during the growing season, they are subject to transplant shock. This shock is brought on because the root–shoot balance has been upset. What plant process is the chief cause of

    dessication (transplant shock)? _____

6. A common practice when transplanting trees in the springtime is to spray the leaves with antidessicants.

    These allow the passage of oxygen and carbon dioxide but prevent passage of what? _____

7. Before the advent of chemical antidessicants, what do you suppose was (and still is) a common method of bringing the mass of leaves, twigs, and branches back into balance with the reduced root mass?

    _____

8. Many people in regions of cold climates start their vegetable gardens early by germinating and growing the seedlings indoors. When these are later transplanted outside in the ground, the recommendation is

    to do this on cloudy days or in the evening. Explain why: _____

    _____

9. If imbibition did not take place when a dry starchy seed is wetted, could water enter the seed by os-

    mosis? _____ Sometime in a test you may be asked to *explain* this.

10. The common kitchen practice of soaking wilted vegetables in cold water to crisp them demonstrates the

    cook's confidence in the process of _____

11. Many shrubs are grown from seedlings in a container of some sort and sold directly to the customer for planting in the oround. Other shrubs are grown in the field, then put temporarily in a container for shipping and sale.

  a. Which shrub has the better chance of survival after planting in the ground? (a) field-grown plant; (b) container-grown plant. Circle (a) or (b).

  b. Which one should be top pruned after planting in the ground? (a) the field grown plant; (b) the container-grown plant. Circle (a) or (b).

12. Explain the basis of your answers to Questions 11.a and 11.b. _____

_____

_____

13. Aside from extra lighting (which is not usually practiced), what is one way that commercial growers of greenhouse tomatoes increase the yield during the winter months? _____

_____

14. A plant wilts or becomes flaccid because some or many of its cells have become _____

15. Most herbaceous plants have very little wood or fibers. What then gives them structural strength and form? _____

16. The first hard frost of autumn kills all the garden vegetables and flowers. This can be explained as due to death of the _____of the cells with the subsequent loss of _____ from the plant.

17. What does a potometer measure? _____

18. What does an osmometer measure? _____

19. Osmosis occurs only in living tissue. Circle one: (a) true; (b) false.

20. Plasmolysis of living cells is caused by the influx of salt. Circle one: (a) true; (b) false.

EXERCISE 18   **STUDENT NAME** _____

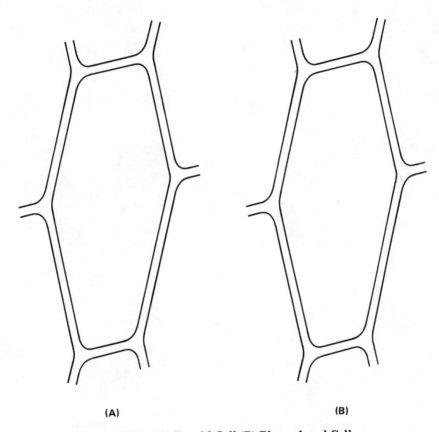

(A)   (B)

Figure 18.2   (A) Turgid Cell (B) Plasmolyzed Cell.

EXERCISE 19

PHOTOSYNTHESIS

The basic framework of organic matter on this planet earth is primarily made of carbon, and green plants are the only ongoing means of capturing carbon from the atmosphere.

Through photosynthesis the green plant

1. Constructs organic compounds from water, minerals, gases, and so on
2. Captures the energy of the sun's light waves and converts it into a biologically usable chemical form of energy
3. Stores this chemical energy in the organic compounds
4. Replenishes the atmosphere of the earth with oxygen

Thus, photosynthesis provides both the structural and the energy resources for all life. Even the photosynthesis of plants of the geological past provides us with energy today in the form of oil, coal, and gas.

I. THE PIGMENTS OF PHOTOSYNTHESIS

Much of plant breeding today is focused on producing plants that are more efficient in photosynthesis. To do this, scientists study, analyze, and experiment with the pigments of photosynthesis. The easiest way to analyze the individual pigments is to separate them from each other. A technique called chromatography may be used to separate the pigments. It has been known for many years and can be done in a variety of ways. Some ways are technologically quite sophisticated.

In today's work you perform the most basic technique of paper chromatography. Paper chromatography works on the principle that different pigments, when dissolved in a solvent (such as water or ether), will move (travel) through a piece of paper at different rates of speed. The rate is determined by (1) how soluble the pigment is in the solvent and (2) the degree of adhesion of the pigment to the surface of the paper.

A. SEPARATION OF THE PIGMENTS IN INK BY PAPER CHROMATOGRAPHY

► ACTIVITY 1   **Proceed** as follows.

1. Obtain a piece of filter paper long and narrow enough to fit inside a test tube [see Figure 19.1(A)].
2. Crease the paper lengthwise to make it more rigid.
3. Ink a straight line across the paper about 2 to 3 cm from one end and let it dry. (Use the ink provided.)
4. Put a little water (about 1 cm or so in depth) in a test tube.
5. Insert the paper in the tube. The inked line should be above the water level. Do not wet the line.
6. Set the tube in a rack or wooden block.
7. After a few minutes, examine the pigment separation on the paper strip.

OBSERVATIONS

1. What color ink did you use to make the line on the paper? _____
2. How many ink colors are visible after separation of the pigments? _____
3. What colors are there? _____
4. The combination of all the pigments leads to some masking of individual colors. Do you agree?   YES   NO
5. Is the distance traveled through the paper the same for all pigments?   YES   NO
6. Are all the pigments, therefore, equally soluble in the water?   YES   NO. Are they equally adsorbed by the paper?   YES   NO
7. The differential adsorption to the paper and the differential solubility of these pigments in water has effectively done what to them? _____

B. SEPARATION OF LEAF PIGMENTS BY PAPER CHROMATOGRAPHY

A leaf extract has been prepared by macerating leaves in a small volume of acetone. This can be done either with a mortar and pestle or in an electric blender. By paper chromatography, you can separate the different pigments out of the leaf extract just as you separated the different pigments out of the ink.

► ACTIVITY 2   **Proceed** as follows.

1. Get a test tube and add to it a small amount of solvent (about 1 cm or so in depth). The solvent is a mixture of 95 parts ether and 5 parts acetone. Plug the tube and set it aside.
2. Obtain a strip of filter paper long and narrow enough to fit inside a test tube [Figure 19.1(A)] and crease it lengthwise.
3. With a medicine dropper, apply a few drops of the leaf extract to the filter paper strip about 2 to 3 cm from one end. Allow the spot to dry. Two or three repeated applications to the same spot may be needed. The spot must be dry before you proceed to step 4.
4. Insert the paper in the test tube. Do not permit the extract on the paper to be directly wet by the solvent. Don't "buckle" the paper. Plug the tube and set it aside.
5. In a few minutes you will see the pigments begin to separate. As the solvent passes through the paper, the various pigments become dissolved in it and move up the paper at different rates.

Examine the chromatograph at frequent intervals because if the pigment separation continues for too long, some of the pigments will be superimposed on each other near the top of the strip.

The four pigments that separate out into their respective colors are

1. Chlorophyll a—Blue green color
2. Chlorophyll b—Pale green color

3. Xanthophyll—Pale yellow
4. Cartoene—Orange yellow

Draw lines on the diagram of the paper strip [Figure 19.1 (B)] to show the position and general appearance of the pigments. **Label** each pigment line.

OBSERVATIONS

1. Chlorophyll b is the least soluble. Which position, therefore, should it occupy on the strip? _____

2. Carotene is the most soluble. Where should it be located? _____

_____

II. THE NECESSITY OF CHLOROPHYLL IN PHOTOSYNTHESIS

The leaves of many plants are naturally variegated, that is, they have margins or patches that are not green but are white or (usually) some shade of yellow or pink. These patches lack the green pigment, chlorophyll. Variegated leaves are considered an attractive feature and are of horticultural value.

For a botany class they are of value because they can be used to demonstrate that where there is no chlorophyll in a leaf, there is no production and storage of the photosynthate, starch.

▶ ACTIVITY 3   **Test** for starch in the variegated leaf. **Proceed** as follows.

1. Obtain a leaf from some variegated variety of plant, such as *Coleus blumei*; Algerian ivy (*Hedera canariensis* v. *variegata*); geranium (*Pelargonium hortorum* v. *marginatum*); etc.
2. Diagram the leaf with the green and nongreen zones in the space provided in Figure 19.2.
3. Boil the leaf in water to kill the tissue and to remove the water-soluble red and blue pigments that mask chlorophyll in some of these plants.
4. Boil the leaf in alcohol for a few minutes to extract the pigments. *Caution: Do not heat the alcohol over an open flame.* Use an enclosed-element electric hot plate or use the beaker of hot water (from step 3) as a means of heating a small beaker of alcohol.
5. Set the leaf in a petri dish and add a few drops of dilute iodine–potassium–iodide (IKI).
6. A brown or dark purple coloration in the leaf indicates the presence of starch.
7. **Label** the leaf diagram you drew as to which area gave a positive starch test.

OBSERVATIONS

1. Did the white or nongreen zones of the leaf give a positive test for starch? YES NO
2. Did the green zones of the leaf give a positive test for starch? YES NO
3. Is chlorophyll present throughout the leaf? YES NO
4. Is the presence of starch a fairly good indication of chlorophyllous tissue in a leaf? YES NO

III. THE NECESSITY OF LIGHT IN PHOTOSYNTHESIS

▶ ACTIVITY 4   **Observe** bean plants (*Phaseolus vulgaris*) or geranium (*Pelargonium* sp.) in which several leaves have squares (about 4 cm² in size) of heavy black paper attached to them (by paper clips). The black paper forms a light-tight screen. The plants were kept in the dark for 72 hours. Then they were exposed to light for 8 to 12 hours just before class.

Remove one of the leaves; remove the black paper and test for starch as follows.

1. Boil the leaf in water.
2. Boil the leaf in alcohol. *Caution: Do not heat the alcohol over an open flame. Take precautions as noted earlier.*

3. Set the leaf in a petri dish and pour some dilute IKI over it.
4. If starch is present, the tissue will turn brown or dark purple.

OBSERVATIONS

1. Did the tissue that had been kept under the light screen give a positive test for starch? YES NO
2. Did the remainder of the leaf show the presence of starch? YES NO
3. What is the conclusion? _____
4. Why were the plants kept in the dark for 72 hours? _____
   _____
5. What would happen to the plants if they were kept in the dark for several weeks at room temperature? _____
6. Commercial greenhouse growers of ornamental plants commonly keep plants in the dark for many weeks when they wish to delay growth for marketing schedules. The plants suffer no setback. What other key environmental factor must be manipulated by the grower to allow for success of this practice? _____

IV. THE NECESSITY OF CARBON DIOXIDE IN PHOTOSYNTHESIS

► ACTIVITY 5   **Observe** the following demonstration.

1. Three test tubes numbered 1 to 3 are each partly filled with water that has previously been heated to remove any dissolved air, then cooled to room temperature.
2. A few drops of phenol red indicator are added to each tube. Phenol red is an indicator here for the presence of carbon dioxide. In the presence of CO_2, phenol red turns yellow. When no CO_2 is present, the indicator is red.
3. Using a soda straw, the instructor will briefly blow his breath into the liquid in tubes 1 and 2 just until the liquid turns yellow. The third tube will remain as the control with no CO_2 in it.
4. A sprig of elodea (*Anacharis canadensis*) or parrots-feather (*Myriophyllum brasiliense*) is placed in tube 1.
5. All three tubes are put under very bright light for a half hour or so.

OBSERVATIONS

1. Have any color changes taken place? YES NO. (The plants may be removed to aid in color comparisons.)
2. Has the control tube 3 changed color? YES NO
3. Has tube 2 changed color? YES NO
4. What does this indicate? _____
5. Has tube 1 with the plant changed color? YES NO
6. What is the color of the liquid in tube 1? _____
7. Is there any CO_2 in the liquid of tube 1 now? YES NO
8. What can you conclude has happened to the CO_2? _____
   _____

V. OXYGEN AS A BY-PRODUCT OF PHOTOSYNTHESIS

► ACTIVITY 6   In Figure 19.3 you see a very simplified summary of the process of photosynthesis. The oxygen produced by photosynthesis, which replenishes the earth's atmosphere, is derived from the water used in the process.

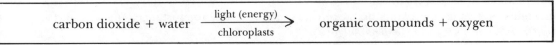

$$\text{carbon dioxide + water} \xrightarrow[\text{chloroplasts}]{\text{light (energy)}} \text{organic compounds + oxygen}$$

Figure 19.3   Simplified Summary of Photosynthesis

Observe the demonstration which shows the release of oxygen by a photosynthesizing aquatic plant, and the influence of light on the rate of photosynthesis.

A few sprigs of elodea (*Anacharis canadensis*) or parrots-feather (*Myriophyllum brasiliense*) have been placed in two jars or beakers with clean tap water. In order to enrich the CO_2 content of the water, several drops of 1 percent $NaHCO_3$ have been added to the water, or the gas has simply been bubbled in by means of blowing through a soda straw.

A funnel trap has been placed over the plants (as shown in Figure 19.4) and a test tube inverted over the stem of the funnel. The gas given off by the plant will be collected in the test tube.

One beaker of plants has been kept in relatively dim light (usually normal classroom light); the other has been kept under fairly bright light.

Note the difference in the amount of gas that has accumulated in each tube. **Watch** what happens when the instructor inserts a glowing splinter in the tubes.

OBSERVATIONS

1. Which plant has produced more gas, (a) the well lighted plant or (b) the dimly lit one? Circle (a) or (b).
2. What happened to the glowing splinter when it was inserted in the test tube?

   _____

3. What is the gas that accumulated in the tube? _____

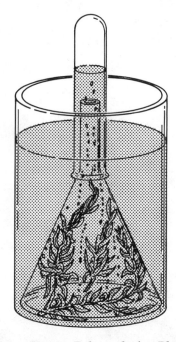

Figure 19.4   Oxygen Release during Photosynthesis

VI. STARCH—A STORAGE PRODUCT OF PHOTOSYNTHESIS

The initial organic compound produced by photosynthesis is a sugar. On a molecular level, sugar molecules occupy more space than starch molecules. Because of this and for other reasons, the plant converts sugars to starch for storage purposes. Starch is more extensively stored than proteins or fats. Some plants do store the sugar sucrose, for instance, sugar cane (*Saccharum officinarum*) and sugar beets (*Beta vulgaris*).

► ACTIVITY 7   **Prepare** a wet mount of starch grains of potato (*Solanum tuberosum*) by lightly scraping the cut surface of a raw white potato with a scalpel. Mount the scrapings in water and cover with a cover glass. Then allow a drop of IKI to run under the cover glass while you hold a piece of paper toweling at the opposite edge of the cover glass to draw off some of the water.

Starch grains often have a laminated appearance much like the growth lines on a clam shell. Starch grains, to some extent, can be used to identify the plant because they are characteristic of the species.

Prepare other wet mounts using flour made from the starch of any of the following plants: sweet potato (*Ipomoea batatas*); rice (*Oryza sativa*); oats (*Avena sativa*); or others. **Examine** these for differences in the starch grains.

EXERCISE 19   STUDENT NAME _____

QUESTIONS

1. Radiant energy is not directly usable by biological systems. What form of energy is usable? _____

   _____

2. Can radiant energy be stored for future use (a) by man; (b) by animals; (c) by plants? Circle as many as apply.

3. Can chemical energy be stored? _____

4. Where is chemical energy stored? Be specific, give the three basic, organic storehouses: (a) _____;

   (b) _____; (c) _____

5. What process converts radiant energy into chemical energy? _____

6. The biological and technological world is like an upside-down pyramid; everything rests on one point.

   What biological reaction occupies that point? _____

7. Although you saw four pigments separate out in the leaf chromatography experiment, only one is the primary pigment of photosynthesis. The others are accessory pigments. Which pigment is it?

   _____

8. In today's exercise you saw that light influences the rate of photosynthesis. Name at least five physical,

   environmental factors that can alter the rate of photosynthesis: (a) _____;

   (b) _____; (c) _____; (d) _____;

   (e) _____

9. Oxygen is a by-product of photosynthesis. From what is it derived? _____

10. The most common storage compound in plants is _____

11. Do you think it would be wasteful or beneficial if a grower in northern Michigan gave extra applications

    of fertilizer to his greenhouse crops during December and January? _____

12. Explain your answer to Question 11. _____

    _____

13. It is usually recommended that plants grown indoors all winter be placed outdoors in a semishady area

    during the summer. Why outside? Why in a semishady place? _____

    _____

14. What element makes up the basic framework of organic matter? _____

15. What, besides green plants, can capture or "fix" this element? _____

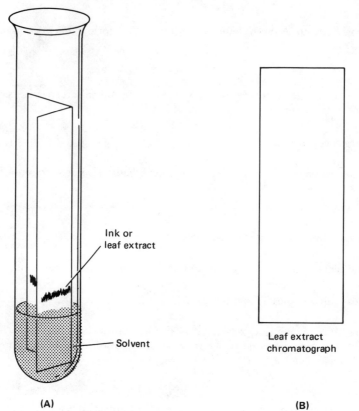

Ink or
leaf extract

Solvent

Leaf extract
chromatograph

Figure 19.1   **(A)**

(B)

Figure 19.2

EXERCISE 20

DIGESTION AND RESPIRATION

Green plants capture the energy of light and put it away for storage in organic molecules. These molecules are mostly starches and fats and (to a lesser degree) proteins. The energy is retrieved through the process of *respiration*.

Respiration does not retrieve the energy directly from the storage molecules. They must first be converted to sugars, and it is from the sugars that respiration then retrieves the energy. The conversion of storage molecules to sugar and other substances is *digestion*. So, digestion and respiration are two closely related functions.

I. DIGESTION

When water is present (and other suitable conditions exist) living systems can convert starches to sugars. The conversion is mediated by enzymes secreted by the living system. Essentially, digestion makes an insoluble substance into a soluble one.

A. PRODUCTION OF STARCH-DIGESTING ENZYMES BY A LIVING SYSTEM

Germinating seeds are excellent sources of starch-digesting enzymes and are easily used to demonstrate the effect of the enzymes.

Starch–Digesting Ability of Germinating Seed

➤ ACTIVITY 1   **Observe** the starch-digesting ability of a germinated seed.

1. Two days before this class, a germinating corn seed was cut lengthwise and placed (with the cut surface down) on a starch-agar plate (petri dish).
2. The instructor now removes the seed from the plate.

3. Dilute iodine-potassium-iodide (IKI), a starch indicator, is poured over the plate and left on for 1 or 2 minutes. Areas of the plate containing starch will react with the IKI and turn a blue-black. Those without starch will not turn color.
4. The IKI is poured off and the plate rinsed with water.

OBSERVATIONS

1. Did all the agar plate turn blue? YES NO
2. Describe the agar zone in the vicinity occupied by the seed. _____
3. What is your conclusion? Is starch present in the vicinity of the seed? YES NO
4. Is starch present in the remainder of the agar plate? YES NO
5. Briefly, what can you conclude about materials exuding from the cut seed? _____
_____

Extraction and Testing of Digestive Enzymes from Germinating Seeds

extraction of the enzyme

► ACTIVITY 2   **Observe** the extraction.

1. One cup of five-to-six-day old germinating wheat seeds is mixed with one to two cups of a very weak buffer solution (0.001 percent sodium acetate). (This controls the acidity of the solution and allows the enzymes to work better than in just plain water.)
2. The seeds and buffered water are mixed in an electric blender for a few minutes.
3. The mixture is filtered with cheesecloth and the filtrate allowed to stand for 15 to 20 minutes to permit the solid residue to settle out.
4. The supernatant should contain digestive enzymes.
5. The supernatant will be referred to as the *extract* in the subsequent tests.

testing the extract for the presence of digestive enzymes   We need to show if starch-digesting enzymes are present in the extract. If starch is added to the extract, and if enzymes are present, the starch should be converted to sugar. If enzymes are not present, the starch will remain starch.

The presence of sugar can be demonstrated by means of an indicator, called Benedict's solution. **Observe** a demonstration of this sugar test. The instructor will add a few drops of Benedict's solution to a test tube containing a sugar solution. The tube is heated to boiling in a water bath. If sugar is present, a yellow to reddish brown precipitate will form.

The presence of starch can be demonstrated by means of the indicator IKI. You have already used this indicator on several occasions. Material that contains starch turns dark when IKI is added to it.

We need to test how long it takes for the enzyme to take effect and to continue being effective. Therefore, sugar and starch tests will be made at the start and at 15-minute intervals for a period of 45 minutes.

► ACTIVITY 3   **Refer** to Figure 20.1 as a guide. **Proceed** as follows.

1. Obtain a 125-ml flask and fill it about one quarter full with the extract (about 50 ml).
2. Add an equal volume of a 1 percent starch suspension to the flask.
3. Shake the flask well.
4. Fill two small test tubes each with about 2 to 3 cm of the starch–extract mixture.
5. Immediately test one tube for starch, using IKI.
6. Immediately test the second tube for sugar, using Benedict's solution as outlined above.
7. Repeat steps 3 through 6 at 15-minute intervals.
8. Record the results and your interpretations and conclusions in the spaces provided in Figure 20.1.

B. THE EFFECT OF HEAT ON DIGESTIVE ENZYMES

Many factors alter or inhibit enzyme activity. These include such things as dehydrated tissues, extremes of acidity or alkalinity, and extremes of temperature. In the following test we will see the effect of a temperature extreme.

► ACTIVITY 4   **Observe** or **proceed** with the following tests. **Refer** to Figure 20.2 as a guide.

1. Fill five test tubes one quarter full with water.
2. Add 15 drops of 1 percent starch suspension to each tube.
3. To tubes 3, 4, and 5 add 15 drops of amylase (a starch-digesting enzyme).
4. Boil tube 3 in a water bath for 15 to 20 minutes.
5. Set all tubes aside for 45 minutes to allow any reactions to proceed.
6. Test each of the tubes as shown in Figure 20.2. Test tubes 1 and 5 for starch with IKI and test tubes 2, 3, and 4 for sugar with Benedict's solution.
7. **Record** the test results and your interpretations and conclusions in the spaces provided.

II. RESPIRATION

Respiration releases the energy stored in glucose (or some glucose product). This energy is then used to drive all the reactions needed to sustain life. Figure 20.3 is a summary equation of the chemical changes that occur in respiration.

$$\text{glucose } + \text{ oxygen} \longrightarrow \text{carbon dioxide } + \text{ water } + \text{ energy}$$

Figure 20.3

A. OXYGEN AS A FACTOR IN SEED GERMINATION

Seed germination is a period of rapid cell division and high respiration rates. Much energy and a great deal of oxygen are consumed in the process.

► ACTIVITY 5   **Observe** the demonstration.

1. The demonstration was started five to six days prior to class.
2. Two wide-mouthed bottles contain water, but one (bottle 2) also has a small vial containing pyrogallate which absorbs the oxygen that is inside the bottle.
3. At the start of the experiment, some thoroughly soaked (but not germinating) seeds were placed in cheesecloth bags and suspended in the bottles over the water.

OBSERVATIONS

1. Did both batches of seed germinate?   YES   NO. Equally so?   YES   NO
2. Which bottle shows more evidence of germination of seeds, (a) the bottle with oxygen or (b) the bottle without oxygen? Circle (a) or (b).
3. Did both bottles have the same amount of moisture?   YES   NO
4. What accounts for the difference in the germination of the seeds? _____
5. **Complete** this statement by filling in the blanks. For most plants, seed germination will not occur in the absence of _____, even though ample amounts of _____ are available.

B. OXYGEN AS A FACTOR IN ROOT GROWTH

Root respiration is extremely critical. Oxygen must be available to roots or the plants die. People and animals destroy air spaces in soil by trampling the ground. Compacted soil like this cannot support plant growth. Witness the grassless shortcut path across a lawn, and the dead or dying trees encircled by paved parking lots.

It's hard to demonstrate soil aeration directly, but root aeration in general can be demonstrated.

► ACTIVITY 6   **Compare** the two sets of cuttings which have been rooted in water for a few weeks. These cuttings may be from coleus (*Coleus blumei*); tomato (*Lycopersicon* sp.); wax begonia (*Begonia semperflorens*); geranium (*Pelargonium hortorum*); wandering Jew (*Zebrina pendula*); purple heart (*Setcreasea purpurea*); or willow (*Salix* sp.).

Set *A* has had no oxygen added to the water other than what diffuses in from the air above the water or from splashing bubbles when the water is replenished.

Set *B* has been growing with the fish tank aerator bubbling air into the water in the container.

OBSERVATIONS

1. Which set has the better root and shoot growth, set *A* or *B*? (Circle one.)
2. If these plants survive growing in water, why won't plants survive in water-logged soil?

_____

3. Some plants (but very few) thrive in flooded soil. Can you name one important food crop that does so? _____

C. RESPIRATION RATES AT DIFFERENT TEMPERATURES

Balancing the rate of respiration with the rate of photosynthesis is a predicament for the plant. To keep living cells functioning, respiration must supply them with energy at all times, day and night. Respiration uses the foods made by photosynthesis. There is a direct positive correlation between respiration and temperature. When temperatures are high, respiration rates are high too (rapidly using up the stored foods).

But, photosynthetic rates depend on both temperature *and* light. If temperature is high but light is low, then the rate of photosynthesis is slow. In this situation, food production may not keep pace with the food losses (of respiration). If this kind of imbalance occurs too often or for too long, not enough food reserves remain for normal growth and development. For this reason, plants kept in heated offices and homes, with usually far less than optimum light for photosynthesis, do not grow as well as in their natural environment.

► ACTIVITY 7   **Observe** the apparatus which has been set up and **refer** to Figure 20.4 as a guide. An easy way to perceive differences in respiration rates is to measure the quantity of CO_2 given off by plants exposed to different temperatures. The apparatus in this demonstration is used to collect the CO_2 produced by seeds respiring at different temperatures. The apparatus (Figure 20.4) and demonstration consists of the following.

Apparatus

1. Four bottles are connected in a series to a hand aspirator (air displacement pump).
2. The system is set up to "wash" air (that is, pump it through) from one bottle to the next. Any CO_2 in the air in these bottles is either absorbed (in bottle 1) or precipitated out as a white precipitate (in bottles 2 and 4).
3. Three different bottles of bottle 3 will be put into the series, one at a time. These bottles contain wheat seeds that have been germinating for several days at different tempera-

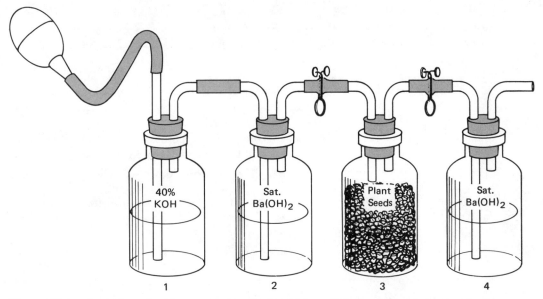

Figure 20.4   Apparatus for Demonstrating the Evolution of Carbon Dioxide in Respiration

tures: one in the refrigerator at 5°C, one at room temperature of 20°C, and one in the incubator kept at 30°C. (Because other classes may have already used these bottles earlier in the day, your demonstration may show only the amount of CO_2 that has accumulated since then.)

4. Bottle 1 contains sodium or potassium hydroxide (NaOH or KOH). When the room air is pumped through the system, the CO_2 in it (the room air) should be removed by the NaOH.
5. Bottles 2 and 4 contain barium hydroxide $(Ba(OH)_2)$. Bottle 2 will precipitate out any remaining CO_2 that washes over from 1.
6. During the germination test period, bottles 3 should be stored with tubing and pinch clamps as shown. After a bottle is connected into the gas-washing series, the clamps are removed.
7. The CO_2 given off by the seeds will be "air washed" into bottle 4, where it will precipitate out as a white precipitate, barium carbonate $(BaCO_3)$.
8. For each bottle 3 used, a new bottle 4 will be used.

Demonstration

Observe the procedure by the instructor.

1. One bottle 3 is put into the series.
2. The hand pump is squeezed 4 to 5 times to force air through the system.
3. **Note** the precipitate that forms in bottle 4 and record in Table 20.1.
4. The same procedure is repeated for the other two bottles 3. The hand pump must be squeezed the same number of times for each analysis. **Record** the results in Table 20.1.
5. The instructor next filters the precipitate and will ovendry it with the filter paper at 80°C for one hour.
6. Each sample will be weighed to the nearest 0.1 gram on a torsion balance.
7. **Record** the weights in Table 20.1.

OBSERVATIONS

1. Does temperature influence the rate of respiration?   YES   NO
2. The temperature difference between 5°C and 20°C is fourfold (assuming that no plant

activity can take place at 0°C or below). Was there a fourfold difference in the respiration rate between these two?   YES   NO

3. How do you explain the result as answered in question 2? _____

_____

_____

4. List at least three other factors or conditions that influence the rate of respiration.
    (a) _____; (b) _____;
    (c) _____

5. The seeds of peas, turnips, and leaf lettuce can be planted outdoors five to six weeks *before* the last frost. If we had used seeds of these plants, would the results have agreed with those obtained with wheat?   YES   NO. Explain: _____

_____

6. Green peppers and watermelon seed should not be planted until two weeks *after* the last frost. If we had used seeds of these plants, would the results have agreed with those obtained with wheat?   YES   NO. Explain: _____

_____

_____

7. In view of the facts given in questions 5 and 6, can you now list another factor (besides the three you listed in question 4) which relates to germination rates?

_____

8. How can we be sure that the precipitate in bottle 4 came from the seeds? Couldn't some CO_2 have been in the air that was pumped from the room into the system? _____

_____

_____

D. ANAEROBIC RESPIRATION

So far, you have been observing and studying various aspects of respiration that involve oxygen, in other words, aerobic respiration. Many bacteria and yeasts normally respire without oxygen. This is anaerobic respiration. CO_2 is again one of the end products, and measuring the amount of it produced is again a method of observing the effect of temperature—this time upon anaerobic respiration.

Figure 20.5 is a simplified summary of the chemical changes that occur in anaerobic respiration.

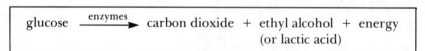

glucose $\xrightarrow{\text{enzymes}}$ carbon dioxide + ethyl alcohol + energy
(or lactic acid)

Figure 20.5   Summary of Chemical Changes in Anaerobic Respiration

Fermentation

The conversion of sugar to ethyl alcohol is called fermentation and is one of the most common types of anaerobic respiration. It is of great commercial value.

► ACTIVITY 8   **Work** in groups of two or four or as directed by the instructor and **proceed** as follows.

1. The instructor will have available a solution in which yeast cells have been actively fermenting sugar for at least a half hour beforehand.
2. Obtain from the instructor two empty fermentation tubes (Figure 20.6) and fill each with the fermenting yeast solution. Tilt the tubes (as demonstrated by the instructor) so as to fill the upright arm with the solution.
3. Set one tube aside on your desk where the room temperature is approximately 20°C.
4. Immediately cool down the second fermentation tube with an ice pack or other procedure as directed by the instructor.

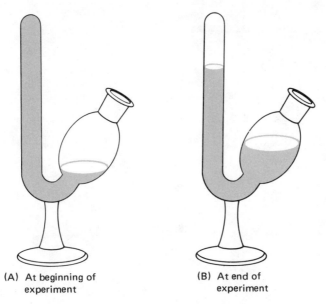

(A) At beginning of
experiment

(B) At end of
experiment

Figure 20.6   Fermentation Tube

5. Place the cooled tube in a refrigerator kept at approximately 5°C.
6. After at least one hour, remove the tube from the refrigerator and compare with the one on your desk.

OBSERVATIONS AND ADDITIONAL TESTS

1. Does the solution still completely fill the arm of both tubes?   YES   NO
2. Which tube shows the greater displacement of the solution from the arm, (a) the 5°C one or (b) the 20°C one? Circle (a) or (b).
3. Measure (with a ruler) the empty part of the arm and record here. (If the tube arm is calibrated, read the amount in milliliters.)
   Refrigerated solution: gas accumulation _____ ml; or _____mm
   Room temperature solution: gas accumulation _____ ml; or _____mm
4. Add 4 to 5 drops of phenol red indicator to the fermentation tube.
5. Without disturbing the gas in the arm, shake gently to distribute the indicator.
6. What is the color of the solution after adding the phenol red? _____
7. You have used this indicator before in previous exercises. This color change is an indicator of the presence of what? _____

CONCLUSIONS

1. Under which conditions did respiration proceed at the faster rate, (a) 5°C, or (b) 20°C? Circle (a) or (b).
2. What was your evidence for this? _____
3. Is carbon dioxide produced during anaerobic respiration? YES NO
4. What was your evidence for this? _____
5. If you had kept one of the fermentation tubes at 50°C, the amount of gas in the arm would have been (a) greater than the 20°C tube; (b) greater than the 5°C tube; (c) less than the 5°C tube; (d) greater than the 5°C tube but less than the 20°C tube. Circle (a); (b); (c); or (d).
6. Explain your answer to question 5. _____

_____

EXERCISE 20   **STUDENT NAME** _____

QUESTIONS

1. The depth of the feeder roots of trees is really quite shallow (20 to 25 cm). Do you think this is a response to the availability of moisture or of oxygen? _____

2. Weed seeds can lie dormant far below the soil surface for many years. A farmer comes along and deep plows the ground. In a few days the area has many weeds germinating. What two factors can you deduce that the weed seed required before germination could occur? (a) _____;

    (b) _____

3. People who grow house plants know that during the winter, night temperatures should be about 10–15 degrees cooler than day temperatures. What immediate effect does this have on what process in the

    plant? (a) _____

    _____

4. What is the purpose of this practice? _____

    _____

5. Two amateur gardeners, Jane and Zane, disagree on whether or not they should soak their seeds before planting them. Jane planted most of her seeds without soaking but did soak the few hard-coated varieties overnight. Zane soaked all of his seeds for four days just to make sure they had plenty of moisture.

    Who will have the smaller gemination percentage, Jane or Zane? _____ Explain: _____

    _____

6. Jane and Zane are also house-plant enthusiasts, but again disagree on cultural practices. Zane's plants are healthier than Jane's, even though the soil in his pots gets dry between waterings. Jane waters her

    plants well every day. What is Jane systematically withholding from her plants? _____

7. One of the main problems that commercial greenhouse operators have is maintaining even heating throughout the house. What two plant processes particularly are affected by temperature differences?

    (a) _____; (b) _____

8. A grower with severe problems of uneven heating in his greenhouses can't sell all his chrysanthemums

    for the same price. Why not? _____

9. Commercial greenhouse growers of ornamental plants face a financial crisis with ever-rising fuel costs. What do you suppose is the focus of many plant-breeding programs in regard to this problem? _____

    _____

10. Do dormant seeds respire? _____

11. Plants that are used for interior design in offices, lobbies, etc., by professional plant care services are used in matched sets of two or three of each kind of plant. They are used in rotation; one is on display while the other two are in a greenhouse. In light of what you have learned in this exercise, explain why this is the best way to manage these display plants. _____

   _____

   _____

12. It's been known for 50 years or more that storing apples at a temperature of 3 to 4°C keeps them firm

   and sweet for five or six months. What is the physiological basis for this practice? _____

   _____

13. More recently the storage life and quality has been doubled to a year by flooding the cold storage vault

   with CO_2. How is this practice similar to that described in question 12? _____

   _____ How is it different? _____

   _____

14. Without herbicides, modern agriculture wouldn't be modern agriculture. Without in-depth knowledge of the normal plant processes of photosynthesis, growth, respiration, and digestion, chemists could not have developed herbicides. From the previous statement, what do you conclude to be the basis of action

   of herbicides in general? _____

   _____

15. Sodium chlorate is an herbicide. If its main *effect* is the depletion of root reserves, what would you con-

   clude to be its *mode of action* on the root tissues? _____

   _____

Table 20.1   Amount of $BaCO_3$ Formed as a Measure of CO_2 Production in the Respiration of Wheat Seedlings Germinating at Different Temperatures

Temperature (°C)	Relative Amount of Precipitate Visible (0, +, ++, etc.)	$BaCO_2$ Production (grams)
5		
20		
30		

EXERCISE 20   **STUDENT NAME** _____

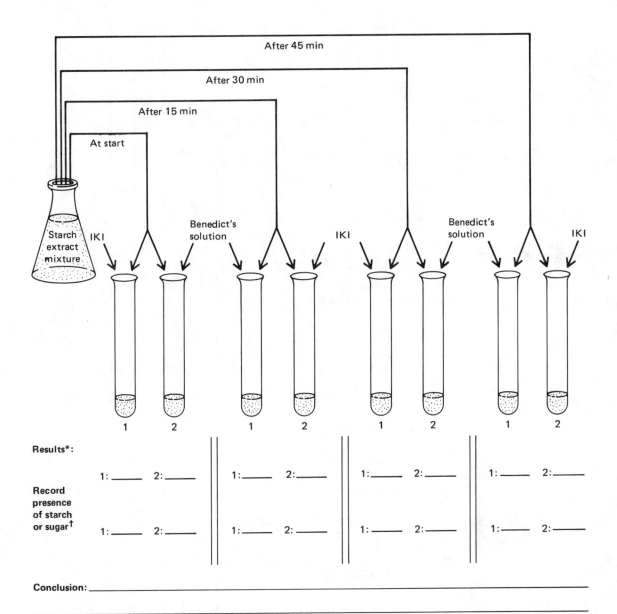

Results*:

Record
presence
of starch
or sugar†

Conclusion: _____

_____

*Use the following symbols: o = no change, +, ++, and
+++ = intensity of positive test.

†Code: St = Starch; Su = Sugar

Figure 20.1   Presence and Extent of Activity of Starch-Digesting Enzymes from Germinating Seeds

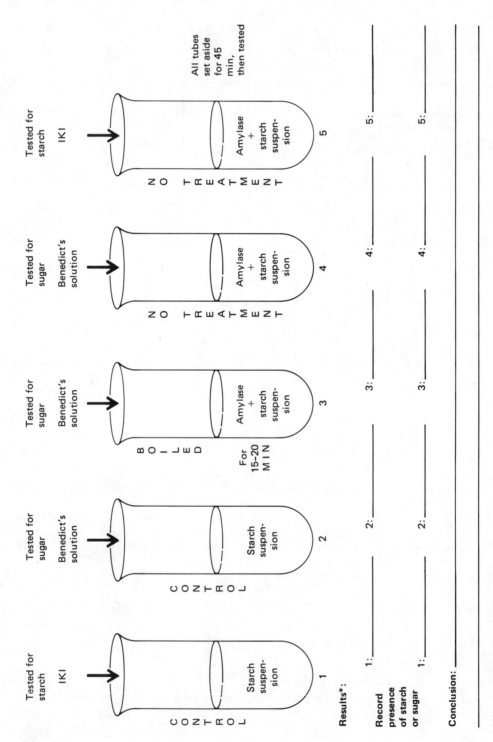

Figure 20.2   The Effect of Heat on Digestive Enzymes

Tube 1 — CONTROL — Starch suspension — Tested for starch — IKI

Tube 2 — CONTROL — Starch suspension — Tested for sugar — Benedict's solution

Tube 3 — BOILED — For 15–20 MIN — Amylase + starch suspension — Tested for sugar — Benedict's solution

Tube 4 — NO TREATMENT — Amylase + starch suspension — Tested for sugar — Benedict's solution

Tube 5 — NO TREATMENT — Amylase + starch suspension — Tested for starch — IKI

All tubes set aside for 45 min, then tested

Results*:

Record presence of starch or sugar

1: _____   2: _____   3: _____   4: _____   5: _____

1: _____   2: _____   3: _____   4: _____   5: _____

Conclusion: _____

*Use the following symbols: + = positive; o = negative

EXERCISE 21

PLANT MOVEMENT AND GROWTH RESPONSES TO STIMULI

Many people think that animals, but not plants, can respond to stimuli from their environment. This idea is wrong. Plants respond to many of the same stimuli that affect animals. It is just that plant response is usually slower and may bring about permanent or semipermanent changes in the growth form of the plant.

Plant growth is controlled by the interplay and balance between several hormones (growth promoting and growth inhibiting). We will study only the auxins, a class of growth-promoting hormones.

In this exercise we will consider just two types of plant responses to stimuli; (1) tropisms and (2) turgor movements. A tropism is an irreversible "bending" or growth curvature induced by auxins. A turgor movement is a reversible, relatively fast movement induced by water pressure changes in the cells.

I. GROWTH IN RESPONSE TO AUXIN

A. THE EFFECT OF AUXIN ON STEM GROWTH CURVATURE

You learned in previous exercises that plant cells grow larger because of the expanding force of internal water pressure. Auxin promotes this enlargement by making the cell wall more plastic, and it then becomes extended by the water pressure.

► ACTIVITY 1   **Work** alone or in groups as directed by the instructor. **Proceed** as follows.

1. **Obtain** from the instructor a small pot or peat cube containing seedlings of oats (*Avena sativa*) or barley (*Hordeum vulgare*) (or other suitable grasses) that are several days old.
2. Examine the tips of the plants and identify the COLEOPTILE, a colorless sheath enclosing the green stem. The coleoptile is easy for you to work with because of its simplicity of structure.

191

3. Obtain several toothpicks and a small quantity of lanolin paste in which auxin is mixed (in the proportion of 1 part auxin per 10,000 parts lanolin). This auxin is indole acetic acid (IAA).
4. Using a tooth pick, apply the auxin paste to the upper 6 to 10 mm of one side of the coleoptile of each seedling. Some seedlings could be left untreated to serve as controls.
5. For a reference marker for later identification, poke a clean toothpick in the soil or peat on the same side of the seedling where the auxin is applied.
6. Set the pots in a warm dark chamber for one or two hours, then remove.
7. **Note** any changes in the stem tips.

OBSERVATIONS

1. Are the auxin-treated stem tips still growing straight up? YES NO
2. How do they look? (a) tip curved; (b) entire stem curved. Circle (a) or (b).
3. Is the curvature toward or away from the side on which the auxin was applied? _____
4. Note the sketch at the right. Which is greater, the distance $A - A'$ or $B - B'$? _____
5. Which side, therefore, has had more cell enlargement? $A - A'$ or $B - B'$? _____
6. Look at your specimens. Which side ($A - A'$ or $B - B'$) corresponds to the side on which you put the auxin paste? _____
7. Why were the seedlings placed in the dark during the experiment? _____
_____
8. Do the untreated seedlings show any curvature? _____
9. Is there a correlation between auxin concentration and differential growth (more growth on one side than on another)? YES NO
10. Did the higher auxin concentration inhibit or promote growth on the side to which it was applied? (Circle one.)
11. In the stems of the untreated seedlings, would you describe the auxin distribution as equal throughout or unequal? (Circle one.)
12. Complete the following statements.
    a. Differential stem growth is induced by differences in _____ of auxin and causes _____ of the stem.
    b. The stem growth assumes a bend in a direction which is _____ the side which has the greater concentration of auxin.

B. THE EFFECT OF THE HERBICIDE 2,4–D ON STEM GROWTH

Natural auxins are very potent hormones. Extremely small quantities can induce curvature and growth. Plant tissues have means of regulating auxin concentrations, otherwise normal growth would be disturbed. Certain herbicides have auxin-like activity, but being synthetics, the plant doesn't always have the ability to degrade (deactivate) them. Hence abnormal growth is induced in the plant, and this eventually kills it. This is the basis of such well-known herbicides as 2,4-D (2,4-dichlorophenoxyacetic acid) and related compounds, silvex, and 2,4,5-T (formerly used to defoliate trees and brush).

► ACTIVITY 2   **Observe** plants that have been sprayed with 2,4-D. Most dicot plants are susceptible to 2,4-D. It is absorbed by the leaves and carried through the phloem. **Note** the abnormal stem bending (the condition is called epinasty). Other symptoms include stem splitting, adventitious root formation, swollen shortened roots, and strap-shaped leaves.

II. PHOTOTROPISM—A GROWTH CURVATURE (BENDING) IN RESPONSE TO LIGHT COMING FROM ONE SIDE

A. POSTURE, PIGMENTATION, AND GROWTH OF PLANTS IN RESPONSE TO LIGHT AND DARKNESS

In the preceding section you have caused the stem to "bend" in the dark simply by applying auxin to one side. You saw also that the bend was in a direction opposite to the side receiving the extra auxin.

This next demonstration shows you the response of plants to light coming from one direction, and the role auxin plays in this response.

► ACTIVITY 3   **Observe** radish seedlings that have been grown in four boxes. Two of the boxes a window made of a glass monochromatic light filter or gelatin paper which allows light of specific wave lengths to pass. One filter allows blue wave lengths of light to pass into the box; another filter allows red wave lengths of light to pass into the box. The third box has clear glass or cellophane, admitting full light (all wave lengths). The fourth box has no window, so the plants are in darkness. A lamp has been placed near the windows of the three boxes. The plants inside these boxes receive light from one direction.

In the probable event that, at this time, it is too early in the class period for you to have seen the results of your experiment in Section IA, we provide you with a bit of information to help you interpret this demonstration.

We know that the stem tip is one of the major centers of auxin synthesis. We also know that under natural conditions (in which the plant receives light from many directions), the auxin moves downward and is distributed evenly across the stem. As a result, growth is even and the young stem grows more or less straight. When auxin is not evenly distributed, growth is uneven; the stem elongates more on one side than the other, resulting in a bend or curve.

OBSERVATIONS

1. In which of the boxes is the distribution of auxin apparently even? _____ _____

2. In which of the boxes is the auxin distribution apparently uneven? _____ _____

3. Which wave length of full light has apparently altered the even distribution of auxin? _____

4. Which wave length of full light apparently has no effect on the distribution of auxin? _____

5. In those seedlings showing curvature, which side on the stems has grown more, the shady side of the illuminated side? _____

6. This means, therefore, that there must be a greater concentration of auxin on which side of these stems? _____

7. Remember that auxin is a mobile substance moving *downward* under the influence of gravity. This particular demonstration shows that auxin also moves _____ under the influence of _____, specifically the _____ wave length.

8. This demonstration shows that the phototropic response of plants is a differential growth rate caused by the _____ (color) wave length of _____

9. What other effect(s) of light is (are) shown in this demonstration? _____ _____

10. Auxin promotes cell elongation. Which plants show the maximum effect of cell elongation? _____

11. Which plants, therefore, show no inhibition of auxin activity? _____

12. You have seen the effect of light on auxin *distribution*. What can you conclude as to the net effect of light on auxin *activity*? _____

13. What wave length is necessary for the synthesis of chlorophyll? _____

Record the results of this experiment in Figure 21.1. Let the symbol ♀ represent a seedling, and sketch the seedlings in the trays, showing their posture. Record color and stem sturdiness.

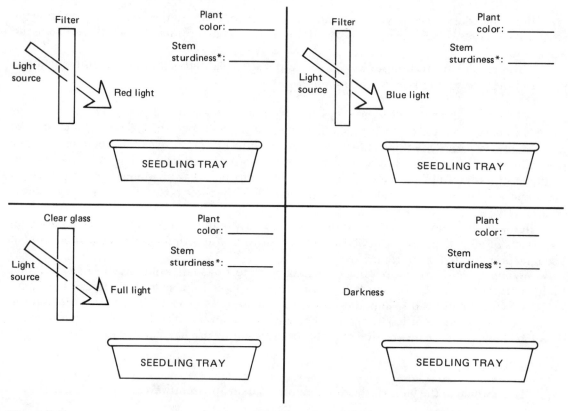

*Use the following symbols for stem sturdiness: o = spindly and weak; +, ++, +++ = increasing degrees of firmness

Figure 21.1   Results of Exposure of Radish Seedlings to Unidirectional Light of Different Wavelengths (color) and Darkness

B. ANOTHER ILLUSTRATION OF THE INFLUENCE OF LIGHT ON AUXIN DISTRIBUTION

➤ ACTIVITY 4   **Observe** two pots of seedlings of oats (*Avena sativa*); barley (*Hordeum vulgare*); or other suitable plants, one set on a rotating platform (clinostat) and the other left stationary. Both plants are illuminated from one direction. Explain the growth response of these two sets of plants in terms of auxin. _____

C. PHOTOTROPIC RESPONSE OF FUNGI TO LIGHT

➤ ACTIVITY 5   **Observe** the culture of the fungus *Phycomyces blakesleeanus* which has been exposed to light coming from one direction.

1. Does the fungus show a phototropic response?   YES   NO.
2. The green plant must have light for photosynthesis and, therefore, its positive phototropic response has survival value. Are fungi green and photosynthetic?   YES   NO.

3. Can you think of any probable value for this response of the fungus? Although photo-synthesis requires light, could it be that normal growth and development also depend to some degree on light? YES NO. Does the fungal response support this? YES NO. Quite so! Many organisms, animals included, depend on light for a variety of biological functions. Organisms of this planet have evolved in close relationship with the spectral energy from our nearest star, the sun.

III. GEOTROPISM

Geotropism is a growth curvature in response to gravity. As in phototropism, it is brought about by the uneven distribution of auxin. Geotropism is described as *negative* if the plant part turns away from the earth and as *positive* if the plant part turns toward the earth.

A. GEOTROPISM IN GERMINATING SEEDS

► ACTIVITY 6   **Observe** corn seeds that have been germinating inside petri dishes and kept moist and stationary by a backing of wet filter paper. The seeds were originally set in the dish with no attempt made to orient them all the same way. The dish has been kept vertical.

OBSERVATIONS

1. In what direction do the roots turn? _____; the shoots? _____
2. Have any roots or shoots made a complete U turn?   YES   NO
3. Of what value to us is this innate geotropic response of the young root and shoot of the seed? _____

B. GEOTROPISM IN THE WHOLE PLANT

► ACTIVITY 7   **Observe** a potted specimen of coleus (*Coleus blumei*); geranium (*Pelargonium hortorum*); *Kalanchoe* sp.; or other suitable herbaceous plant that has been set on its side for several days or (if class time permits) for an hour or so. The plant should be knocked out of its pot so that the roots can be seen. If the plants have been grown in peat cubes, they make even better speci-mens for this illustration because the roots are more readily seen.

OBSERVATIONS

1. In what direction are the root tips turned? _____; the shoots and leaves? _____
2. If auxin moves in response to gravity, where is the higher concentration of auxin in the root? _____; in the shoot? _____
3. In order for the root to "bend" downward, which of its sides (upper or lower) would have had to grow more? _____
4. In order for the shoot to "bend" upward, which of its sides (lower or upper) would have had to grow more? _____
5. Approximately the same distribution (concentration) of auxin exists on the lower side of the horizontal roots and shoots of this plant. How can you account then for their opposite growth response? _____
_____

IV. TURGOR MOVEMENTS

Turgor movements result from rapid changes in cell size resulting from loss of turgor pres-sure. Turgor pressure drops because of loss of water. When water pressure is regained by the cells, the movement of the plant parts affected is reversed.

► ACTIVITY 8

1. **Observe** the reaction of the leaves and leaflets of the sensitive-plant (*Mimosa pudica*) to various stimuli and intensity of stimuli, such as touching them with a pencil or holding a lighted match near (but not *too* near).

2. **Observe** specimens (if available) of Venus's flytrap (*Dionaea muscipula*) or sundew (*Drosera rotundifolia*). The reactions of these plants are in response to the pressure of the weight of insects who light on them. These plants are insectivorous. Changes in leaf position form traps for insects which are then digested by enzymes secreted by the leaves.

3. **Examine** demonstration microscopes with slides of cross sections of leaves of corn (*Zea mays*) showing the leaf in the folded and open positions. **Locate** the bulliform cells whose collapse due to loss of water pressure causes the corn leaf to fold lengthwise. Under what environmental conditions would the leaf fold? _____.
This folding is beneficial because it reduces water loss. How is this reduction accomplished? _____

EXERCISE 21   **STUDENT NAME** _____

QUESTIONS

1. What are two environmental factors to which plants respond by a growth curvature?

   (a) _____; (b) _____

2. Which wave length(s) of full light is (are) responsible for plant bending toward light?_____

   _____

3. Is light necessary for the synthesis of chlorophyll? _____ Any specific wave length? _____

4. The word auxin is derived from the Greek, *auxein*, to increase. Give the specific action exerted on the

   cell by auxin (which is the basis for its name). _____

   _____

5. The synthesis and development of the 2,4-D family of herbicides grew out of basic botanical and chemi-

   cal studies of _____ in plants.

6. If you were a chemist charged with the research project to develop a synthetic growth hormone to be
   used as an herbicide to interfere with weed seed germination, *what growth response* of the seed would you

   target in on? _____ (There *is* such an herbicide. It is naptalam; it blocks IAA action

   and the seed loses its normal behavior to this growth response.)

7. The same concentration of auxin occurs on the lower side of both stem and root of a plant placed horizontally. On this lower side; (a) is root cell elongation inhibited or promoted? Circle one. (b) is shoot
   cell elongation inhibited or promoted? Circle one.

8. From your observations of the root and shoot growth response of the horizontally placed plant, it can be

   concluded that the _____concentration of _____that inhibits _____

   growth, also promotes _____ growth.

9. Suppose that during one spring a cold snap occurs in a Michigan apple orchard. It kills many of the
   pollinating bees but not the flowers. The grower knows that without pollination, no seeds will be formed
   and that without seeds, no fruit will develop. What do you suppose the seeds produce and without which

   no fruit develops? _____

10. What could this apple grower do (and in fact it is common practice) in order to get his trees to set fruit?

    _____

PART II

Survey of the Plant Kingdom

Exercise 22

BACTERIA

Today you will look at some of the simplest organisms we know—the bacteria. They are single-celled organisms. Although structurally simple, their metabolism is complex. Many authorities consider them plants chiefly because they have a cell wall. Other specialists think bacteria are so different that they belong in a separate subkingdom, or even kingdom, distinct from both plants and animals.

Only the blue-green algae bear any resemblance to bacteria. They are commonly placed in the same division (Schizophyta) with the bacteria because of certain similarities in internal cell structure and in their reproduction by simple binary fission.

There is tremendous variety within the bacteria. We find modes of life among them more varied than all other plant types together. There are unicellular and colony-like forms. Many resemble fungi, slime molds, protozoans, or algae. Some are aquatic, some terrestrial. A few are autotrophic. Most are heterotrophic, and of these, more than 99 percent are harmless saprophytes.

We have time to study only the most common bacteria which also happen to be among those with the simplest forms.

I. CHARACTERISTICS OF THE MOST COMMON BACTERIA

1. Unicellular and microscopic
2. Nonphotosynthetic
3. Saprophytic
4. Rigid cell walls
5. Reproduction by simple fission

II. MICROSCOPIC APPEARANCE OF BACTERIA

There are three cell shapes.

1. COCCUS   A spherical shaped cell (plural: cocci).
2. BACILLUS   A rod-shaped cell (plural: bacilli).
3. SPIRILLUM   A rod-shaped cell with one or more curves (plural: spirilli).

► ACTIVITY 1   **Examine** a prepared slide showing the three cell shapes found in different species of bacteria. Do the cells always occur as single isolated cells?   YES   NO. Do you see any chains of cocci?   YES   NO. These chains are considered colonies. They are called STREPTOCOCCI. Certain species of bacteria characteristically grow in chains of cells. Do you see any chains of bacillus forms?   YES   NO. Are all bacilli the same thickness?   YES   NO.

How many cells would form a DIPLOCOCCUS colony? _____ Are there any on your slide?   YES   NO. Cocci that grow in clusters forming flat, irregular plates are called STAPHYLOCOCCI.

III. SIMPLE STAINING

► ACTIVITY 2   **Prepare** a simple stain of a bacterial smear. It is difficult to see bacterial cells, even with high magnification of the microscope. Dyes are used to make them more visible and to distinguish among different species. We will do a simple stain which involves a single dye. You will be provided with agar slant cultures of two organisms, *Bacillus megaterium* (a rod-shaped species) and *Micrococcus luteus* (a spherical bacterium).

Proceed as follows.

Preparing the Slide

1. Thoroughly **clean** a slide with scouring powder. (Good smears **require** clean slides! Otherwise you won't get even staining and even spread of the cells.) Rinse and wipe dry, being careful to hold the slide at its edges. (Fingerprints are oily!)
2. Gently flame the slide to remove any residual grease. This promotes an even smear.
3. Set the slide down with the clean, flamed side up.

Making the Smear

In the following steps, use the sterile technique demonstrated by the instructor.

1. With a bacteriological inoculating loop, put one drop of tap water near one end of the slide (but not too close) and another drop near the other end.
2. Hold the culture tube (agar slant containing the bacterium) in your left hand, parallel to and on top of your fingers, holding it in place with your thumb.
3. Hold the inoculating loop in your right hand and flame the wire to redness in the flame of your burner. Gently flame the lower part of the handle also. Do not set the loop down.
4. Grasp the closure (cap or cotton plug) of the culture tube with the little finger of your right hand and twist off the closure.
5. Flame the lip of the tube gently for a second or so.
6. Still holding the closure, insert the sterile loop into the tube and gently touch it to the layer of bacterial growth.
7. Remove the loop from the tube and replace the closure. Set the tube down in a block or carton.
8. Emulsify the bacteria on the loop in one drop of water on the slide. Do not get the suspension too heavy. It should be barely cloudy. (Have the instructor check it.) Smear it around evenly to cover an area the size of a nickel (2 cm diameter).
9. Reflame the loop to sterilize it and set it down.

10. Repeat the procedures (steps 1 through 9) for the second bacterial culture.
11. Allow the slide to air dry (without heating which may distort wet cells).
12. Pass the slide (smear side up) back and forth through the flame of your burner until it is warm to the back of your hand, but *not hot*. Gentle heating causes the cells to adhere to the slide. Excessive heat distorts the cells.

Staining the Slide

1. Place the slide on a staining rack or across the top of a beaker that is three quarters full with tap water.
2. Cover each smear with 3 to 4 drops of crystal-violet dye and let it stay on for one minute. Other dyes can be used, but crystal-violet is best for this simple staining.
3. *Gently* rinse off the dye in *slowly* running tap water or rinse it in the beaker, pouring off, and adding clean water a couple of times.
4. Drain off the excess water from the slide on a paper towel.
5. Blot the slide dry (don't rub or wipe!) between folds of paper towel.

Examining the Slide

Examine the slide under low or medium magnification to locate the bacteria. Find an area where the cells are separated so you can see them clearly. Shift to high power or to the oil immersion lens if your microscope has one.

Do all the cells look the same? YES NO. Is there much variation in their size? YES NO. Is the staining uniform or irregular (that is, some places more heavily stained than others)? _____. Do you have many dense clusters of cells? YES NO. Does most of your smear have a fairly even scattering of cells? YES NO. Probably you see large, dense clusters and uneven staining. As a trainee bacteriologist, you couldn't be expected to make an evenly smeared and stained slide the first time around. But, perhaps you did! Good!

IV. MOTILITY IN BACTERIA

➤ ACTIVITY 3   **Examine** a bacterium that shows motility, such as *Rhodospirillum rubrum*. **Follow** this procedure.

1. Using a sterile bacteriological loop, transfer several loopsful of a broth culture of *Rhodospirillum rubrum* to a clean glass slide. (It is not necessary to make a hanging drop.)
2. Put a little vaseline in the palm of your left hand. Touch the four edges of a cover glass to the vaseline. Then lower the cover glass on the liquid on the slide. The vaseline will seal your slide and prevent differential drying and other problems while you are trying to find the bacterium.
3. Examine on high magnification and cut down on your light in order to see the organism better. Use oil immersion if it's available.

You probably won't be able to see the flagella. The bacterium has tufts of flagella at each end. What is the cell shape of *R. rubrum*? _____. How can you tell that the organism is really moving and not just showing Brownian movement? _____

V. RESISTANCE AND SENSITIVITY TO BACTERIOSTATIC AGENTS

Compounds that inhibit bacterial growth are said to be bacteriostatic. Some of these agents are relatively *simple* compounds, for instance, detergents, disinfectants, and antiseptics. Antibiotics are totally different kinds of bacteriostatic agents. They are the products of metabolism of certain fungi and soil bacteria. Examples of antibiotics are penicillin, tetracycline, and aureomycin. Bacterial species differ in their response to different bacteriostatic agents. We speak of this as their resistance or sensitivity to these agents.

The following procedure is a common method for determining bacterial response to bacteriostatic agents.

▶ ACTIVITY 4   **Work** in pairs, each student using a different organism.

1. Obtain a sterile agar petri dish and broth cultures of *Bacillus megaterium* and *Escherichia coli*.
2. With a wax pencil, draw a line on the glass of the bottom (agar-filled) half to divide the dish into two equal areas.
3. Each student of a pair inoculates one half of the agar plate with one organism. Do not use the same organism on both halves of the plate. Do this by the following method:
   a. Dip a cotton swab into a broth culture of the organism.
   b. Gently streak the surface of the agar plate with the swab as directed by the instructor. *Do not* dig into the agar with the swab.
   c. Keep the cover on the agar plate at all times except when introducing material into it.
4. Obtain paper disks that have been impregnated with a solution containing some bacteriostatic agent. Choose three agents. Some suggestions are; iodine; methiolate; vinegar; Listerine or other mouth wash; Lysol concentrate; liquid bleach (sodium hypochlorite —a 5 percent aqueous solution); sodium chloride (20 percent solution); and various antibiotics as are available.
5. You need to keep track of which bacteriostatic agent is in which disk. Either the instructor will have precoded the disks before autoclaving, or you can put a code number on each with pencil. Another method is to write a number on the bottom of the agar plate with a grease pencil and then place the disks over the number. Record which bacteriostatic agent each number represents.
6. Use the same three agents on both organisms on the plate.
7. With forceps, gently lay the disks on the agar in the area of your streaks. *Do not place the three disks close to each other.* Each student in a pair places three disks on her (his) half of the petri dish.
8. Cover the petri dish, invert it, and give it to the instructor to be incubated until the next laboratory session.
9. During the next laboratory session, examine the plates to determine the resistance or the sensitivity of the organisms to the bacteriostatic agents. Clear agar around the disk shows that growth has been suppressed.
10. Record this information in Table 22.1. Additional spaces are allotted for organisms other than *E. coli* and *B. megaterium*.

VI. NITROGEN-FIXING BACTERIA

Nitrogen-fixing bacteria convert (fix) nitrogen gas (N_2), which is in the atmosphere and in soil air, to organic nitrogen compounds essential to plant growth. In modern farming, standard practice employs abundant amounts of nitrogen fertilizer. Without nitrogen, plants (and ultimately, animals) cannot make proteins. Bacteria also play an important role in the cycling of other minerals and making them available to plants. Some of these other minerals are calcium, sulfur, iron, magnesium, and molybdenum.

The natural, and only, resevoir of nitrogen is not rock strata, but air—both atmospheric and soil air. By means of chemical technology, a variety of nitrogen fertilizers are synthesized, the process starting with atmospheric nitrogen. However, this chemical synthesis requires fuel; bacterial synthesis does not. (*Note*: it *does* require energy—the respiratory energy of the bacterial cell.)

Rhizobium sp. is the most important soil organism for the fixation of atmospheric nitrogen. However, the bacterium cannot fix the nitrogen unless it is living in the roots of plants, especially

Table 22.1   Response of Certain Bacterial Species to Some Bacteriostatic agents

Bacteriostatic Agent	*E. coli*	*B. megaterium*		

Code: R—resistance; S—sensitivity.

the legume plants, for example, alfalfa (*Medicago* sp.); sweet clover (*Melilotus* sp.); clover (*Trifolium* sp.); *Lespedeza* sp.; birdsfoot trefoil (*Lotus* sp.); soybean (*Glycine* sp.); pea (*Pisum* sp.); and vetch (*Vicia* sp.).

Species of *Rhizobium* invade the root hairs of these plants (and a few other, nonleguminous plants). Their presence stimulates the plant to form small tumor-like growths, called NODULES, around them. Inside the nodules, the bacteria consume food compounds made by the plant. They also fix nitrogen and synthesize organic nitrogen compounds which the plant uses. The association is MUTUALISTIC, both plant and bacteria deriving substantial benefits from it.

After harvest or plowing under of these crops, the soil is enriched by the nitrogen compounds released from the dead plant debris and the bacterial cells that eventually die. This is the basis for rotating (alternating) legume crops with other crops. The introduction of clover into customary crop rotations in the 16th century revolutionized agricultural practices throughout the world.

► ACTIVITY 5   **Examine** the roots of leguminous plants and note the nodules. If time permits, crush open a nodule in a drop of water on a slide. Transfer some of the liquid to another slide and prepare a stained slide for examination as you did earlier in this exercise.

EXERCISE 22   **STUDENT NAME** _____

QUESTIONS

1. What are the two general features of similarity between bacteria and blue-green algae?

   (a) _____

   (b) _____

2. In general, why are bacteria considered more plant-like than animal-like? _____

   _____

3. What is the term for a spherical-shaped cell? _____

4. What is the plural for this term? _____

5. When we speak of a single bacterial cell, it is called: (a) a bacteria; (b) a bacterium. Circle (a) or (b).

6. What is the term for a rod-shaped bacterial cell? _____

7. When making a stained smear, what are two undesirable results if the slide is not absolutely clean?

   (a) _____; (b) _____

8. Define a bacteriostatic agent. _____

9. Briefly, in one short sentence, define a nitrogen-fixing bacterium. _____

   _____

10. Crop rotation with legumes has been practiced (by western civilization, at least) since when? _____

    _____

11. Briefly, in a short *phrase*, what does the farmer *do* when he rotates crops? _____

    _____

12. Corn shows dramatic response to the addition of nitrogen fertilizer. Soybeans show a slight response, at

    best. Can you explain this? _____

    _____

13. Name the genus that is the most important soil organism for fixing nitrogen. _____

14. The three major nutrients used by plants are nitrogen, potassium, and phosphorus. What is the world's

    only natural storehouse for nitrogen? _____ What is the source

    material for potassium and phosphorus? _____

15. If you were to dig up a clover plant (roots, nodules, and all), in *what living tissues* would you find (by

proper assay) the organic nitrogen compounds synthesized by the bacteria? _____

_____

16. If a field has no leguminous (or other suitable) plants growing in it, will the *Rhizobium* fix nitrogen?

_____

17. The source of free nitrogen for bacterial synthesis of organic nitrogen compounds is the air. What is the

source of free nitrogen forthis synthesis by industrial chemical technology? _____

18. Less than 1 percent of known bacteria are harmful. Circle one: (a) true; (b) false.

19. Since the common bacteria all *look* alike (there are only three different shapes for the thousands of species), how then do you suppose a bacteriologist can make a positive identification of a bacterium?

_____

_____

20. The name Schizophyta is made up of the Greek words "phyta," meaning plant, and "schiza" meaning to splinter. What characteristic of bacteria is the basis for the composition of the name Schizophyta?

_____

EXERCISE 23

FUNGI I:
INTRODUCTION TO FUNGI,
PRIMITIVE FUNGI, AND SLIME MOLDS

Fungi are considered to be plants. But they don't look much like plants. They don't have any of the familiar roots, stems, leaves, or tissues that you've seen in your earlier studies. In fact, they are so different that we have to use a different vocabulary to describe them.

Fungi lack chlorophyll. So, instead of synthesizing their own food like other plants, they decompose and consume organic matter made by other organisms. The vast majority are beneficial. They decompose dead organic matter that otherwise would accumulate in the environment. These fungi are called saprophytes. A minority live on living material, such as other plants or the flesh of animals. These are the parasites. As with all "culprits," the destructive activity of parasites receives far more attention from people than the less conspicuous, but beneficial, action of the saprophytes.

Some terms used for fungi.

FUNGUS	Term for a single plant (plural: fungi).
MOLD	A commonly used term for fungus.
THALLUS	Name for the entire plant body, both vegetative parts and reproductive parts. (This term is also used for algae and bryophytes.)
MYCELIUM	Vegetative ·or assimilative phase of the thallus. Consists of white masses of microscopic, branching filaments, called hyphae. A mycelium is perennial. It lives for several years and usually isn't seen because it grows under the soil surface or inside the host material.
HYPHA	A branching, threadlike tubule which, in huge numbers, comprises the mycelium (plural: hyphae).
FRUITING BODY	Reproductive part of the thallus that bears the sexually produced spores. In some species it is a microscopic or extremely small single sac. In other species it is a conspicuous, fleshy organ. It may live for a few days or several years.
COENOCYTE	A structure, such as the mycelium, with numerous nuclei and no cell walls separating them. The adjective is coenocytic.
NONSEPTATE	Means the same as coenocytic—without cross walls; refers specifically to filaments or tubules that lack cross walls.
SEPATE	With cross walls.
PERFECT STAGE	The phase of the life cycle in which sexual reproduction occurs. Takes place only once a year in most fungi. Brings about genetic variation in the population.
IMPERFECT STAGE	The phase of the life cycle in which asexual reproduction occurs. Takes place many times during the season. It is the chief means for increasing the population size.

I. THE BIG PICTURE OF THE WORLD OF TRUE FUNGI

Before diving into details, let's get a general idea—a "big picture" of the world of the true fungi. The variety of types is a reflection of their evolution. In this evolution we can see that four major trends or lines of advance have occurred:

1. A trend from aquatic life to terrestrial life.
2. A trend from small inconspicuous size to large conspicuous size (especially in the fruiting body).
3. A trend from a simple sac (that merely releases spores) to complex discharge and dispersal mechanisms (that spread spores more efficiently).
4. A trend from a haploid thallus to a diploid thallus.

Figure 23.1 and Table 23.1 illustrate the general differences between the classes of true fungi (Eumycophyta). These classes are

1. Class Oomycetes—Water molds and their relatives. Examples: downy mildews, potato blight fungus, white rusts, pythium.
2. Class Zygomycetes—Primitive terrestrial molds without fruiting bodies. Examples: bread mold, fly fungi.
3. Class Ascomycetes—Terrestrial molds most of which have fruiting bodies that are spherical, flask-, or cup-shaped. Examples: morel, cup fungi, dutch elm disease fungus, apple scab fungus, ergot.
4. Class Basidiomycetes—Terrestrial molds most of which have fruiting bodies generally larger than those of ascomycetes and also very diverse in form. Examples: mushrooms, puffballs, rusts, smuts, stinkhorns.

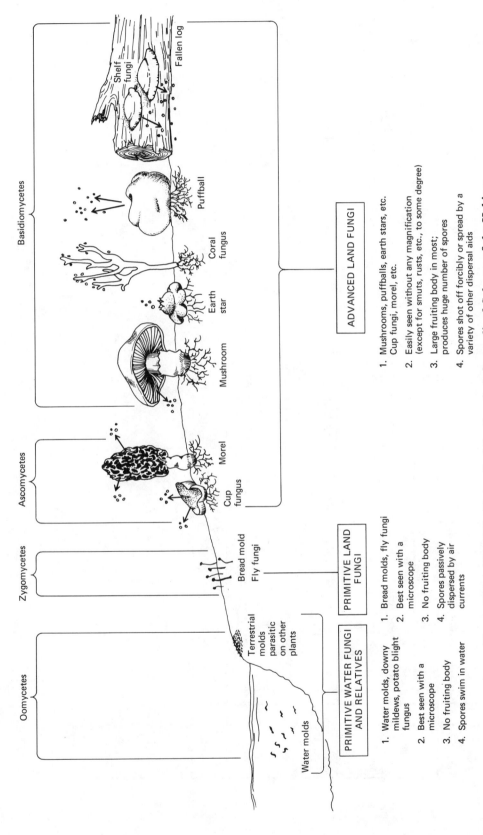

Oomycetes

Zygomycetes

Ascomycetes

Basidiomycetes

Shelf
fungi

Fallen log

Puffball

Coral
fungus

Earth
star

Mushroom

Morel

Cup
fungus

Bread mold
Fly fungi

Terrestrial
molds
parasitic
on other
plants

Water molds

PRIMITIVE WATER FUNGI
AND RELATIVES

1. Water molds, downy
   mildews, potato blight
   fungus
2. Best seen with a
   microscope
3. No fruiting body
4. Spores swim in water

PRIMITIVE LAND
FUNGI

1. Bread molds, fly fungi
2. Best seen with a
   microscope
3. No fruiting body
4. Spores passively
   dispersed by air
   currents

ADVANCED LAND FUNGI

1. Mushrooms, puffballs, earth stars, etc.
   Cup fungi, morel, etc.
2. Easily seen without any magnification
   (except for smuts, rusts, etc., to some degree)
3. Large fruiting body in most;
   produces huge number of spores
4. Spores shot off forcibly or spread by a
   variety of other dispersal aids

Figure 23.1   The "Big Picture" of the Fungi. A Generalized Scheme of the Habitat,
Relative Size, and Evolutionary Status of the Various Groups

Table 23.1   Differences among the True Fungi (Eumycophyta)

Class	Oomycetes	Zygomycetes	Ascomycetes	Basidiomycetes	Fungi Imperfecti
Habitat	Some aquatic, some terrestrial	Mostly terrestrial	Mostly terrestrial	Mostly terrestrial	Mostly terrestrial
Mycelium	Nonseptate, growth not extensive	Nonseptate, growth not extensive	Septate, extensive in growth	Septate, extensive in growth	Septate, extensive in growth
Sexual spore stage	Thick-walled resting stage, called OOSPORE, not in a fruiting body	Thick-walled resting stage, called ZYGOSPORE, not in a fruiting body	ASCOSPORES (usually 8) borne inside a sac (ascus), usually in a fruiting body; many lack the sexual stage	BASIDIOSPORES (usually 4) borne on the outside of a club-like cell (basidium), usually in a fruiting body	None
Asexual spore stage	Motile spores (ZOOSPORES) produced inside a sporangium	Nonmotile spores (APLANOSPORES) produced inside a sporangium	Nonmotile spores (CONIDIOSPORES) formed on the tips of specialized filaments (CONIDIOPHORES); some lack an asexual stage	Nonmotile spores (CONIDIOSPORES) formed on the tips of specialized filaments (CONIDIOPHORES); many lack an asexual stage	Nonmotile spores (CONIDIOSPORES) formed on the tips of specialized filaments (CONIDIOPHORES)
Motile cells	Present	None	None	None	None

II. MAKING SENSE OUT OF PLANT LIFE CYCLES

In this exercise and in those that follow you will study the details of the life cycles of representative plant groups. These details can be confusing and often a colossal bore to many beginning students. But basically all life cycles are the same. They're all involved with survival of the group.

Any plant group has four basic requisites for survival, which it meets as follows.

Requisite	Accomplished by
1. It must increase its numbers 2. It must disperse its offspring 3. It must obtain and assimilate foods for energy 4. It must put some genetic variety into its populations	1. Asexual reproduction (primarily) 2. Motile cells or air- or animal-carried cells or seeds, etc. 3. Decomposing and consuming organic matter—if a fungus; photosynthesis—if a green plant 4. Sexual fusion and meiosis

These can be restated as *four survival phases*:

1. Multiplication
2. Dispersal
3. Assimilation
4. Genetic variation

The details of every life cycle fall into one of these four phases. Variety is the "spice of life," and every group has its own style of going through these phases. The style may differ but the essence of each phase is the same.

Genetic variation is very important in life cycles and needs a bit more explanation here. Meiosis and sexual fusion work together. Just as a card dealer shuffles the cards between each game so that the players can receive different combinations of cards, so MEIOSIS rearranges genes and gene groups so that the resulting haploid nuclei receive new gene combinations. These new gene combinations contribute to genetic variety. Meiosis also segregates two chromosome sets so that each daughter cell gets one set.

SEXUAL FUSION brings together two haploid nuclei, both bearing somewhat different gene combinations in their otherwise matching chromosome sets.

III. DIVISION EUMYCOPHYTA—CLASS OOMYCETES

A. AQUATIC OOMYCETES

Mycelium—The Vegetative Part of the Thallus

▶ ACTIVITY 1   **Examine** water cultures containing living water molds, such as species of *Saprolegnia* or *Achlya*. The mold will be growing on dead hemp seeds or dead insects in the water. **Examine** the MYCELIUM under the microscope. What are the filaments or tubes that compose this mycelium? _____. Can you see any cross walls in these tubes?   YES   NO. Give the adjective used to describe this type of mycelium. _____. Is this mycelium haploid or diploid? _____. (See Figure 23.2.) What survival phase does the mycelium represent? _____. In Table 23.2, **enter** the name mycelium in the box opposite the survival phase it represents.

The Reproductive Parts of the Thallus

Examine a prepared slide showing the asexual and sexual organs of *Saprolegnia* sp. or *Achlya* sp. **Refer** to Figure 23.2 as a guide.

asexual structures   **Look** for oval, cigar-shaped sacs at the tip of a hypha. Each such sac is a SPORANGIUM. It contains many cells that are released and swim about or are carried about by water currents. These motile spores are called ZOOSPORES. These settle down and encyst for a while. Later the encysted cells germinate into secondary zoospores. Each secondary zoospore can grow into a new thallus. What two survival phases are represented here? (a) _____; (b) _____. In Table 23.2 **enter** the names of these structures in the boxes opposite the survival phases they represent.

sexual structures   **Look** for round bodies attached along the sides of the hyphae. These are the female sex organs, called OOGONIA (singular: OOGONIUM). MEIOSIS occurs within the oogonium and 5 to 10 haploid EGG NUCLEI are formed. Male sex organs, called ANTHE-RIDIA (singular: ANTHERIDIUM) look more like an ordinary hypha. MEIOSIS occurs within the antheridia, producing haploid SPERM NUCLEI. When an antheridium comes in contact with an oogonium, it puts out finger-like extensions, called FERTILIZATION TUBES. These can pierce the oogonium, and each tube enters one egg and delivers a sperm nucleus directly to the egg. If fertilization has occurred, you will see four to five thick-walled, opaque zygotes inside the oogonium. These zygotes, termed OOSPORES, are diploid; they have two sets of genes— one from the sperm and one from the egg. After a period of dormancy (which typically coincides with adverse, droughty conditions or the cold season), each OOSPORE germinates into a short tube with a terminal sporangium, the GERM SPORANGIUM. Several germ sporangia may be formed. The germ sporangium produces cells which, when released, are motile ZOOSPORES. These are carried about by water and eventually can grow into a new thallus.

These structures studied in this section fall into what two survival phases? (a) _____; (b) _____. **Enter** the names of these structures in the appropriate boxes in Table 23.2. **Complete** the labeling of the life cycle of *Saprolegnia* in Figure 23.2, using the capitalized terms defined in the preceding description. **Label** the four survival phases in Figure 23.2

B. TERRESTRIAL OOMYCETES

Certain of the water molds cause serious plant diseases. *Phytophthora* causes blight in potatoes and brought on the potato famine in Ireland in the 1840s; *Pythium* causes root rot in many greenhouse and field crops; and downy mildew disease is caused by a variety of water molds one of which almost wiped out the grape-wine industry in France in 1882.

IV. DIVISION EUMYCOPHYTA—CLASS ZYGOMYCETES

Zygomycetes are primitive terrestrial molds without fruiting bodies. Some of these fungi are saprophytes, such as the bread molds, and others are parasitic on insects, such as the fly fungi.

▶ ACTIVITY 2   **Examine** living specimens of *Rhizopus stolonifera.* **Refer** to Figure 23.3 as a guide. **Obtain** an intact portion of the mycelium that has been grown on 1-cm² sections of cellophane paper over agar. **Mount** the cellophane section with the mycelium on a *dry* slide and examine *uncovered* under medium power (100×) on your microscope.

The Mycelium and the Asexual Structures

Note that the mycelium is a white or grayish white mass of branched filaments not very different from those of *Saprolegnia.* Is this mycelium haploid or diploid? _____. (See Figure 23.3.) What survival phase does this mycelium represent? _____. **Enter** the name mycelium in Table 23.2 in the appropriate box opposite the survival phase it represents.

If the culture is old enough, you will be able to see tiny black spherical bodies, the SPORAN-GIA (singular: sporangium) attached to erect hyphae, called SPORANGIOPHORES. Sporangia contain haploid cells, SPORANGIOSPORES. These have been formed from simple mitotic di-vision of the haploid nuclei of the thallus. Do they contain the same gene combinations as the parent thallus?   YES   NO. When released, they are dispersed about by air currents and later each can germinate into a new thallus. Will this new thallus be haploid or diploid? _____ _____. What survival requisite does asexual reproduction accomplish? _____ _____.

Add a drop of water to the mycelium on the slide and cover with a cover glass. **Examine** the sporangia and spores under the microscope. **Note** the cluster of root-like hyphae at the base of each sporangiophore. These are called RHIZOIDS. They anchor the hyphae. Coming off laterally from the base of the sporangiophore are horizontal-growing hyphae, known as STOLONS. What survival phase(s) are these structures involved in? _____ _____. **Enter** the names of these structures in the appropriate boxes in Table 23.2.

Sexual Structures and the Process of Conjugation

Refer to Figure 23.3 as a guide again. **Examine** a prepared slide of *Rhizopus stolonifera* or a living culture of the organism showing sexual reproduction. When two different strains of mycelia (we call them "plus" and "minus" strains) come in contact, every HYPHA that touches another sends out a side bulge at the point of contact. This bulge is called a PROGAMETANGIUM. Each progametangium grows larger and pushes against the other. They form a bridge between the hyphae. Soon the nuclear and cytoplasmic contents of each progametangium is walled off from the rest of its hypha. This walled-off part is now called a GAMETANGIUM. This word simply means a "sac of gametes," and that's what it is—a sac of nuclei that will behave as gametes, the sperms or eggs.

When gametangia are mature, their intervening walls break down, and the nuclear contents of the two fuse, that is, fertilization, CONJUGATION, takes place. **Look** to see if you can find a dark (dense) sphere in the "bridge" formed by the gametangia. This sphere is the ZYGOSPORE. It contains the diploid nuclei formed by nuclear fusions inside the gametangia. This zygospore remains dormant for a while, and then its nuclei undergo meiosis to form haploid nuclei. All these haploid nuclei abort except one. This one nucleus and the accompanying cytoplasm develop into a GERM SPORANGIUM which produces SPORANGIOSPORES all of one nuclear makeup. Sporangiospores germinate into the thallus. Most of these structures described here represent what survival phase? _____. In Table 23.2 **enter** the names of these structures in the box opposite the appropriate survival phase they represent.

Complete the labeling of the life cycle of *Rhizopus stolonifera* in Figure 23.3, using the capitalized terms defined in the preceding description.

► ACTIVITY 3   If available, **examine** living cultures of "plus" and "minus" strains of *Phycomyces blakesleeanus* which have sexually crossed and produced zygospores. **Note** that the zygospores are formed only along the line of contact of the two strains.

V. MYXOMYCOPHYTA—THE SLIME MOLDS

Slime molds are a small, insignificant group of very little economic importance. Botanists call them fungi because, like fungi, they produce sporangia. Zoologists think they are protozoans because the thallus looks and behaves like a giant amoeba. Since they are not totally fungus-like or animal-like, opinions differ about their identity. So they are classified as a group totally separate of either plants or animals. This group is placed in the division or phylum Myxomycophyta.

► ACTIVITY 4   **Examine** a living slime mold such as *Physarum polycephalum* growing on agar or filter paper in a petri dish. **Use** 100 × magnification with the microscope. *do not* expose the plasmodium to bright light, or within 30 minutes it will go into a sporangial stage. The thallus or assimilative phase is a broad, flat-lying mass of naked protoplasm known as the PLASMODIUM. Portions of it fan out wider than others and you can see "veins" of streaming nuclei and particles of cytoplasm.

The plasmodium is a coenocyte. It flows over the surface by protoplasmic streaming and engulfs bacteria and other minute particles in its path. During its reproductive phase, it stops moving and produces erect, sac-like structures, called SPORANGIA. Cells inside these divide by meiosis to form HAPLOID SPORES. These haploid spores are released, blown around, and lie about for a while going through a period of DORMANCY. Later they germinate and two kinds of cells are formed, flagellated SWARM CELLS and amoeba-like MYXAMOEBA. These behave as gametes, two swarm cells unite or two myxamoeba unite. Their protoplasms fuse but not their nuclei. This fusion of protoplasm is known as PLASMOGAMY. Later the nuclei fuse. This is known as KARYOGAMY. Subsequently a new plasmodium develops.

Enter the names of the structures described here in the appropriate boxes in Table 23.2 aside the corresponding survival phase each represents.

Complete the labeling of the life cycle of *Physarum polycephalum* (Figure 23.4), using the capitalized terms defined in the preceding description.

EXERCISE 23   **STUDENT NAME** _____

QUESTIONS

1. What does asexual reproduction achieve for any plant group? _____

   _____

2. What is the general term for the entire plant body of a fungus? _____

3. What is the term for the vegetative part of the fungus plant body? _____

4. What does meiosis achieve for any plant group? _____

   _____

5. If a fungus has no motile cells, how then is it dispersed? _____

   _____

6. There are more beneficial fungi than disease-causing fungi. Circle one: (a) true; (b) false.

7. Which are more primitive, aquatic or terrestrial fungi? _____

8. Which is the more common event, sexual or asexual reproduction? _____

9. What is the term used to describe a tube that has cross walls? _____

10. What is the term used to describe a multinucleate structure without any cell walls? _____

    _____

11. What is the term used to describe the gametic union of protoplasm but not the accompanying union of

    the nuclei? _____

12. What is the term used for the fusion of two nuclei? _____

13. Every life cycle can be reduced to four survival phases. List them:

    (a) _____; (b) _____;

    (c) _____; (d) _____

14. Which plays the greater role in adding to the population size, sexual reproduction or asexual reproduc-

    tion? _____

15. Why isn't a slime mold considered to be a 100 percent fungus? _____

    _____

16. Name the divison to which all true fungi belong: _____

17. Name the division for the slime molds: _____

18. Do Zygomycetes have large, conspicuous fruiting bodies? _____

19. List the four evolutionary trends seen in the fungi:

(a) _____ ;

(b) _____ ;

(c) _____ ;

(d) _____

20. Using general terms, name the cells and /or events of a fungus life cycle which are involved in bringing

about genetic variation. _____

_____

21. What structure in the life cycle of a fungus is equivalent to the plasmodium of a slime mold?

_____

Table 23.2   Life Cycle Cells, Stages, or Events Related to Methods of Species Survival

Survival Phase	Accomplished in *Saprolegnia* sp. by	Accomplished in *Rhizopus* sp. by	Accomplished in *Physarum* sp. by
Multiplication			
Dispersal			
Assimilation			
Genetic variation			

STUDENT NAME _____

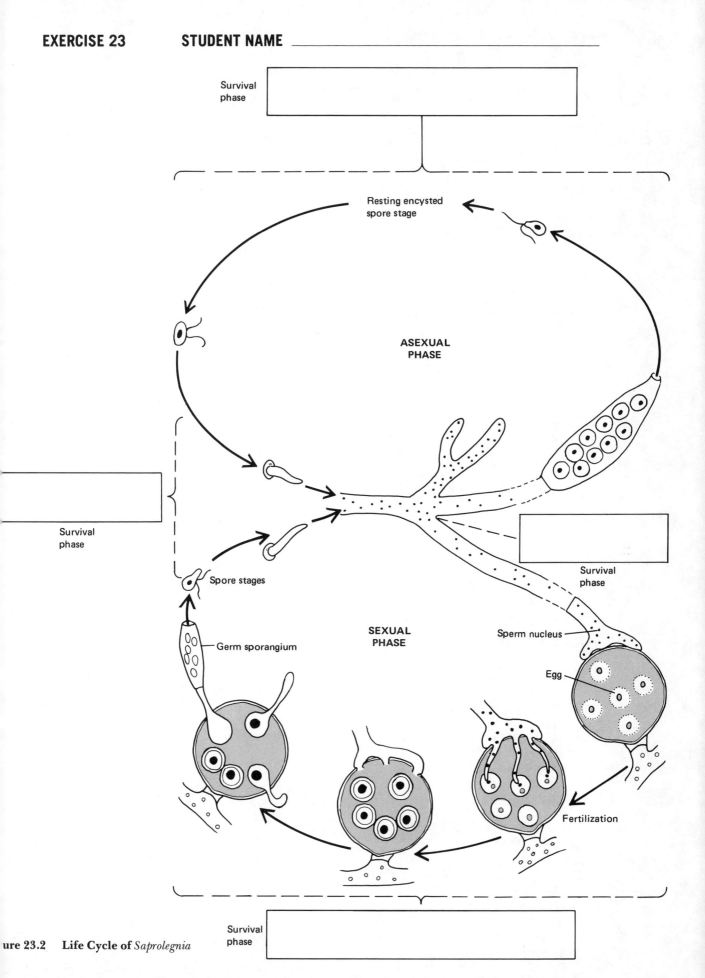

Survival phase

Resting encysted spore stage

ASEXUAL PHASE

Survival phase

Survival phase

Spore stages

Germ sporangium

SEXUAL PHASE

Sperm nucleus

Egg

Fertilization

Survival phase

ure 23.2   Life Cycle of *Saprolegnia*

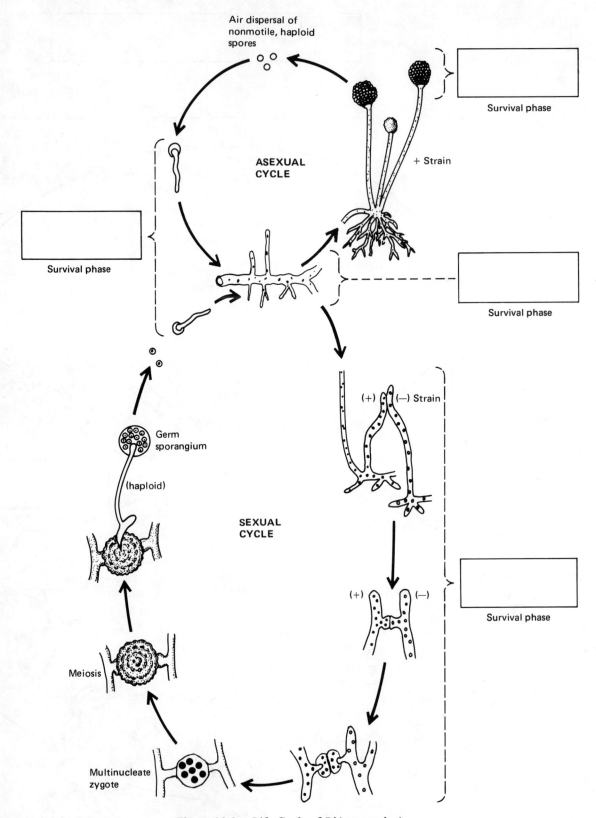

Air dispersal of
nonmotile, haploid
spores

ASEXUAL
CYCLE

+ Strain

Survival phase

Survival phase

Survival phase

(+) (—) Strain

(+) (—)

Survival phase

Germ
sporangium

(haploid)

SEXUAL
CYCLE

Meiosis

Multinucleate
zygote

Figure 23.3   Life Cycle of *Rhizopus stoloniera*

STUDENT NAME _____

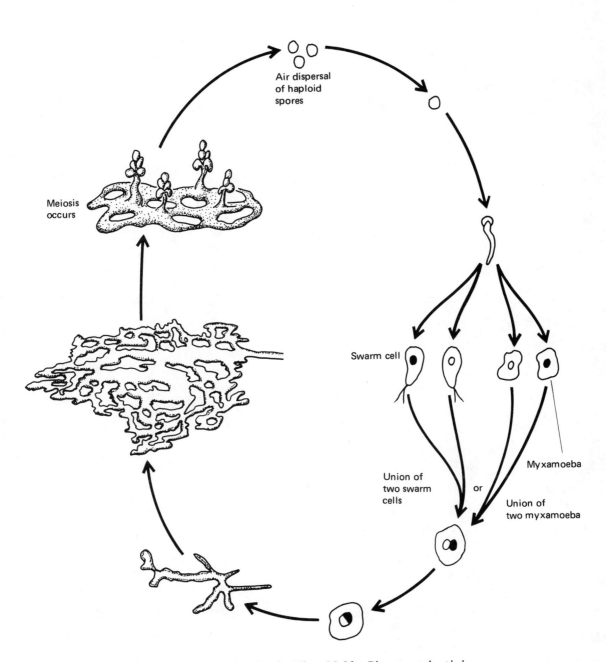

Figure 23.4   Life Cycle of a Slime Mold—*Physarum polycephalum*

EXERCISE 24

FUNGI II:
THE HIGHER FUNGI

The higher fungi consist of two biologically distinct groups, the ascomycetes and the basidiomycetes (of the division Eumycophyta). A third, artificial group known as the Fungi Imperfecti (or Deuteromycetes) has been created as a classification pigeonhole for all those higher fungi in which only asexual reproduction has been seen. It turns out that most species falling into the category of Fungi Imperfecti are in reality ascomycetes; many others are basidiomycetes. They have either lost their ability to reproduce sexually (through evolutionary reduction), or scientists have just missed finding the sexual stages.

I. CLASS ASCOMYCETES—THE SAC FUNGI

Ascomycetes are terrestrial molds most of which have fruiting bodies that are spherical, flask-, or cup-shaped, and whose sexually formed spores are borne in a sac, called an ASCUS.

Ascomycetes are the largest group of fungi. Economically they are very important. Many, such as yeasts (*Saccharomyces* sp.) and species of *Penicillium*, are valuable in industry and medicine. Some, such as the truffles (*Tuber* sp.) and morel (*Morchella esculenta*) are prized as food delicacies. A very large number of them cause tremendous damage to field and orchard crops, timber, and ornamental plants.

Their life cycle has a SEXUAL or ASCUS STAGE and an ASEXUAL or CONIDIAL STAGE. Many ascomycetes don't seem to have a sexual stage in their life cycle. They have either lost this stage during their evolution, or they do really have it but we've just not ever seen it. Species without a known sexual stage are very numerous and cause many plant diseases. As noted earlier, we classify them all in one group known as the fungi imperfecti (because they lack the so-called "perfect" or sexual stage). Comparative features of ascomycetes with other major fungal groups are given in Table 23.1 in the preceding exercise.

A. THE ASCOCARP—THE SEXUAL FRUITING BODY OF THE ASCOMYCETES

If you look back at Figure 23.1, the evolutionary advance of the fungi, you will see that the sexually formed fruiting body of the ascomycete has become a large, fleshy organ. You can easily see it and can recognize the species by it. It is called an ASCOCARP. Many ascocarps forcibly "shoot" the spores out with a puffing action. This is repeated many times until all the spores are ejected. The puffing action is brought on by changes in moisture conditions in the ascocarp. Not all ascocarps are built exactly the same. Let's look at a few.

Open-Type Ascocarp—Apothecium

► ACTIVITY 1   **Examine** dried or pickled specimens of the fungus *Peziza* which has a simple, open ascocarp, called an APOTHECIUM (Figure 24.1). The apothecium is cup-shaped and wide open across the top. Any fungus having this type is called a "cup fungus." Below the CUP part is a slender, supporting STALK. When the fungus is alive, the ascocarp is often brightly colored (and not a dull, ugly black or brown color like the preserved ones). It also may be fleshy, leathery, or gelatinous. Within the shallow cup are tiny sacs, each called an ASCUS. These contain the spores, called ASCOSPORES.

Examine a prepared slide of a cross section of the Ascocarp of *Peziza* sp. under the microscope. **Find** the ASCI (singular: ascus). These are vertically elongate, tube-like sacs. Each contains eight haploid spores, called ASCOSPORES. When mature, the ascus ruptures and releases the spores. **Note** the hyphae, called STERILE HYPHAE, between the asci. These are thought to flex and move and so help in releasing the spores from the surface of the ascocarp. Asci and sterile hyphae together form the fertile layer, called the HYMENIUM. A hymenium is present in all ascocarp types. **Label** the *Peziza* apothecium in Figure 24.1, using the capitalized terms defined in the preceding descriptions.

► ACTIVITY 2   **Examine** morels (*Morchella esculenta*). These are fungi that superficially resemble mushrooms, though they are more slender and columnar. The surface has ridges, so the morel appears spongy. The depressions between the ridges are lined with the HYMENIUM. Each depression is the same basic OPEN-TYPE ASCOCARP or APOTHECIUM as in *Peziza*.

Closed-Type Ascocarp—Cleistothecium

► ACTIVITY 3   **Examine** a lilac (*Syringa* sp.) leaf infected with powdery mildew—a disease caused by a variety of ascomycete fungi. **Find** the minute, pepper-grain-size black bodies. These black bodies are CLEISTOTHECIA, closed-type ascocarps. They are ball-like objects that are hollow inside. This hollow cavity is lined by the hymenium consisting of the asci and sterile hyphae. **Note** that the cleistothecia have RIGID APPENDAGES with branched tips. These aid in dispersal. **Put** some cleistothecia on a slide with water and cover glass. **Press down** on the cover glass in order to rupture the cleistothecium, ASCI will squeeze out of each CLEISTOTHECIUM. Powdery mildew fungi attack many ornamental plants such as roses, chrysanthemums, snapdragons, African violets, and turfgrasses, including the bluegrasses. Whitish gray patches of mycelia on leaves, flowers, or stems are evidence of powdery mildew fungi. If sections of your lawn are heavily shaded by trees, they are very likely to develop powdery mildew infection.

Label the powdery mildew cleistothecium in Figure 24.2, using the capitalized terms defined in the preceding description.

Partially Open-Type Ascocarp—Perithecium

► ACTIVITY 4   **Examine** living cultures of *Sordaria fimicola* growing on an agar plate. These are ascomycetes whose ascocarp is a PERITHECIUM, a closed sphere-shaped organ with a single pore at the top. **Mount** a portion of the fungus on a slide, add some water, cover, and examine. **Press down** on the cover slip, and you should be able to split some of the perithecia. ASCI and STERILE HYPHAE are inside the perithecium.

Label the perithecium of Figure 24.3, using the capitalized terms defined in the preceding description.

B. LIFE CYCLE OF AN ASCOMYCETE

Life cycles of ascomycetes are very complicated and many are not well understood. A generalized cycle is shown in Figure 24.4. **Refer** to this figure in the description that follows.

Sexual Phase

Evidence indicates that the site of both fertilization and meiosis is the YOUNG ASCUS (**A** in Figure 24.4). When sexual reproduction occurs (about once a year), the mature mycelium forms an ASCOCARP. Each young ascus within the ascocarp contains two HAPLOID NUCLEI. These fuse to form a DIPLOID ZYGOTE NUCLEUS. The zygote nucleus divides by meiosis. Eventually, each of the eight HAPLOID NUCLEI in the ascus (**E**) becomes invested by protoplasm and a cell wall, and thus eight ASCOSPORES are formed in the ASCUS (**F**).

When mature, the ascus opens at its tip, and each ASCOSPORE is released (and often forcibly ejected by the ascocarp). It is carried away by air currents or other dispersal agents. Eventually the GERMINATION of the ASCOSPORE occurs and a YOUNG MYCELIUM takes form.

The growing mycelium digests and assimilates food materials from the substratum upon which it lives. It becomes an extensive, cotton-like mass of white hyphae permeating the substratum. When mycelia are mature and food reserves have accumulated, they reproduce, either sexually (as just described) or asexually (as more often happens).

Asexual Phase

The MATURE MYCELIUM produces erect hyphae, called CONIDIOPHORES. These form at their tips CONIDIOSPORES (by mitotic cell divisions). Before splitting off, conidiospores commonly remain together in chains. The conidiophores often form branching patterns characteristic of the species (see Figure 24.5). Conidiospores are dispersed by air currents. The GERMINATION of a CONIDIOSPORE forms a young mycelium . . . and the cycle cranks on again many times more.

► ACTIVITY 5   **Complete** the labeling of the ascomycete life cycle in Figure 24.4, using the capitalized terms defined in the preceding description. **Enter** in Figure 24.4 each of the four survival phases. **Enter** in Table 24.1 the names of the structures involved in each of the four survival phases of an ascomycete.

C. YEAST

Yeasts are ascomycetes in which the entire thallus is just one cell. The unicellular condition of yeasts appears to have developed in connection with life in sugary solutions. Certain other fungi that are not ascomycetes will assume a "yeast" condition if grown in sugary solutions.

► ACTIVITY 6   **Examine** *Saccharomyces cerevisiae*, the common brewer's yeast, by placing a small drop of yeast culture on a slide. **Cover** with a cover glass and examine under high power.

Asexual reproduction takes place by budding. A NEW CELL occurs as a BUD which later breaks off of the PARENT CELL. Commonly though, new cells themselves start to bud before they break off the parent cell.

You can't see the nucleus in this preparation, but a large central vacuole is usually visible.

We will not consider sexual reproduction of yeast.

Label Figure 24.6, using the capitalized terms defined in the preceding description.

II. CLASS FUNGI IMPERFECTI (DEUTEROMYCETES)

Most of the members of this group are ascomycetes. A minority are basidiomycetes. They reproduce only by CONIDIA, as far as scientists have been able to determine. It is theorized that they have either lost their sexual stages during their evolution, or that we've just never found these stages.

Some of the worst fungal pests are in this group. To mention a few, we have *Phomopsis* sp. which cause serious dieback in conifers, and canker in gardenias. *Fusarium* sp. cause a wilt of asters and carnations (both of which are important commercial greenhouse crops); *Fusarium* sp. also cause a serious blight in turfgrass, flax, cabbage, and tomato. *Botrytis* sp. infect the leaves of chrysanthemums, begonias, carnations, and other commercially valuable greenhouse crops. *Helminthosporum* sp. cause blight and leaf spot diseases of turfgrasses. *Rhizoctonia* sp. are a common cause of root rot disease in greenhouse plants.

In contrast a *Penicillium* sp., the blue-green mold commonly seen on decaying citrus fruit, is the source of penicillin (the first antibiotic). Some species are used in the making of Roquefort, Camembert, and blue cheese. *Aspergillus* is another extremely common fungus. It will grow on almost anything, provided there's a little moisture.

► **ACTIVITY 7**   **Obtain** from an agar plate culture a strip of cellophane paper containing the mycelium of *Penicillium* or *Aspergillus*. **Mount** it on a dry slide and examine under medium magnification. **Note** the CONIDIOPHORES and the CHAINS of CONIDIA. Are the conidiophores of *Aspergillus* and of *Penicillium* arranged the same way?   YES   NO.

Make a wet mount of the mycelium by adding a drop of 70 percent ethanol and then a drop of 5 percent potassium hydroxide (KOH). **Cover** with a cover glass and examine under the microscope. **Note** the branching (if any) of the conidiophore and the arrangement of the conidia.

Label the conidiophores and conidia of *Penicillium* and *Aspergillus* in Figure 24.5. Also **label** each drawing for the plant it represents.

III. CLASS BASIDIOMYCETES

Basidiomycetes are the second largest group of fungi. Most of the large and conspicuous fungi seen in fields and woods are basidiomycetes. These include mushrooms, stinkhorns, puffballs, jelly fungi, and shelf fungi. In this class there are also numerous microfungi, the most important of which are the smuts and rusts. These are extremely virulent plant pathogens causing extensive damage to many crops and ornamental plants.

Distinguishing characteristics of the basidiomycetes are outlined in Table 23.1 of the preceding exercise.

A. THE BASIDIOCARP—THE SEXUAL FRUITING
BODY OF THE BASIDIOMYCETES

The fruiting body or BASIDIOCARP is the part of the fungus that people see and recognize—not as a basidiocarp, but as a mushroom, toadstool, puffball, stinkhorn, jelly fungus, or shelf fungus on the side of an old tree stump, a dead log, or ground.

► **ACTIVITY 8**   **Examine** a mushroom specimen—a BASIDIOCARP. The basidiocarp can be compared to an umbrella. The "handle" is the STIPE and the cap-like portion is the PILEUS. On the underside of the pileus you will see things that look like, and are called, GILLS. The gills are plate-like. On their surface is the HYMENIUM, the spore-forming layer. You can't see anything smaller than the gills without going to cut sections on slides and using the microscope.

Label all the parts of the mushroom shown in Figure 24.7, using the capitalized terms defined in the preceding description.

▶ ACTIVITY 9   If available, **examine** spore prints that the instructor may provide or which you could make yourself. A spore print shows you the arrangement of the spore-bearing surfaces of which there is a great variety among the basidiomycetes. The spore print of a mushroom can be made in the following way.

1. Set the pileus of a ripe mushroom (with its gill side down) on a piece of white paper.
2. Cover it with a glass or bowl so that air currents won't blow away the spores as they drop from the gills.
3. Let the fungus remain for an hour or perhaps longer on the paper.
4. Remove the cover and examine the spore pattern (print).

B. LIFE CYCLE OF A BASIDIOMYCETE

Sexual Phase

Refer to Figure 24.7 as a guide in this description. The vegetative or assimilative phase is (as with other fungi) the MYCELIUM. The PRIMARY MYCELIUM is the first stage, a mass of HYPHAE formed by the germination of HAPLOID SPORES. Commonly there are different strains of mycelia which, for convenience, we will call "plus" and "minus" strains. Hyphae of the primary mycelium are septate and uninucleate.

When hyphae of primary mycelia of different strains come in contact, they merge. The result of this is the formation of a SECONDARY MYCELIUM.

Cells of the hyphae of the secondary mycelium have two nuclei, one from each of the two primary mycelia. The secondary mycelium is very long-lived, compared to the primary. Eventually, the growth of several hyphae becomes compactly organized into the BASIDIOCARP. Spores are released from the basidiocarp. They germinate into primary mycelia and the cycle starts over.

Spore Formation

Spore formation takes place on the surface of the GILLS. **Refer** to Figure 24.7 again as a guide in this description. A terminal cell of a hypha starts to become a YOUNG BASIDIUM (plural: basidia). Its two nuclei fuse, that is, fertilization takes place. This finally brings together the two sets of genes from the original two spores that produced the primary mycelia.

The diploid nucleus then divides by MEIOSIS. Four HAPLOID NUCLEI are formed.

The YOUNG BASIDIUM cell enlarges and produces four knob-like protuberances. As the knobs enlarge, the haploid nuclei move up into them, one nucleus into each. These four enlarged structures, each with one nucleus, are the BASIDIOSPORES. The number of spores is characteristically four. Each spore is attached to the *outside* of the basidium by a slender stalk. This differs from ascomycetes which produce spores *inside* the ascus.

When a spore is mature, it is forcibly ejected from the surface of the gill. (Remember, this spore formation has been taking place on the gills on the underside of the basidiocarp.) The catapulting of each spore helps to launch it farther into the air for better dispersal. A little droplet of water forms at the base of the spore. Changes in surface tension of this droplet of water is the triggering mechanism that catapults the spore off the basidium and the gill.

▶ ACTIVITY 10   **Using** the microscope, examine a prepared slide showing a section through the cap of a gill-type mushroom such as *Coprinus* sp. Under low magnification, **note** the central STIPE surrounded by the radiating GILLS. **Change** to high magnification and examine the HYMENIUM with the BASIDIA and BASIDIOSPORES. The mature spores are much larger than the basidia. They will also be scattered about and not necessarily all attached to the basidia.

Label the enlarged gill sections shown in Figure 24.7, using the capitalized terms here described. **Complete** the labeling of the life cycle of a basidiomycete in Figure 24.7 using the capitalized terms defined in the preceding description. **Enter** in Figure 24.7 each of the four survival phases. In Table 24.1 **enter** the names of the structures involved in each of the four survival phases of a basidiomycete.

Asexual Reproduction

Basidiomycetes reproduce asexually by a variety of methods. Many seem to have no asexual cycle. One method is by the production of conidia as is found in the ascomycetes. We will not go into the details of this.

C. SMUTS AND RUSTS

These are basidiomycetes that are microfungi. They all parasitize plants and cause tremendous damage, particularly to fruits and cereal grains. Their fruiting bodies are microscopic but occur in such large masses that their presence is revealed by rust-colored lesions or black, sooty-looking malformed tissues of the infected plants.

► ACTIVITY 11   **Examine** plants infected with rusts or smuts.

EXERCISE 24   STUDENT NAME _____

QUESTIONS

1. All the fungi for which no sexual reproduction has been observed are put into one group known as the

   _____

2. Which class of fungi is the largest? _____

3. Nuclear fusion is termed _____

4. Which group exhibits the greater variety of fruiting bodies, (a) the ascomycetes or (b) the basidiomy-

   cetes? Circle (a) or (b).

5. Basidiomycetes and ascomycetes are all terrestrial fungi. Circle one: (a) true; (b) false.

6. In ascomycetes, which is the more common method of reproduction, by conidia or by ascospores?

   _____

7. Complete the following: The effect of meiosis is twofold: (a) it segregates a double set of chromosomes

   into two single sets which go into two daughter cells, and (b) _____

   _____

   _____

8. Name the assimilative phase of fungi. _____

9. Name a cell in the dispersal phase of an ascomycete. _____

10. Name a cell in the dispersal phase of a basidiomycete. _____

11. Conidia are the only known cells for dispersal in the Fungi Imperfecti. Circle one: (a) true; (b) false.

12. What is the typical number of ascospores formed in an ascus? _____

13. What is the typical number of basidiospores formed on a basidium? _____

14. Many ascomycetes forcibly eject the spores. How is ejection accomplished? _____

    _____

15. Basidiospores are "shot" off by surface tension changes in a drop of _____ located at the

    base of the _____

16. Smuts and rusts belong to what class of fungi? _____

17. Fungi Imperfecti consist mostly of species belonging to which class of fungi? _____

    _____

18. Conidia and conidiospores are found only in ascomycetes. Circle one: (a) true; (b) false.

19. Name two common examples of basidiomycetes: (a) _____; (b) _____

20. Name the *fungus class* for each of the following.

    a. Spores of the sexual life cycle are commonly formed in groups of eight and borne *inside* the spore-forming structure: _____

    b. Spores of the sexual life cycle are formed in groups of four and borne *outside* the spore-forming structure: _____

Table 24.1   Life Cycle Cells, Stages, or Events Related to Methods of Species Survival

Survival Phase	Accomplished in an Ascomycete by	Accomplished in a Basidiomycete by
Multiplication		
Dispersal		
Assimilation		
Genetic variation		

EXERCISE 24   **STUDENT NAME** _____

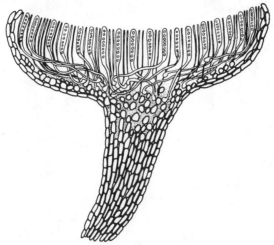

Figure 24.1   Apothecium

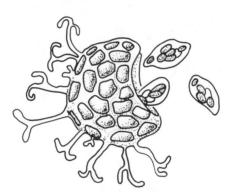

Figure 24.2   Cleistothecium

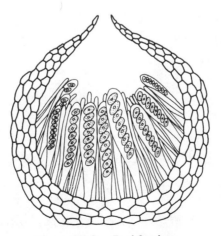

Figure 24.3   Perithecium

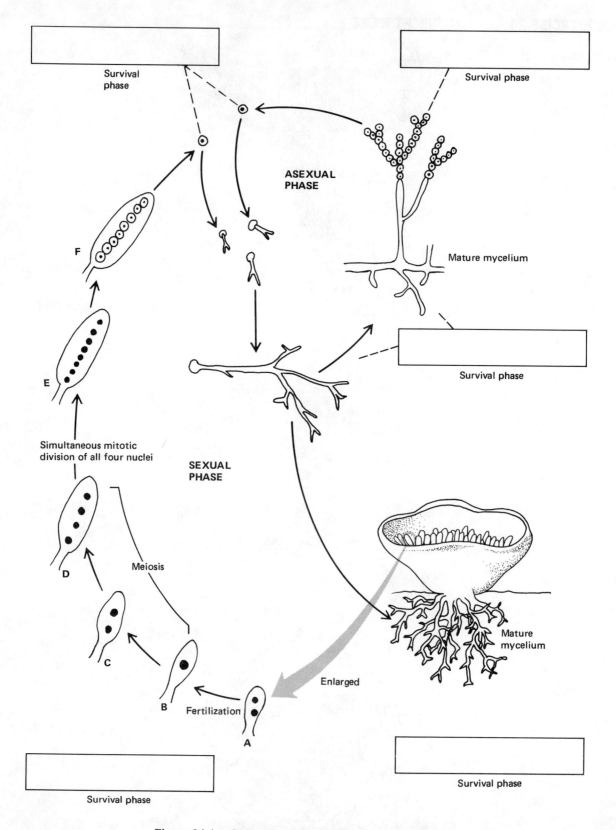

Survival phase

Survival phase

ASEXUAL PHASE

Mature mycelium

F

E

Simultaneous mitotic division of all four nuclei

SEXUAL PHASE

Survival phase

D

Meiosis

C

B

Fertilization

A

Enlarged

Mature mycelium

Survival phase

Survival phase

Figure 24.4   Generalized Life Cycle of an Ascomycete

EXERCISE 24   **STUDENT NAME** _____

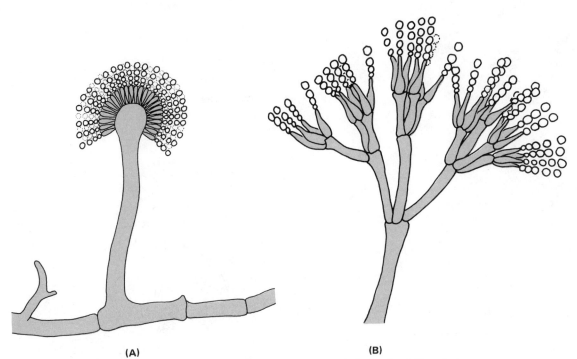

(A)   (B)

**Figure 24.5   Two members of the Fungi Imperfecti Showing
Two Types of Conidiophore Branching Patterns**

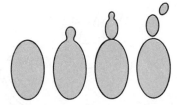

Figure 24.6   Asexual Reproduction of Brewer's Yeast (Saccharomyces cerevisiae)

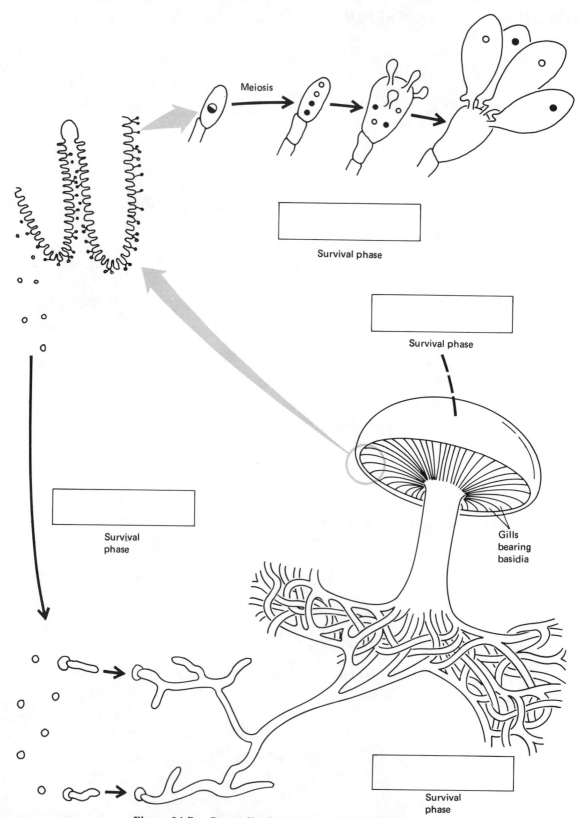

Meiosis

Survival phase

Survival phase

Gills bearing basidia

Survival phase

Survival phase

Figure 24.7   Generalized Life Cycle of a Basidiomycete

Exercise 25

ALGAE I: GREEN ALGAE AND EUGLENOIDS

Algae are a huge, diverse assemblage of fairly simple, photosynthetic plants whose origins date back over 4 billion years. Scientists recognize several distinct groups (divisions). Differences among the groups are based on size, habitat, structural complexities, reproduction, physiology, and biochemistry of the plants.

Some terms used for algae:

ANTHERIDIUM — In algae, a single cell that contains sperm. (Plural: antheridia.)

CHLOROPLAST — A specialized cytoplasmic body containing chlorophyll.

CILIUM — Short, hair-like process on a cell whose vibrating or lashing movements serve to propel the cell through water. (Plural: cilia.)

COLONY — A thallus type composed of a group of closely associated but independently functioning cells or unicellular organisms among which there are no marked structural differences and little or no division of function.

EYESPOT — A small pigmented structure thought to be sensitive to light.

FILAMENT — A thallus type in which the cells occur in a thread-like row.

FLAGELLUM — A very long whip-like process on a cell whose vibrating or lashing movements serve to propel the cell through water. (Plural: flagella.)

GAMETANGIUM — A general term applied to any cell within which gametes are formed.

HETEROGAMY — A type of sexual reproduction in which male and female gametes differ in size, structure, and behavior.

ISOGAMY — A type of sexual reproduction in which all gametes are alike in size and structure and often in behavior; considered primitive. (Adjective: isogamous.)

OOGAMY	A type of sexual reproduction involving a larger, nonmotile egg and a smaller, motile sperm. (Adjective: oogamous.)
OOGONIUM	A single cell that contains one or several eggs. (Plural: oogonia.)
PYRENOID	A center of starch formation on certain chloroplasts.
THALLUS	The name for the entire plant body.
ZOOSPORE	A flagellated motile cell.
ZYGOSPORE	A diploid spore formed by the fusion of isogametes (or isogametangia); often goes into a resting or dormant stage.

I. DIVISION CHLOROPHYTA—GREEN ALGAE

Next to the diatoms, the green algae are the second largest group of algae. They are the most diverse of all algae. Some approach the structural complexity of simple land plants. Individual plants are microscopic, but colonies of them can form masses visible to the eye.

The function, structure, and chemical makeup of their cells are similar, and in many cases, identical, to those of higher plants. Taking all these features into account, it is generally agreed that the green algae are probably ancestral to all land plants.

General facts about green algae are

Usual color	Green.
Distribution	Mostly in fresh waters and on land.
Habitat and niches	Most species are floaters in the plankton of rivers, lakes, reservoirs, and creeks. Also in soil, snow, on rocks, tree bark, flower pots, and fish tanks. A few species are floaters in plankton of the neritic zone (continental shelves) and littoral (coastal) zones.
Motility	By flagella.
Body types	Single cells, filaments, and colonies.
Value	Important source of oxygen and food for aquatic organisms; some are eaten by humans.

A. LINES OF EVOLUTION IN THE GREEN ALGAE

Two general trends are recognizable in the evolution of green algae. These are

1. Increase in plant size, from a single-cell type to a filamentous or colonial type.
2. Change in sexual reproduction, from isogamy to oogamy.

The increase in size is seen in the range of growth forms found today. These growth forms or body types illustrate some of the probable stages of evolution. Each is a successful way of life which, within certain limits, has survived to the present.

Unicellular organisms are the most primitive. Filamentous and colonial forms probably arose from them by cell divisions occurring in different planes as indicated in Figure 25.1.

Study the generalized life cycle shown in Figure 25.2. **Refer back** to this cycle as you examine the various plants.

B. REPRESENTATIVE GREEN ALGAE

Single-Cell-Type Isogamous Green Alga—*Chlamydomonas* or *Carteria*

Chlamydomonas (as well as *Carteria*) is a single-celled plant. Thus the thallus is one cell. Its sexual reproduction is isogamous.

thallus

➤ ACTIVITY 1   **Get** a clean slide and add to it a drop of methyl cellulose and a drop of water from the *Chlamydomonas* or *Carteria* culture. **Stir** the two drops together gently, using the edge of a cover

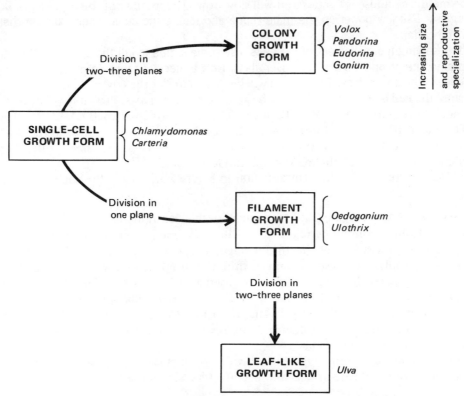

Figure 25.1   Some of the Different Forms of Green Algae Illustrating
Some of the Probable Paths and Stages of Evolution

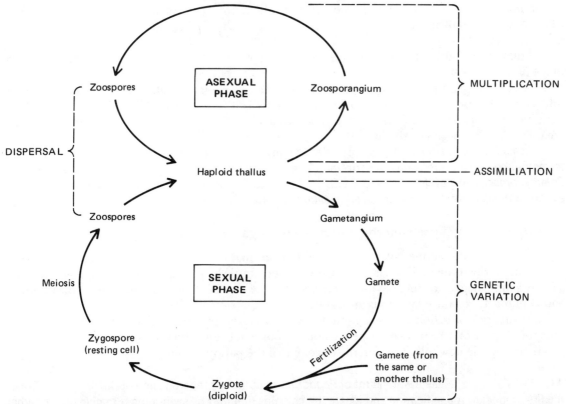

Figure 25.2   Generalized Life Cycle of Green Algae. Each Part Is Concerned with One of the Four Survival Phases of the Population—Dispersal, Multiplication, Assimilation, and Genetic Variation

glass. The methyl cellulose is viscous and will slow down the motility of the organisms. **Cover** with the cover glass. **Use** low power on the microscope and locate the cells. Then **examine** them under higher magnification.

After observing the movement of the cells, **use** some iodine to kill the cells and stain the parts. **Put** a drop of iodine on one edge of the cover glass. **Hold** a piece of rough-torn paper toweling on the opposite edge of the cover glass. The iodine will be drawn under the cover glass.

Because the cell is small, you may not be able to see all its parts. Probably you can see at least some of the following: the CELL WALL, the NUCLEUS, the cup-shaped CHLOROPLAST, the EYESPOT (an orange or red spot), the PYRENOID (starch body in the lower part of the chloroplast), and two anterior FLAGELLA (four in *Carteria*). If you cannot see the flagella, reduce the light coming through the microscope and look again.

Label the mature thallus (the central figure) in Figure 25.3, using the capitalized terms here described.

reproduction   **Refer** to Figure 25.3 for a guide for this description.

The sexual phase commonly takes place when adverse conditions arise, for example, the pond dries up or cold weather sets in. In this stage, a vegetative cell will subdivide its protoplast into 8 to 32 protoplasts. These remain within the original cell wall. This cell is now a GAMETANGIUM. Later the protoplasts are released as separate cells. They develop two flagella (four in *Carteria*). These cells are ISOGAMETES. When the flagella of two gametes come in contact, they intertwine and then the two cells fuse to form a ZYGOTE. The zygote develops a thick wall and goes into a period of dormancy. This resting cell is the ZYGOSPORE. It can resist droughty or other adverse conditions.

When dormancy ends, the zygospore divides by meiosis and forms four ZOOSPORES. Zoospores swim about, increase in size, and become typical adult cells.

▶ ACTIVITY 2   If material is available, **prepare** a slide with a drop of water from a culture of one strain of *Chlamydomonas*. **Observe** for a minute to check for the presence of organisms. To this slide add a drop of water from a culture of a different strain of *Chlamydomonas*. **Observe** quickly the behavior of cells. **Describe:** _____

Label the the sexual phase of the life cycle in Figure 25.3, using the capitalized terms defined in the preceding description.

Asexual reproduction is a common occurrence. A vegetative cell subdivides its protoplast into two, four, or eight daughter protoplasts. This cell is now a ZOOSPORANGIUM. Each daughter protoplast, when released, becomes a flagellated ZOOSPORE. These swim about, increase in size, and become adult cells.

Complete the labeling of the life cycle of *Chlamydomonas* in Figure 25.3, using the capitalized terms defined in the preceding description. **Enter** the terms designating the survival phases in the *Chlamydomonas* life cycle in Figure 25.3. In Table 25.1 **enter** the names of the cells, phases, or events of this life cycle that are involved in each of the four survival phases.

Simple-Colony-Type Heterogamous Green Alga—*Pandorina*

Pandorina is a simple colony of *Chlamydomonas*-like cells. *Pandorina* represents a stage in evolution in which the thallus has been made bigger simply by the aggregation of a few cells with a fairly simple degree of coordination between them. 4 or 32 cells are arranged in a compact ball and kept together by a gelatinuous envelope.

Sexual reproduction is heterogamous, and the zygote produced divides by meiosis to form zoospores, each of which forms a new colony. In asexual reproduction, each cell in the colony produces a whole new colony within the envelope of the parent colony.

▶ ACTIVITY 3   **Make** a wet mount of *Pandorina* or obtain a prepared slide of the organism. **Examine** first under low magnification and locate the colonies that may be swimming. **Examine** under higher magnification and note the cellular details such as eyespot, chloroplast, and flagella.

Complex-Colony-Type Oogamous Green Alga — *Volvox*

Volvox represents the peak of evolutionary development of colony-type green algae. The number of *Chlamydomonas*-like cells in the colonies of vario/s species of *Volvox* ranges from 500 to 50,000. Sexual reproduction is oogamous. Male gametes are formed in large numbers by multiple subdivisions of the protoplasts of certain adult vegetative cells. These cells are the ANTHERIDIA (singular: antheridium). The female gamete, the egg, is large and nonmotile. Only one or a very few of them are formed within any one parent cell. The parent cells are the OOGONIA (singular: oogonium).

▶ ACTIVITY 4   **Make** a wet mount from a culture of *Volvox*. **Note** that each cell of the colony is a *Chlamydomonas*–like cell.

The coordinated movement of the flagella of all the cells allows the entire colony to behave as one individual organism, moving forward as it also spins on a vertical axis. The colony varies from spherical to ovoid. All the cells are held together by a gelatinous envelope and by thin strands of protoplasm between neighboring cells. In addition, each cell is enveloped by an individual sheath.

Volvox shows some specialization and division of labor among the cells. Reproductive cells occur at the base or hind end of the colony, while the forward or leading end has only vegetative cells. Eyespots in the vegetative cells are larger than those of the reproductive cells. Small daughter colonies formed as a result of either sexual or asexual reproduction are usually seen floating in the center of the larger parent colony. When the parent colony disintegrates, the daughter colonies are released and become mature colonies.

Simple-Filament-Type Isogamous Green Alga—*Ulothrix*

thallus   The thallus of *Ulothrix* is a simple, unbranched (unbranched is considered primitive) filament with only the basal cell, the holdfast cell, differentiated from the rest of the filament. It attaches the filament to some rock, pebble, or other object.

▶ ACTIVITY 5   **Examine** a prepared slide or living specimens of *Ulothrix*. **Note** that the cells are all alike. The CHLOROPLAST in each cell is C-shaped, and may have one or more PYRENOIDS. No eyespot is present. Why not? _____
_____The NUCLEUS may or may not be visible.

Label the thallus of *Ulothrix* in Figure 25.4, using the capitalized terms defined in the preceding description.

reproduction   Any cell except the holdfast cell in the filament is capable of reproducing the plant, either asexually or sexually. **Refer** to Figure 25.4 as a guide in this description. In the asexual phase the protoplast of a cell divides by mitosis to form four or eight daughter protoplasts. These are released and become flagellated ZOOSPORES. The parent cell that produced these zoospores is called a ZOOSPORANGIUM. After swimming about, each dispersed zoospore settles down, loses its flagella, and undergoes mitotic divisions to form a new FILAMENT.

In the sexual phase the protoplast of a cell divides many times to form 32 to 64 daughter protoplasts. When freed from the parent cell (the GAMETANGIUM), these daughter protoplasts function as GAMETES. They are ISOGAMETES. Each has two flagella, not four as in zoospores. They are also smaller than zoospores.

Eventually two isogametes from different filaments fuse to form a quadriflagellate ZYGOTE. The zygote goes through a dormant period as a ZYGOSPORE. Later it divides by meiosis to form four haploid ZOOSPORES. Each of these will grow into a new filament.

Is there any basic difference in the cyclic events of this life cycle and of *Chlamydomonas*? YES   NO. Is there any real difference in individual cell behavior? YES   NO. What, then, is the only difference between the *Chlamydomonas* plant and the plant *Ulothrix*? _____

Complete the labeling of the life cycle of *Ulothrix* in Figure 25.4, using the capitalized terms defined in the preceding description. **Enter** the terms designating the four survival phases in the *Ulothrix* life cycle in Figure 25.4. In Table 25.1 **enter** the names of the cells, phases, or events of this life cycle that are involved in each of the four survival phases.

Simple-Filament-Type Oogamous Green Alga—*Oedogonium*

thallus

▶ ACTIVITY 6   **Examine** a prepared slide or living specimens of *Oedogonium*. *Oedogonium* is, like *Ulothrix*, a simple, unbranched filament. But it reproduces by oogamy, an advanced method compared to the isogamy of Ulothrix. Cells in the filament are mostly elongate. The CHLOROPLAST is reticulate, that is, in a delicate meshwork. PYRENOIDS are located at the intersections of the reticulum, and you see them as scattered dark bodies in the cell. The NUCLEUS is usually centrally placed. A HOLDFAST cell is present in some species.

reproduction   **Refer** to Figure 25.5 as a guide in this description.

▶ ACTIVITY 7   To study the sexual phase, **examine** prepared slides with filaments of *Oedogonium* showing antheridia and oogonia. Here and there among the elongate vegetative cells in the filament are groups of much smaller disk-like cells. Each of the short disk-like cells is an ANTHERIDIUM. Each forms two PROTOPLASTS. After release, these protoplasts become SPERM. They are small, ovoid cells with a ring of flagella at one end.

Locate an oogonium. This will be a single cell whose entire protoplast has become a large, globose, opaque EGG. The oogonium has a FERTILIZATION PORE through which a sperm cell can enter to fertilize the egg.

After fertilization, the ZYGOTE that is formed remains within the oogonium. It develops a thick wall, goes dormant, and is known as an OOSPORE. As the filament decays, the oospore is released but remains dormant for several months. Later the oospore divides by meiosis to form four haploid ZOOSPORES. These have a ring of flagella at one end. They resemble sperm but are larger cells. They swim about and later each grows into a new filament.

The asexual phase can occur in two ways, (1) by fragmentation of the filament, each fragment growing into a new filament; or (2) by the transformation of the protoplast of a vegetative cell into a ZOOSPORE as shown in Figure 25.5. After release from the old cell wall, the zoospore swims about and then grows into a new filament.

Complete the labeling of the life cycle of *Oedogonium* in Figure 25.5, using the capitalized terms defined in the preceding description. **Enter** the terms designating the four survival phases in the *Oedogonium* life cycle in Figure 25.5. In Table 25.1 **enter** the names of the cells, phases, or events of this life cycle that are involved in each of the four survival phases.

Leafy Type Green Alga—*Ulva*

Ulva (sea lettuce) is leafy. It exemplifies another evolutionary line in green algae. The thallus is a thin flat sheet of cells instead of a chain or a cluster. It grows attached to other objects and has a holdfast with rhizoid- or finger-like extensions.

▶ ACTIVITY 8   If available, **examine** preserved specimens or dried herbarium specimens of *Ulva*.

II. DIVISION EUGLENOPHYTA—EUGLENOIDS

Euglenoids are unicellular organisms whose classification as plants or animals can be debated equally either way. They exhibit both plant and animal characteristics. They are not related to any algae, but because they occupy the same kinds of habitats as algae and are of comparable simplicity, they are included in this study of algae. Zoologists include them in studies of Protozoa.

General facts about euglenoids are

Usual color	Green; a few are colorless.
Habitat	Fresh water, predominantly.
Body types	Single cell.
Cell wall	Variable; often a flexible pellicle.
Motility	By one long whip-like flagellum.
Value	A food source for fish and other aquatic animals.

► ACTIVITY 9   **Put** a drop of methyl cellulose on a clean slide and add to it a drop of water from a culture of *Euglena*. **Using** the edge of the cover glass, stir gently. **Cover** with a cover glass and under low magnification, locate the organisms.

Note the swimming motion. **Reduce** the light and you may be able to see a long flagellum at one end of the cell. Near the base of the flagellum you may be able to see a GULLET. Some euglenoids are thought to ingest food into the gullet.

Note that the cell is elongate with a pointed end and a blunt end. The blunt end is the leading end of the cell that is propelled through the water by the flagellum. A light-sensitive, pigmented body, called the EYESPOT, is near the blunt end. Can you see it? (YES) NO. Do you see CHLOROPLASTS? (YES) NO. Is a NUCLEUS present? (YES) NO. Does the organism change shape? (YES) NO. Does it have a rigid wall, then? YES (NO.) The changes in cell shape are possible because the protoplast is bounded by a flexible PELLICLE.

vacuum cleaner hose

EXERCISE 25   STUDENT NAME _____

QUESTIONS

1. Do the life cycles of green algae contain the same four survival phases as fungi? _____

2. List these phases. Use just one word or phrase to identify each. (a) _____;

   (b) _____; (c) _____;

   (d) _____

3. What is the term for the most primitive level of gamete specialization? _____

4. What are the three body types or growth forms found among green algae? (a) _____;

   (b) _____; (c) _____

5. Which growth form or body type is considered the most primitive? _____

6. Give at least two reasons why green algae are considered ancestral to land plants.

   (a) _____

   (b) _____

7. Are green algae ordinarily visible to the naked eye? _____

8. Are the sperm and eggs of green algae produced by (a) multicellular organs or (b) single cells? Circle (a)

   or (b).

9. What component of the cell has a distinct and recognizable shape for each genus of the green algae

   studied? _____

10. In which stage is there a rearrangement of chromosomes and gene combinations, in (a) mitosis or in (b)

    meiosis? Circle (a) or (b).

11. Which cells, therefore, carry new genetic "blueprints" for making the new filament, the zoospores of the

    (a) sexual phase or (b) asexual phase? Circle (a) or (b).

12. Name a colonial green alga with oogamous reproduction. _____

13. Why wouldn't you expect to find eyespots in cells in filamentous colonies? _____

    _____

14. Name the two cell types in the green alga life cycle that function in dispersal of the population.

    (a) _____; (b) _____

15. Briefly, what are the two broad, general evolutionary trends recognizable in green algae?

(a) _____; (b) _____

16. What does the thallus of *Chlamydomonas* consist of? _____

17. Would you expect to find many species of green algae in the ocean? _____

18. Are euglenoids algae? _____

19. Which phase of the life cycle of the green alga is mainly responsible for maintaining and increasing the

number of individuals in the population? _____

20. What are the two events in the life cycle that introduce genetic variation into the population?

(a) _____; (b) _____

Table 25.1   Life Cycle Cells, Stages, or Events Related to Methods of Species Survival

Survival Phase	Accomplished in *Chlamydomonas* sp. by	Accomplished in *Ulothrix* sp. by	Accomplished in *Oedogonium* sp. by
Multiplication			
Dispersal			
Assimilation			
Genetic variation			

EXERCISE 25   **STUDENT NAME** _____

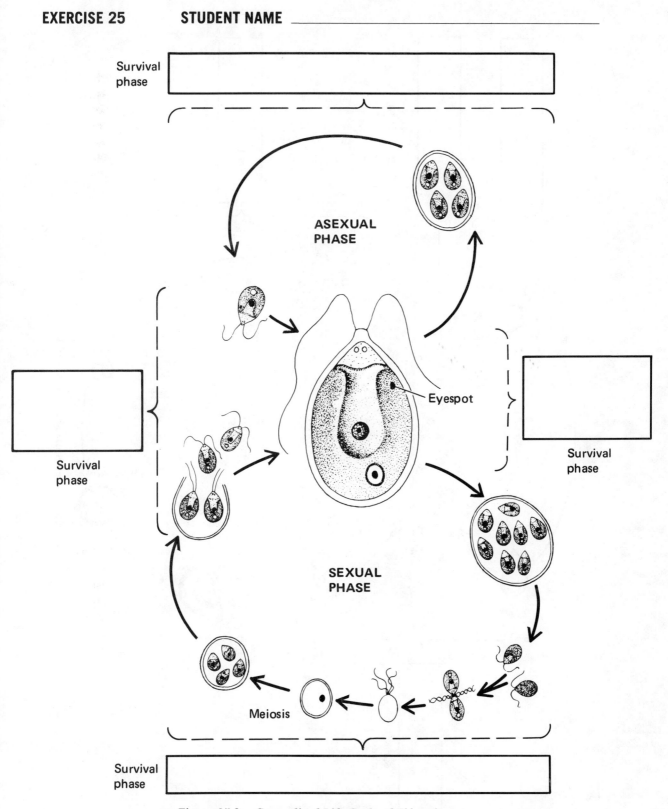

Figure 25.3   **Generalized Life Cycle of** *Chlamydomonas*

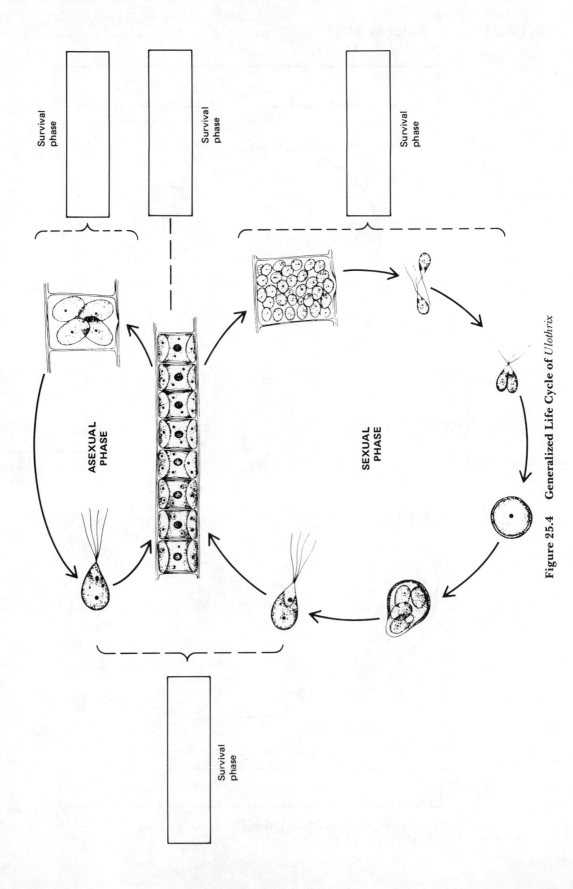

Figure 25.4   Generalized Life Cycle of *Ulothrix*

EXERCISE 25   **STUDENT NAME** _____

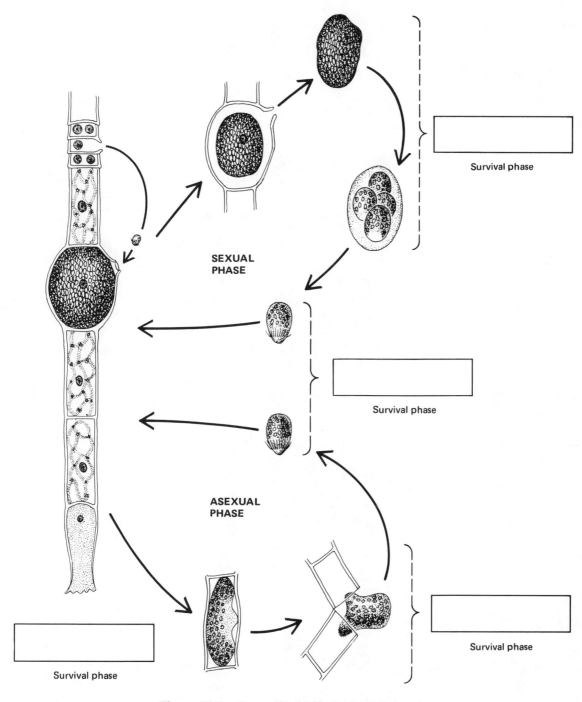

Figure 25.5   **Generalized Life Cycle of** *Oedogonium*

EXERCISE 26

ALGAE II:
BLUE-GREEN ALGAE AND GOLDEN-BROWN ALGAE

Today's work covers two groups of algae, the division Cyanophyta, the blue-green algae, and the Division Chrysophyta of which you will study only the diatoms.

I. DIVISION CYANOPHYTA—BLUE-GREEN ALGAE

The Cyanophyta are the blue-green algae and are the most primitive of the chlorophyll-bearing plants. They are more similar to bacteria than to algae or other plants. So, more commonly now they are classified with the bacteria. For your introduction to them, however, it is more convenient to study them with the algae.

General facts about blue-green algae are

Usual color	Blue-green, but many are red, brown, or purple.
Distribution	Mostly in ocean and brackish waters.
Habitat and niches	Most species grow in tidal marshes and mud flats; many occur in upper 1–3 cm of the soil surface. A few species are in caves, hot springs, geyser basins, wet cliffs, and waterfall ledges, and as plankton of ponds, lakes, rivers, etc.
Motility	No.
Body types	Most are filaments; some are loose colonies; some are unicellular.
Value	Have both beneficial and detrimental effects. Excess growth in fresh waters creates nuisance and depletes oxygen content.

There isn't much cellular detail visible in a blue-green alga. Nor are there any complicated life cycles. Reproduction is asexual only. It occurs by cell divisions, spores, and fragmentation of the thallus. Let's take a look at a few plants and see what is visible to you.

A. *GLEOCAPSA*

▶ ACTIVITY 1   **Get** a clean slide and make a wet mount from a culture of *Gleocapsa*. **Cover** with a cover slip. **Examine** under the microscope. What body type does *Gleocapsa* have? _____. What shape are the CELLS, rectangular or spherical? _____. Each cell is surrounded by a mucilaginous capsule. All the cells together in a group are also embedded in one common capsule. How many cells do you see in a group? Too many to count?   YES   NO. How many? Circle one: (a) 2 to 8 cells; (b) more than 8; (c) more than 20. Can you see any nuclei?   YES   NO. Any chloroplasts?   YES   NO. Are any cells swimming?   YES   NO. **Put** a drop of methylene blue stain on one edge of the cover glass. By holding a bit of paper toweling at one edge of the cover glass, you can draw water out and pull the stain under the cover glass. The stain will be absorbed by the SHEATH. Can you see individual cell sheaths?   YES   NO. Is the entire COLONY surrounded by a sheath?   YES   NO. **Label** the colony of *Gleocapsa* in Figure 26.1, indicating the cells, sheaths, and colony.

B. *OSCILLATORIA*

▶ ACTIVITY 2   **Make** a wet mount of *Oscillatoria*. What do you see, colonies or filaments? (Circle one.) Is there a SHEATH surrounding the plant?   YES   NO. Which is the bigger dimension, the length of each CELL or the width of each? (Circle one.) How would you describe the cell contents, (a) evenly granular or (b) with various different inclusions? Circle (a) or (b). Do the filaments move back and forth?   YES   NO. What does the name *Oscillatoria* imply? _____
_____.

   Oscillatoria reproduces simply by fragmenting the filament. It happens this way: frequently a cell in the filament dies and the dead cell becomes a weak, breaking point or SEPARATION DISK in the filament. The short chain of cells between any two disks is called a HORMOGONIUM (plural: hormogonia). When a filament breaks, the hormogonia (in some inexplicable way) are able to move through the water, thereby accomplishing dispersal. Cell divisions within the hormogonium lengthen the chain and produce a new filament. (Hormogonium is a name coined from two Greek words roughly meaning a "reproducing necklace.")
   Label the filament of *Oscillatoria* in Figure 26.2, using the capitalized terms defined in the preceding description.

C. *ANABAENA*

▶ ACTIVITY 3   **Make** a wet mount of living specimens of *Anabaena* and cover with a cover slip. Is this a filament or a colony? (Circle one.) Do all the cells in the plant look alike as in *Oscillatoria*?   YES   NO. Are the cells disk-shaped?   YES   NO. Look for large, oval cells that are clear or transparent. These are HETEROCYSTS. They are typical of many blue-green algae. When a filament breaks into smaller fragments, it always breaks at a heterocyst. In some species, heterocysts are at the end of the filament (terminal heterocysts). In others they are interspersed (intercalary heterocysts) among the VEGETATIVE CELLS. The position varies with different species and is helpful in classifying some of the blue-green algae.
   Label a heterocyst, a vegetative cell, and the sheath in the segment of filament of *Anabaena* in Figure 26.3.

D. *NOSTOC*

   Nostoc is a filamentous alga whose filaments grow extensively intertwined with each other, forming larger and larger colonies that are ball-shaped. They have a very thick firm gelatinous sheath. Some colonies become large enough to be seen with the naked eye.

▶ ACTIVITY 4   **Make** a wet mount of living *Nostoc* and examine the individual filaments. Do you see any heterocysts?   YES   NO. **Observe** a few macroscopic, jelly-like colonies.

E. SUMMARY OF BLUE-GREEN ALGAE

In the *summary* of your study of the blue-green algae, did you see evidence of any of the following features? Circle YES or NO.

Cell nucleus	YES	NO
Chloroplasts	YES	NO
Flagella	YES	NO
Motile plants or cells	YES	NO
Gametangia or zoosporangia	YES	NO
Pyrenoids	YES	NO
Eyespots	YES	NO

II. DIVISION CHRYSOPHYTA — GOLDEN-BROWN ALGAE

The Chrysophyta include three different, but related, algal groups. Time permits study of only one group, namely, the diatoms.

General facts about diatoms are

Usual color	Golden brown or greenish brown.
Distribution	Most are aquatic; freshwater and marine species approximately equal in numbers. Some are terrestrial.
Habitat and niches	As plankton of the seas and fresh waters. They are one of the two most abundant algae of the plankton of midocean; also on soil surface (as the most abundant of all algae), as well as in hot springs, wet rock, etc.
Motility	A gliding motion in some forms.
Body types	Mainly unicellular, a few are weakly colonial. Cell walls composed of two overlapping valves that fit together in a way similar to the parts of a candy box. Cells may be circular, triangular, or rectangular in shape.
Value	Extremely valuable. They are the principal primary food producers of the sea; the basic link in all marine food chains; the major single world source of photosynthetically produced food and oxygen. Their skeletal fossil remains are mined as diatomaceous earth which has many industrial uses.

Diatoms are single-celled plants with extremely novel walls. The living protoplast of the cell is enclosed by two quite glassy walls that fit one over the other like the two halves of a candy box. These two wall halves are called VALVES. Valves typically have markings in a wide variety of patterns. After diatoms die, their glassy valves remain intact, and over the past millennia have accumulated in vast rock deposits known as diatomaceous earth.

Looking down at a scattering of diatoms, you often see two basic shapes, round ones and rectangular ones. Triangular-shaped ones are also present, but are basically round in symmetry. The rectangular-shaped ones are termed PENNATE and the round and triangular-shaped ones are called CENTRIC.

➤ ACTIVITY 5   **Make** a wet mount from a culture of living soil diatoms. (A good place to obtain diatoms is from the surface soil of a potted plant.) **Examine** the slide under low, and then high magnification. Do you see any pennate diatoms? YES NO. Are there any centric forms? YES NO. Look at the illustrations of Exercise 27, Algae Key. Do any of your specimens resemble the illustrated ones? YES NO. Can you make any tentative identification of your specimens? (a) _____; (b) _____; (c) _____ _____; (d) _____.

Pennate forms move by a gliding motion. **Observe** a few cells for a few minutes and you will be able to see this movement in some of them.

Make a wet mount, this time using some diatomaceous earth mixed in a little water. Do you see any pennate forms?   YES   NO. Are there any centric forms?   YES   NO. If you see centric forms, then your sample of diatomaceous earth is of marine origin, because centric forms occur almost exclusively in the sea. From your observations of these two samples, what would you say about the distribution of pennate forms? Are they organisms of (a) the sea only, (b) fresh water only, or (c) both seas and fresh waters? Circle (a), (b), or (c).

EXERCISE 26   STUDENT NAME _____

QUESTIONS

1. What group of algae constitute the basic link in all marine food chains? (common name) _____

   _____

2. Which group of algae is so primitive that we consider them more properly classified with bacteria than

   with the algae? (division name) _____

3. List two features of this group which "brands" them as primitive. (a) _____;

   (b) _____

4. In a brief sentence define diatomaceous earth. _____

   _____

5. Suppose you are a geologist studying sedimentary deposits. In these you discover strata of diatomaceous
   earth in which you find only centric type diatoms. You know right away that these deposits were laid

   down in (a) a body of fresh water; (b) the sea. Circle (a) or (b).

6. Phosphorous is a mineral utilized in plant growth. If we saturate our streams and other waters with soap

   washings high in phosphates, what happens to the populations of algae in these waters? _____

   _____

7. Fish and other animal life in fresh waters often die when there is an overgrowth of algae. If we rule
   out toxins (poisons) released by some algae, exactly what is it then that brings on death to these animals?

   _____

8. Visitors to Yellowstone National Park are always impressed by the colored, massive, rock ledges
   and terraces deposited around the hot springs by the action of algae living in the waters. Which algal

   group is responsible for this? (common name) _____

9. Do blue-green algae reproduce sexually? _____

10. A characteristic feature of blue-green algae is the presence of a _____

   enveloping the cells or colony.

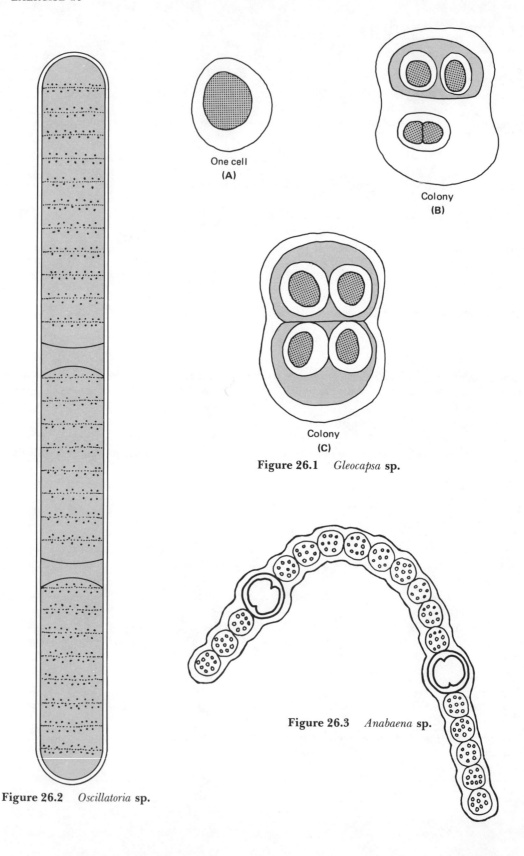

One cell
(A)

Colony
(B)

Colony
(C)

Figure 26.1   *Gleocapsa* **sp.**

Figure 26.2   *Oscillatoria* **sp.**

Figure 26.3   *Anabaena* **sp.**

EXERCISE 27

ALGAE III:
IDENTIFICATION OF SOME COMMON
FRESHWATER ALGAE

This exercise consists of a key designed to enable you to make an approximate identification of a variety of the most common freshwater and terrestrial algae. Several hundred genera have been deliberately left out to simplify the key. Except in a few cases, the majority of algae you'll find in any freshwater or soil sample will belong to the genera in this key.

You may discover in your sample numerous "little round green things" (LRGTs) which, even under the highest magnification of your microscope, are still LRGTs. They defy identification by visual appearance alone. Even experts on algae (phycologists) often need to culture LRGTs for months and study their life cycles before the genus and species can be identified.

Section IV shows examples of desmids, diatoms, and other algae. These are so variable that the key would lose its simplicity if we were to include them. If you have trouble going through the key, then look at Section IV, and you may discover that your unknown is a diatom or a desmid. Remember that there are thousands of species of desmids and diatoms, and Section IV illustrates just generalized representatives.

I. KEY TO SOME COMMON FRESHWATER ALGAE[1,2]

1. Cells blue-green, sometimes olive green or purplish; pigmentation evenly spread throughout the cell; no nucleus or other small bodies visible within the protoplast; cell wall thin and usually with a mucilaginous sheath......................**Blue-green algae**
    (Section II)

1′. Cells green to yellow green; pigments contained within definite plastids; nucleus and plastids visible; cell wall distinct and typically without a sheath................**Green algae**
    (Section III)

[1]Courtesy of Dr. Richard L. Smith, Eastern Illinois University, Charleston, IL.

[2]*Directions for Using Key*: This key consists of pairs or couplets of choices, 1–1′, 2–2′, and so on. Compare your plant specimen with the descriptions given in couplet 1–1′. Choose whichever statement (1 or 1′) is more applicable to your specimen and go on to the next statement as indicated by the number at the extreme right. Continue through the key, each time choosing between a pair of choices until you terminate at the name of a plant. In case you get lost and forget which choices you've previously made, the numbers written in parentheses indicate your previous position in the key.

II. BLUE-GREEN ALGAE

1.	Unicellular or colonial	2
1'.	Filamentous	6
2(1).	Cells ellipsoidal or cylindrical	**Gloeothece**
2'.	Cells spherical or hemispherical	3
3(2').	Cells arranged in rows, forming a more or less rectangular sheet	**Merismopedia**
3'.	Cells solitary or united in round to irregular colonies	4
4(3').	Numerous small cells in a colony; sheath watery, often indistinct	**Polycystis**
4'.	Few (less than 50) cells in a colony; sheath thick, often in layers around cells and/or colony	5
5(4').	Sheath colored	**Gloeocapsa**
5'.	Sheath colorless	**Chroococcus**
6(1').	Without heterocysts	7
6'.	With heterocysts	10
7(6).	Without an evident mucilaginous sheath	8
7'.	With an evident mucilaginous sheath	9
8(7).	Filaments straight, with cross walls	**Oscillatoria**
8'.	Filaments spiral, cross walls indistinct	**Spirulina**
9(7').	One filament within a sheath	**Lyngbya**
9'.	Many filaments in a sheath	**Microcoleus**
10(6').	Heterocysts terminal	11
10'.	Heterocysts intercalary	12
11(10).	Filaments tapering	**Rivularia**
11'.	Filaments not tapering	**Cylindrospermum**
12(10').	Filaments showing apparent branching at the heterocysts	**Tolypothrix**
12'.	Filaments unbranched	13
13(12').	Heterocysts round; filaments within globular mucilaginous mass	**Nostoc**
13'.	Heterocysts oval; filaments not in a globular mucilaginous mass	**Anabaena**

III. GREEN ALGAE AND EUGLENOIDS

1.	Plants unicellular or colonial	2
1'.	Plants filamentous	14
2(1).	Motile by flagella	3
2'.	Nonmotile	8
3(2).	Unicellular	4
3'.	Colonial	5
4(3).	Cells long and tapering when swimming, round when resting; one flagellum and a red eyespot	**Euglena**
4'.	Cells oval; 2 flagella; chloroplast cup-shaped	**Chlamydomonas**
5(3').	Colony a flat plate, 4–8–16 cells	**Gonium**
5'.	Colony spherical	6
6(5').	Cells close together, usually 16	**Pandorina**
6'.	Cells remote from each other	7
7(6').	32 (usually) to 128 cells	**Eudorina**

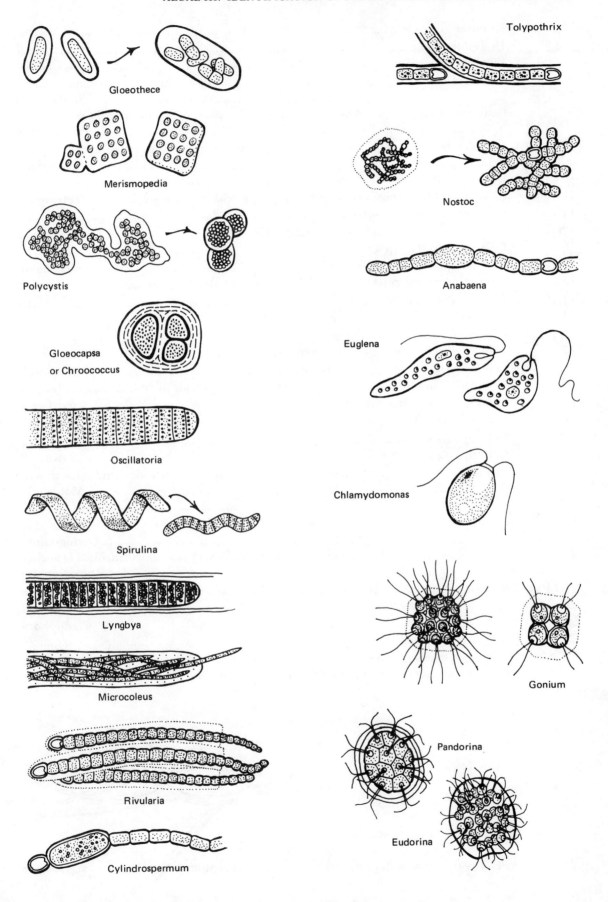

Gloeothece

Tolypothrix

Merismopedia

Nostoc

Polycystis

Anabaena

Gloeocapsa
or Chroococcus

Euglena

Oscillatoria

Chlamydomonas

Spirulina

Gonium

Lyngbya

Microcoleus

Pandorina

Rivularia

Eudorina

Cylindrospermum

7'.	Hundreds of cells	**Volvox**
8(2').	Unicellular	9
8'.	Colonial	10
9(8).	Cells in two symmetrical halves connected by a narrow isthmus (see Section IV)	**Desmids**
9'.	Cells spherical to oval, sometimes in irregular masses; on wood or moist soil	**Protococcus**
10(8').	Colonies with four cells (sometimes eight) in a row; spines often on end cells	**Scenedesmus**
10'.	Colonies with more than four cells	11
11(10').	Colonies mucilaginous; cells in groups of four within mucilage	**Tetraspora**
11'.	Colonies not mucilaginous	12
12(11').	Cells forming a net often visible to the unaided eye	**Hydrodictyon**
12'.	Cells forming a flat plate	13
13(12').	Plate irregular; some cells with long, sheathed bristles	**Coleochaete**
13'.	Plate regular; marginal cells with lobes, horns, or short spines	**Pediastrum**
14(1').	Filaments unbranched	15
14'.	Filaments branched	21
15(14).	Cells short, usually not twice as long as broad	16
15'.	Cells long, at least twice as long as broad	17
16(15).	Cells with thick walls; chloroplast diffuse	**Microspora**
16'.	Cells with thin walls; chloroplast in a ring around interior of cell	**Ulothrix**
17(15').	Chloroplasts star-shaped, usually two per cell	**Zygnema**
17'.	Chloroplasts not star-shaped	18
18(17').	Chloroplasts spiral	**Spirogyra**
18'.	Chloroplasts not spiral	19
19(18').	Cell wall thin; chloroplast a flat plate which is broad in surface view and appears as a thin line in side view	**Mougeotia**
19'.	Cell wall thick	20
20(19').	Chloroplast more or less uniform; some cells with apical caps; swollen oogonia	**Oedogonium**
20'.	Chloroplast dense and granular; cells large, with a very few short rhizoidal branches	**Rhizoclonium**
21(14').	Branches short, with a bulb-like base tapering into a long spine	**Bulbochaete**
21'.	Branches relatively long	22
22(21').	Cells with thick walls	**Cladophora**
22'.	Cells with thin walls	23
23(22').	Plant body showing marked differentiation between a single row of large cells forming the main axis and numerous tufts of short lateral branches with small cells **Draparnaldia**	

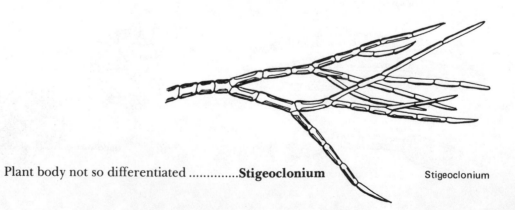

23'.	Plant body not so differentiated .... **Stigeoclonium**

Stigeoclonium

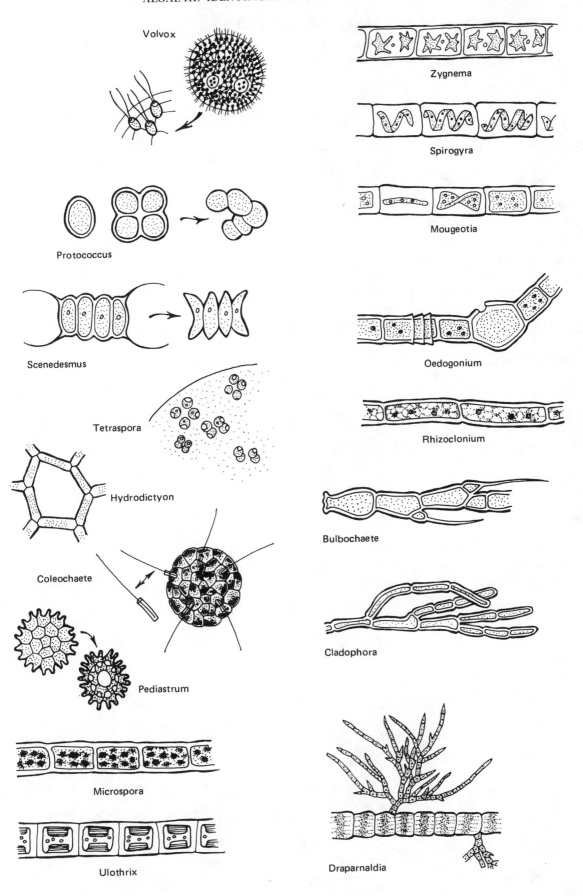

Volvox

Zygnema

Spirogyra

Mougeotia

Protococcus

Scenedesmus

Oedogonium

Tetraspora

Rhizoclonium

Hydrodictyon

Bulbochaete

Coleochaete

Cladophora

Pediastrum

Microspora

Draparnaldia

Ulothrix

IV. OTHER ALGAE NOT INCLUDED IN KEY

A. GREEN ALGAE—DESMIDS

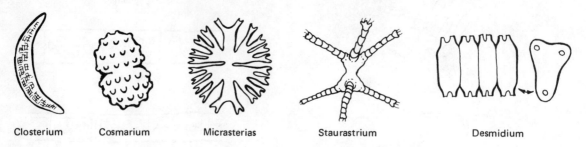

Closterium   Cosmarium   Micrasterias   Staurastrium   Desmidium

B. CHARA

A large alga easily seen with the unaided eye. Plants attached to the soil beneath the water and often forming extensive meadows. Plants may be 0.3 to 1 meter long. Plant body consists of a cylindrical axis bearing a whorl of branches at its several "nodes." Internode lengths up to 5 cm long.

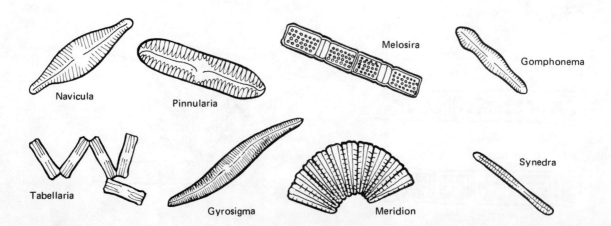

Chara

C. DIATOMS

Often yellow or brown in mass. Usually single-celled motile plants with glassy silicon walls showing striae (lines). A large family of many forms.

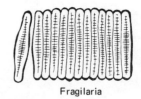

Fragilaria

Navicula   Pinnularia   Melosira   Gomphonema

Tabellaria   Gyrosigma   Meridion   Synedra

D. OTHER ALGAE

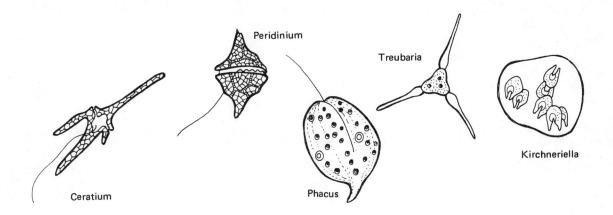

Ceratium

Peridinium

Phacus

Treubaria

Kirchneriella

EXERCISE 28

ALGAE IV: BROWN ALGAE, RED ALGAE, AND LICHENS

Those of you who live along coastal areas will be familiar with some of the brown algae of today's exercise. They are what most people recognize as seaweeds. Red algae, because of their smaller size, are not so commonly known. Lichens are basically land plants that many of you at one time or another have seen growing on tree trunks or rocky places.

Life cycles of brown algae and of red algae are often complex, and we need not go into their details. We can, however, get a general idea of the structure of these algae. Lichens are interesting in structure because each plant is composed of an alga and a fungus.

I. DIVISION PHAEOPHYTA—BROWN ALGAE

The Phaeophyta are the brown seaweeds. They range from microscopic filamentous species to big, shrub-like plants. Some of the large ones have an internal anatomy similar in some ways to land plants, though no one suggests they are related.

General facts about brown algae are

Usual color	Brown
Distribution	Marine, predominantly; mostly in the cooler to very cold oceanic waters or their cold currents in warmer latitudes.
Habitat and niches	Majority are in rough waters of the tidal zone. Attached to rocks of the shoreline or to the ocean bottom at 15- to 20-meter depths. Some are microscopic in the plankton; some are large floating masses.
Motility	Flagellated cells in reproduction stages
Body types	Range from microscopic filaments to large, shrub-like plants with broad, leathery blades.
Value	Of some value. Are the source of alginates which have wide industrial uses; also used as supplementary feed for livestock and as soil fertilizer.

A. THE ROCKWEEDS

Rockweeds live in the rough waters of the tidal zone. Their tough, leathery thallus is well adapted to survive the battering waves.

► ACTIVITY 1   **Examine** specimens of *Fucus*, a common rockweed. **Note** that the THALLUS has a HOLDFAST, a root-like disk that firmly attaches the thallus to rocks. Otherwise the plants would be torn away by the rough water. Above the holdfast is a short stem-like part, the STIPE. The remainder of the plant consists of a much branched BLADE. The branching is a simple, equal-forking pattern. **Note** the AIR BLADDERS along the blade. These keep the plant afloat at high tide. At the tips of the branches are swollen, pointed sacs, the RECEPTACLES. These are covered with small pores that lead into cavities that contain GAMETANGIA. **Feel** how slippery the plants are. This is due to mucilage which protects the plant from dessication when it is exposed to air at low tide.

Label the *Fucus* thallus in Figure 28.1, using the capitalized terms defined in the preceding description.

B. THE KELPS

► ACTIVITY 2   **Examine** specimens of *Nereocystis* or *Laminaria* if available. Kelps are much larger plants than rockweeds. Many of the same basic plant parts are identifiable in the kelps as you saw in the rockweeds. You should be able to recognize the HOLDFAST, STIPE, BLADE, and AIR BLADDER. Again, **note** the tough, leathery, slimy feel to the plants.

Label Figures 28.2 and 28.3, using the capitalized terms noted in the preceding description.

C. SARGASSUM

► ACTIVITY 3   If available, **examine** *Sargassum*, a floating brown alga in tropical waters. **Note** its stem-like STIPE, leaf-like BLADES, and berry-like AIR BLADDERS. These plants float in large numbers, often extending for hundreds and thousands of hectares over the sea's surface. The ships of Christopher Columbus sailed right into one such area, much to the crew's dismay. They feared their ships would be dragged under by it. The Sargasso Sea is named after these plants.

Label Figures 28.4, using the capitalized terms defined in the preceding description.

II. DIVISION RHODOPHYTA—RED ALGAE

Red algae are small, fragile, delicate plants, rarely more than 0.7 meter long. They look like feathery, or lacy tufts of plants. Most of the species are filamentous. They have complex life cycles which we will not consider.

General facts about red algae are

Usual color	Bright red or pink, but many are also bright green or brown
Distribution	Marine; the oceans of the temperate zone, subtropics, and tropics.
Habitats and niches	In calm waters; attached to rocks or other plants in the calmer, deeper waters beyond the tidal zone; a few at shallower depths.
Motility	No.
Body types	Filaments or leaf-like parts that form delicate, branched, tufted, membranous or somewhat fleshy plants.
Value	Of some value. Source of carrageenin widely used commercially for its gelling properties; source of agar used in a variety of manufactured goods and processed foods.

► ACTIVITY 4   **Examine** the various specimens of red algae available, including *Stenogramma* (note the method of branching); *Gelidium* (a source of agar); and *Rhodymenia* (one of the larger species).

III. LICHENS

A lichen is a plant made up of a fungus and an alga living together as one unit. The association is mutualistic, each organism deriving substantial benefits from it. The alga synthesizes carbohydrates and other food materials. The fungus absorbs and retains moisture. Most of the lichen is composed of fungal hyphae. The algal cells are nested among the hyphae and are permeated by root-like growths from them. The fungus is usually an Ascomycete and the alga is usually one of the single-cell-type blue-green (Cyanophyta) or green (Chlorophyta) algae.

There are three growth forms or body types of lichens:

1. CRUSTOSE   The lichen forms just a crust on the surface on which it is growing.
2. FOLIOSE   The lichen is more or less leafy in appearance.
3. FRUTICOSE   The lichen is shrub-like with multibranching and intertwined fibrouslike parts.

General facts about lichens are

Usual color	Variable. Can be gray-green, white, orange, yellow, brown, or black. Many bright colors, especially in the reproductive parts.
Distribution	Terrestrial, worldwide
Habitat and niches	Wide variety of habitats and climatic conditions; tree trunks, rocks, any unpainted, weather-exposed wood, and certain types of poor soils.
Body types	1. Crust-like—crustose 2. Leafy—foliose 3. Shrubby—fruticose
Value	Pasturage for reindeer and caribou in the arctic; human consumption in the Orient; variety of medicinal and industrial uses; used for dying cloth in primitive cultures.

► ACTIVITY 5   **Examine** species of lichens which show the crustose, foliose and fruticose growth forms. **Note** the brightly colored, usually cup-shaped, fruiting bodies. These fruiting bodies are ASCOCARPS (refer back to Exercise 24 if you've forgotten what ascocarps are). They are functionless for the lichen but useful to us in identifying the plant.

If fresh specimens are available, **examine** the lichens with a dissecting microscope.

► ACTIVITY 6   **Examine** a prepared slide of a section of the foliose-type lichen *Physcia* (or other available material) and note the internal organization. **Refer** to Figure 28.5 as a guide in this description while you examine the slide.

The Thallus or Body

The body or thallus of the lichen has four layers:

1. UPPER CORTEX   The upper surface layer, composed of fungal hyphae so highly twisted and compact that the cortex appears cellular.
2. ALGAL LAYER   Just below the upper cortex, composed of clumps of algal cells lying among loosely interwoven fungal hyphae.
3. MEDULLA   A thick layer of loosely packed fungal hyphae.
4. LOWER CORTEX   Similar to the upper cortex but with RHIZOIDS (root-like hyphal growths).

The Ascocarps

Examine the slide again and this time look for saucer-like structures growing up from the top of the thallus. These structures (you may see only one) are the fruiting bodies or ASCOCARPS. In the depression on the top side of the ascocarp is the HYMENIUM, a layer of ASCI and

STERILE HYPHAE. Although ascocarps are present, they play no role in reproducing the lichen. Instead, lichens reproduce by fragments breaking off the plant or by special, regenerative clumps of hyphae and algae, called soredia. Fragments and soredia get blown about and grow into new lichens.

Label the diagram of the lichen *Physcia* in Figure 28.5, using the capitalized terms defined in the preceding description.

EXERCISE 28   **STUDENT NAME** _____

QUESTIONS

1. Rough tidal water is a common habitat for the kelps and rockweeds. Give two adaptations (structural and otherwise) found in these plants which help them survive in this habitat.

   (a) _____

   (b) _____

2. Red algae characteristically live in quiet, calm water depths below wave action. What effect has this had

   on their growth forms? _____

3. Lichens typically have ascocarps. They, therefore, reproduce by cells, called ascospores. Circle one:

   (a) true; (b) false.

4. Which layer of a lichen thallus is photosynthetic? _____

5. Kelps are pretty big, weighty plants. What keeps them afloat? _____

   _____

6. There are no delicate, filamentous brown algae. Circle one: (a) true; (b) false.

7. The fungus portion of most lichens belongs to which class of fungi? _____

8. Lichens are very sensitive to air pollution. If you live in the city or a suburb, think for a minute. Do you

   see many trees with their barks encrusted with lichens? _____

9. The bulk of a lichen plant is formed by which of its component members, (a) the alga; (b) the fungus?

   Circle (a) or (b).

10. Red algae are predominantly marine. Circle one: (a) true; (b) false.

11. Suppose you're hired as the nature counselor at a summer camp near a big lake.

    a. Should you get out your old botany notes and "bone up" on *all* the algae? _____

    b. Would your fishing line get snagged in rockweeds? _____

    c. Could you find any diatoms to show the campers? _____. You'd expect to find (a) pennate

       types; (b) centric types; (c) both types; (d) neither. Circle (a); (b); (c); or (d).

    d. As you lead the way under a waterfall, what group of algae will you know enough to point out grow-

       ing on the wet ledges? (division name) _____

e. On one of the trails you come across some lichens. One of the campers asks you which lichen it is.

What part of the plant will be of particular value to you when you key it out? _____

_____

f. The camp has a microscope and you've brought your algae key from this manual. You collect some

lake water. Would *most* of the key be of use to you? _____

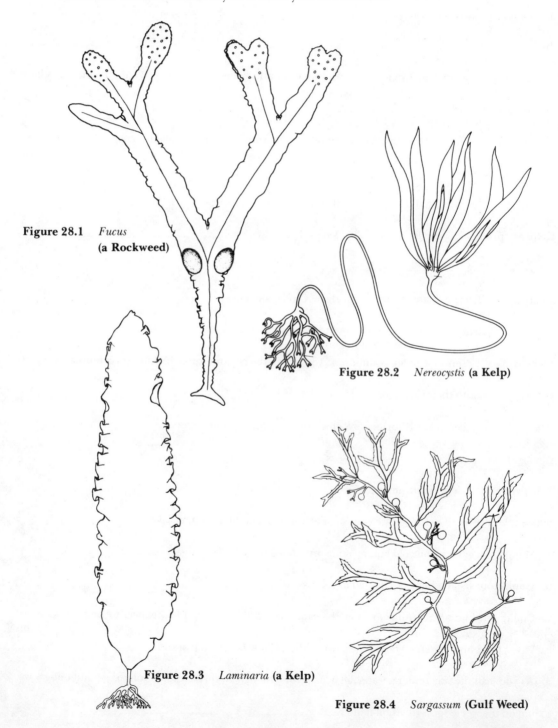

Figure 28.1 *Fucus*
(a Rockweed)

Figure 28.2 *Nereocystis* **(a Kelp)**

Figure 28.3 *Laminaria* **(a Kelp)**

Figure 28.4 *Sargassum* **(Gulf Weed)**

STUDENT NAME _____

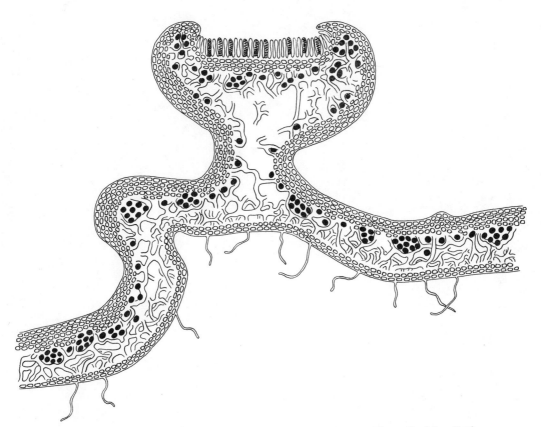

Figure 28.5   Cross Section of a Lichen with Cup-Fungus Type Fruiting Body

Exercise 29

MOSSES AND LIVERWORTS
(AND INTRODUCTION TO LAND PLANTS)

I. INTRODUCTION TO LAND PLANTS

Conditions on land are extremely variable compared to those found in oceans, seas, lakes, and rivers. Temperature, moisture, wind, exposure—these vary from place to place, and so we find on land a wide variety of habitats. The life of a photosynthetic land plant involves the same four basic requisites for survival (see Exercise 23) as you saw in algae and fungi. But land plants can meet these requisites in many different ways.

How did this variety come about? It came about through the evolution of a way to substantially increase genetic variation. This increase in genetic variation is due largely to the increase in the number of meiotic events in the life cycle.

The more meiotic events (that is, number of cells dividing by meiosis), the more the chromosomes and genes are shuffled into new gene combinations. New gene combinations in individuals lead to genetic variation in a population. Thus, ultimately, very different organisms with different genetic makeup have come to occupy and survive in the varied land habitats.

Compare the generalized life cycles in Figures 29.1 and 29.2. In the land plant (Figure 29.1) you can see that for each zygote formed, there are many meiotic events generated. Compare this to the one meiotic event generated for each zygote formed in the life cycle of the basic alga or fungus (Figure 29.2).

The thallus (or plant body) that produces the spores in the land plant is called *sporophyte* or *sporophyte generation*. The thallus (or plant body) that produces the gametes (eggs and sperm) is called *gametophyte* or *gametophyte generation*. So, in the life cycle of the land plant we find an *alternation of two generations*: a sporophyte and a gametophyte.

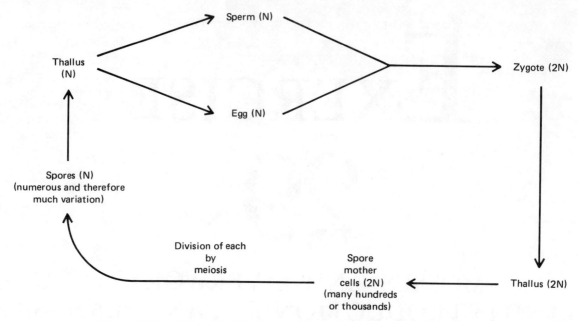

Figure 29.1   Generalized Basic Life Cycle of the Land Plant. The Egg and Sperm Do Not Necessarily Come from the Same Thallus

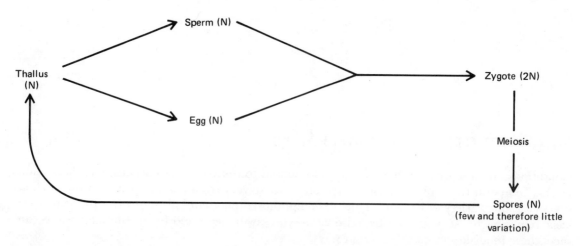

Figure 29.2   Generalized Life Cycle of the Basic Alga and Fungus. The Egg and Sperm Do Not Necessarily Come from the Same Thallus (Some algae and fungi have more complex cycles.)

II. BRYOPHYTES—MOSSES AND LIVERWORTS

Mosses and liverworts make up most of the bryophytes (division Bryophyta), which are tiny plants averaging less than 10 cm in height. Their small size is largely due to their lack of tissues for support and for internal conduction (xylem and phloem). The absence of these tissues also limits them to relatively few types of habitats. Most grow in moist or wet shady places. Some mosses are adapted to dry, sunny, exposed sites.

In bryophytes, the gametophyte thallus is the chief photosynthetic body in the life cycle. But it is also the thallus that produces the gametes. Its small, ground-hugging growth form favors fertilization because a surface film of water is necessary for sperm transport to the egg. But its small size automatically limits the photosynthetic surface. Also, the general lack of a cuticle makes it liable to dehydration.

The sporophyte thallus is totally supported, and partially nourished, by the tiny gametophyte thallus. Thus the sporophyte can never grow any larger than the gametophyte. This combination of features has led to a blind alley as regards evolution in bryophytes (see Table 29.1). The first bryophytes may have evolved from certain highly advanced green algae. In time, many became rather elaborate little plants which thrived long before the dinosaurs. However, they were probably always suited to only a limited range of terrestrial habitats, and as time went on they have declined into the very simple, uncomplicated types we see today.

In this exercise you will study representatives of two classes of bryophytes, the mosses (class Musci) and the liverworts (class Hepaticae).

Table 29.1   Summary of the Progressive and Primitive Features of the Bryophytes as they Relate to Successful Life on Land

Progressive Features over the Algae	Primitive Features
1. Eggs and sperm enclosed by protective organs, the archegonia and antheridia. These prevent dehydration and death of the gametes 2. Embryo protected within the archegonium and nourished by the female parent plant 3. Introduction of a sporophyte generation into the life cycle leads to more genetic variation through meiotic divisions of numerous spore mother cells	1. The incorporation of photosynthesis and sexual reproduction in one plant (the gametophyte) is not a good combination of functions. Necessity of water for fertilization keeps the gametophyte low to facilitate sperm transfer by rain and water droplets. This limits the potential for assimilative functions and growth 2. No vascular or supporting tissue 3. Little or not cuticle; few stomates 4. Water necessary for fertilization

III. GENERALIZED LIFE CYCLE OF BRYOPHYTES

Study Figure 29.3, which is a generalized life cycle of bryophytes, and refer to it as you proceed through the exercise. Also **study** Table 29.2 which shows how the life cycle stages, structures, and cells are involved in species survival.

Table 29.2   Life Cycle Stages, Structures, and Cells Related to the Four Survival Phases in Species of Bryophytes

Survival Phase	Accomplished by
Multiplication	1. Asexual reproduction (gemmae, shoot fragments, and new branches from the gametophyte) 2. Spore production by the sporophyte
Dispersal	1. Spores from the sporophyte 2. Gemmae and shoot fragments from the gametophyte (to some extent)
Assimilation	Gametophyte thallus
Genetic variation	1. Meiosis by spore mother cells in the sporangium of the sporophyte 2. Sexual fusion in the gametophyte generation

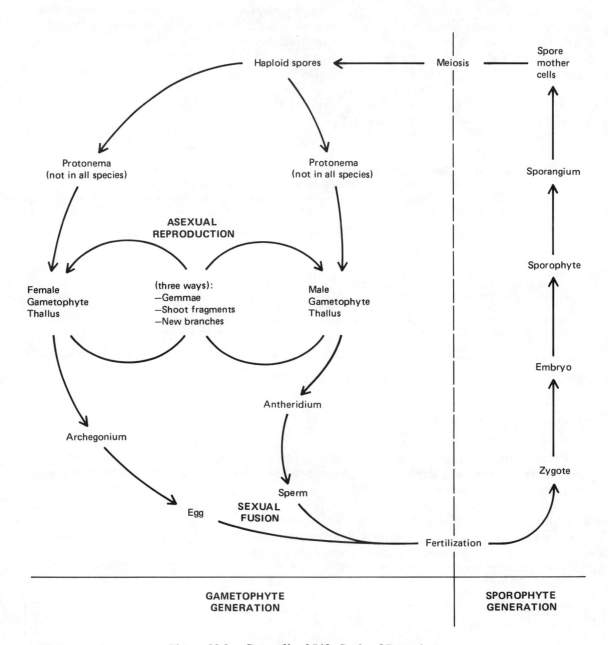

Figure 29.3   Generalized Life Cycle of Bryophytes

IV. THE MOSSES—CLASS MUSCI

A. THE GAMETOPHYTE THALLUS

The gametophyte is multifunctional. It serves in asexual and sexual reproduction; it photosynthesizes, manufacturing foods for its own growth and that of the embryo and sporophyte; and it supports the sporophyte.

Vegetative Parts

► ACTIVITY 1   **Examine** the leafy gametophyte plants of species of *Mnium*, *Polytrichum*, or *Funaria*. If live plants are provided, you can see that numerous tiny gametophytes form a dense, cushiony, carpet-like mat. The mat has been collected from a forest floor or other such shady ground.

With forceps, **pull up** one of the gametophytes. *Do not* pull up any plant that has a sporophyte on it (that is, any plant having a thin, leafless filament and capsule growing at the tip). Pick only those plants that are leafy all the way to the top.

The plant typically has a slender, erect, stem-like axis crowded with leaf-like appendages. The root-like, multicellular filaments at the base are RHIZOIDS. (You may not see any rhizoids because they usually break off when the plant is pulled up). **Remove** a "LEAF" and mount it on a slide for microscopic examination. How many cells thick is the leaf? _____. Are there stomates? YES NO. Is there a midrib? YES NO. Can you see xylem cells? YES NO. Is there a vein system? YES NO. Is this a true leaf? YES NO. Does it carry on photosynthesis? YES NO. **Look** at a rhizoid on a slide under your microscope. Is the rhizoid made up of one or of several cells? (Circle one.) Is there any xylem? YES NO. Any root hairs? YES NO. Is the rhizoid a true root? YES NO.

Label A and **B** of the moss life cycle in Figure 29.4, using the capitalized terms defined in the preceding descriptions.

Reproductive Parts

Examine the tip of a leafy plant. If the tip has a flat, "flower-like" cluster of leaves, the plant is a MALE GAMETOPHYTE, and the male reproductive organs are located in this flat circlet of leaves. If the leaf arrangement at the tip is no different than the rest of the stem, the plant is a FEMALE GAMETOPHYTE, and the female reproductive organs occur hidden among the leaves at the tip.

Rain drops splattering over the plants carry sperm from male plants to the female. As a droplet of rain sits in the cup-like circlet of leaves on the male head, sperm are released into it. Subsequent ricocheting or splattering of these droplets results in some of them landing on female plants. The motile SPERM them swim through the water toward the female organs. **Label** the male plant and the female plant in Figure 29.4.

Reproductive Organs

antheridia

► ACTIVITY 2   **Examine** prepared slides of the tip of a male gametophyte showing the ANTHERIDIA (plural), the male reproductive organs that produce the sperm. An ANTHERIDIUM (singular) is broadly or narrowly oval with a one-layered outer coat, the JACKET LAYER. Virtually all the antheridia that are intact and recognizable on your slide contain tightly packed, angular cells that are in some stage of forming sperm. These are the SPERM-FORMING CELLS. Intermingled among the antheridia are "leaves" and special slender filaments known as paraphyses (singular; paraphysis) or sterile hairs. These are erect and hair-like with expanded, club-like tips. What function do you suppose the "leaves" and paraphyses serve? _____
_____What is the function of the jacket layer? _____.

Label the antheridium and its parts in Figure 29.4, using the capitalized terms defined in the preceding description.

sperm

▶ ACTIVITY 3   **Obtain** a male gametophyte plant of *Mnium* or *Polytrichum* or other moss and note the characteristic cup-shaped circlet of leaves at the tip. This tip is designed to catch and hold a droplet of rain or moisture. The collected water will eventually cause the antheridia to split open and release sperm. When mature, sperm develop two flagella by which they maneuver through the watery film or droplets covering the plants. **Place** the plant on a slide and, using a needle, gently press the base of the head. Squeeze out a few antheridia and examine them under the microscope. After a while the antheridia should probably rupture and discharge sperm. **Keep** the slide and check it periodically to see if sperm are released.

archegonia

▶ ACTIVITY 4   **Examine** a prepared slide of a moss gametophyte showing ARCHEGONIA (plural). Find an ARCHEGONIUM (singular). This is shaped something like a bowling pin or a long necked vase. It consists of (1) an oval, or globe-shaped base, the VENTER, within which is one EGG, and (2) a long, slender chimney of cells, called the NECK. The archegonium exudes a chemical that attracts sperm cells which swim down the neck which in turn channels them toward the egg. One sperm fertilizes the egg and a ZYGOTE is formed. The zygote develops by many cell divisions into the EMBRYO. The embryo continues growth, developing into the sporophyte. All this growth occurs within the archegonium. Because of the minute size of the archegonium, your slide section may not have archegonia that are all complete, and you may see necks without venters attached, and vice versa.

Label the archegonium and its parts, both before and after fertilization, in Figure 29.4, using the capitalized terms defined in the preceding description.

B. THE SPOROPHYTE THALLUS

The sporophyte produces and discharges the spores. These are then carried by air currents (in some cases, water currents) away from the parent plant. A benefit to this is that the new gametophyte plants developing from spores are thus less likely to crowd and compete with the old parent population for light and nutrients.

▶ ACTIVITY 5   **Examine** plants bearing sporophytes. The SPOROPHYTE plant is designed well for its function of spore discharge. There is a little CAPSULE or SPORANGIUM borne at the top of an upright stalk, the SETA. The base of the seta is called the FOOT. The foot is not usually visible to you. It is embedded in an archegonium at the tip of a gametophyte. The elevated capsule is a definite advantage for spore dispersal by air. The capsule serves three functions; the basal part is photosynthetic; the central core produces spore mother cells that divide by meiosis to form the spores; and the tip is designed to discharge spores.

Dissect a capsule. With forceps, gently pinch the very tip of the capsule and pull off the loosely fitting hood, the CALYPTRA. The calyptra is the remnant of the archegonium that was carried upward by the growth of the sporophyte. Next you can see a trim little, tight-fitting lid, the OPERCULUM. **Remove** the operculum with forceps. Transfer the capsule to a small blob of Vaseline on a slide and, using the low power of the microscope, orient the capsule so that you are looking down into the mouth of the capsule. Or, cut off the upper end of the capsule with a sharp razor blade and mount it in water on a slide. Look at the tooth-like projections encircling the rim of the mouth. These projections are the PERISTOME. In many mosses, like *Mnium*, the teeth forming the peristome are hygroscopic (sensitive to changes in air moisture), and they bend in and out in response to these changes. This bending in and out of the teeth flips the spores out of the capsule. In other mosses, like *Polytrichum*, the teeth do not bend in and out. Instead, they are all joined to a membrane that stretches across the mouth, and the spores sift out through tiny holes in the membrane between the teeth tips. In this case, the teeth serve to slow down the escape of spores, thus prolonging the period of dispersal. What is the advantage of this? _____

_____

Test the hygroscopic mechanism yourself. Take two capsules from sporophytes. Moisten one in a wet paper towel; allow the other one to dry in the warmth of a lamp (or just let it air dry). Don't lose it! After a few minutes, look at the curvature of the peristome teeth in each capsule. **Describe** what you see. _____.

Would high- or low-moisture conditions favor release of the spores? (Circle one.) **Label** the sporophyte and the parts of the sporanangium in Figure 29.4, using the capitalized terms defined in the preceding description.

C. SPORES AND PROTONEMA

Spores

▶ ACTIVITY 6   **Crush** a capsule of a moss with a needle and force out some spores into a drop of water on a slide. **Cover** with a cover glass and **examine** under the microscope. Are the spores green? YES NO. Are they all alike? YES NO. Spores are the haploid daughter cells of SPORE MOTHER CELLS that divide by MEIOSIS within the capsule. The reshuffling of chromosome pairs and rearranging of genes that occurs during meiosis produces haploid spores with many different genetic combinations. This serves to ensure adaptive variability among the gametophyte plants that result from spore germination.

Complete the labeling of the sporangium and its contents in Figure 29.4, using the capitalized terms defined in the preceding description.

The Protonema

▶ ACTIVITY 7   **Examine** living spores that are germinating on a nutrient medium. Or, if these are not available, examine prepared slides of moss protonema. A protonema is formed by SPORE GERMINATION. The spore divides and the succeeding cells keep on dividing to form a cylindrical, much-branched, green algal-looking stage of the moss, called the PROTONEMA. The protonema lives for a long time. As it matures, it produces small leafy BUDS. Growth of each bud produces an erect leafy gametophyte thallus.

Label the protonema and its parts in Figure 29.4, using the capitalized terms defined in the preceding description.

D. PEAT MOSS—SPHAGNUM

Mosses are biologically and ecologically important. However, in terms of direct-use economic value, the only one of importance is the peat moss, *Sphagnum*. This is because it is one of the most absorbent of all materials. When a *Sphagnum* plant is completely dry, it can absorb and hold up to 200 times its own weight in water. This property makes it useful for many purposes. The plant also contains compounds that inhibit bacterial growth. In the past, some American Indians used it for diapering infants; up to the American Civil War doctors used it for dressing wounds. Today it is most commonly used for a variety of purposes in the ornamental plant industry. Commercial growers pack the roots of shrubs and trees in wet peat moss so they won't dry out during shipment. Soils containing high quantities of *Sphagnum* are known as peat. Packaged peat is widely sold for use in commercial greenhouses and home gardening.

The Sphagnum Plant

▶ ACTIVITY 8   **Examine** living, preserved, or dried *Sphagnum* plants. Note the "stems," "leaves," and "branches."

Determine the water-holding capacity of *Sphagnum*. Weigh a small plastic drinking glass or beaker and record the weight to the nearest 0.1 gram. Fill the glass with dry *Sphagnum* and reweigh.

Record the combined weight of glass and *Sphagnum*. Fill the glass with water and allow the *Sphagnum* to soak for 10 to 15 minutes. Invert and allow the excess water to drain off for 5 minutes. Reweigh the beaker and wet *Sphagnum*. From the increased weight, you can find the amount of water absorbed by the *Sphagnum*. Comparing this to the dry weight of the *Sphagnum*, you can determine the water-holding capacity of the plant.

Weight of beaker + dry *Sphagnum*:	_____ grams
Weight of empty beaker:	_____ grams
Weight of dry *Sphagnum*:	_____ grams
Weight of beaker + wet *Sphagnum*:	_____ grams
Amount of water held by *Sphagnum*:	_____ grams
Water-holding capacity of *Sphagnum*:	_____ grams

Internal Structure of *Sphagnum* "Leaves"

► ACTIVITY 9   The *Sphagnum* leaf has *special* large, water-holding cells. **Mount** a leaf of living or preserved *Sphagnum* in water and examine it under the microscope. **Note** that there are two cell types. Thin, tube-like cells connect with each other in an open meshwork. These cells contain numerous small bodies which you should recognize as the _____. The larger, colorless cells between the tube-shaped cells have small pores in their walls. The walls also show spiral glassy-looking thickenings. Which of the two cells types forms the greater portion of the leaf? _____. What is the obvious function of these cells?

_____.

V. THE LIVERWORTS—CLASS HEPATICAE

Less commonly encountered, and even less commonly recognized, are the liverworts. In many ways they closely parallel the mosses, but there are essential differences between them. Most liverwort species (more than 80 percent) have leafy foliose gametophytes, and this makes it difficult to readily distinguish them from mosses. A minority of liverworts have what is termed a thallose gametophyte. Thallose gametophytes form broad, branching ribbons that are conspicuously lobed or ruffled looking. They are generally much larger than leafy foliose gametophytes. Both thallose and foliose gametophytes are primarily prostrate or creeping and not erect as are most moss gametophytes. **Refer** to Figure 29.5 as a guide and examine specimens of liverworts.

A. THALLOSE TYPE GAMETOPHYTE

► ACTIVITY 10   **Examine** the gametophytes of species of *Marchantia* or *Conocephalum*. Note that the plant has a dark green, RIBBON-LIKE THALLUS and branches in an equal forking pattern. The upper surface has a central groove running its length and it also has diamond-shaped markings which outline the underlying internal air chambers. Photosynthetic cells occupy the air chambers. Each area has a stomate-like pore for air passage. **Examine** the underside of the gametophyte and look at the dense growth of fine, hair-like RHIZOIDS.

Reproductive Parts

Examine gametophytes of *Marchantia*. You should easily find three kinds of special structures on the top side of the plants. One kind is a little cup, GEMMAE CUPS. These serve for asexual reproduction. The other two kinds look like miniature plants growing upright on the surface of the ribbon. These are the male ANTHERIDIOPHORES and the female ARCHEGONIOPHORES. They occur on separate gametophyte plants.

antheridiophore **Examine** an ANTHERIDIOPHORE. This consists of a small, flat plate or disk, the ANTHERIDIAL RECEPTACLE borne at the tip of a short, slender stalk. ANTHERIDIA are embedded in this disk and release sperm through pores at the top.

archegoniophore   **Examine** an ARCHEGONIOPHORE. This is a tiny, umbrella-like structure with several free-hanging, finger-like lobes, called RAYS. ARCHEGONIA hang suspended, neck down, on the underside of each ray. Sperm are carried from antheridiophores by surface moisture to the archegoniophores, and there they swim up the necks of the archegonia. Fertilization and the development of the sporophyte takes place inside the archegonium as in mosses.

gemmae cups   Gemmae cups are small, round, bowl-shaped structures. If you look down directly into the open gemmae cup and use a hand lens or low power of the microscope, you can see the many little disk-shaped GEMMAE. Gemmae are washed or flicked out of the cup by rain and carried away from the parent plant. Under favorable conditions, a gemma will develop RHIZOIDS, and a YOUNG NEW GAMETOPHYTE is produced from it.

Label all the parts of the gametophyte thallus of the liverwort in Figure 29.5, using the capitalized terms defined in the preceding descriptions.

B. THE SPOROPHYTE

In most cases, the mature sporophyte is simply a small sack, the CAPSULE or SPORANGIUM, suspended by a short stalk and hanging out of an archegonium on the underside of a ray of an archegoniophore. When the capsule matures, it splits open. Special spindle-shaped cells, known as ELATERS, occur among the SPORES. Their spiral banded walls are hygroscopic, and the elaters perform squirming movements that help to loosen the spore mass and throw spores out of the capsule.

Label all the parts of the sporophyte of the liverwort in Figure 29.5, using the capitalized terms defined in the preceding description.

EXERCISE 29   **STUDENT NAME** _____

QUESTIONS

1. What two completely different functions are performed by the gametophyte? (a) _____

_____; (b) _____

2. If moss gametophytes were much taller plants, what problems would they probably have, as regards

   fertilization? _____

3. In the absence of xylem, how does the gametophyte take up soil moisture? _____

   _____

4. Give at least two reasons why the gametophyte plant is not the best designed structure for efficiency in

   photosynthesis. (a) _____; (b) _____

   _____

5. What prevents the sporophyte from developing into a taller plant so as to achieve better spore trajec-

   tory? _____

6. What is an advantage of dispersing spores? _____

7. Describe the habitat best suited to the majority of mosses. _____

   _____

8. Does the spore develop into the embryo? _____

9. Is the embryo haploid or diploid? _____

10. During what event of the life cycle of moss are several new gene combinations created by a reshuffling of

    chromosome pairs and rearrangement of genes? _____

11. Which generation develops from the growth of the embryo? _____

12. Does meiosis lead to the formation of gametes or to spores? _____

13. Which of the following is (are) a haploid: (a) spore(s); (b) egg; (c) sperm. Circle your choice(s).

14. The zygote divides to form many cells that become organized into the _____

    which then matures into the _____ generation.

15. Which group of algae are thought to be ancestral to bryophytes? _____

16. Cite at least three different reasons as to why the structural design of the bryophytes has made them inadequately adapted for successful land life.

   (a) _____ ;

   (b) _____ ;

   (c) _____

17. Increase in genetic variation in land plants is essentially due to _____

   _____

18. For each zygote formed in the life cycle of most algae and fungi, how many chances are there for the

   creation of new gene combinations? (a) one; (b) a few; (c) numerous. Choose one: _____

20. In most algae the cell that divides by meiosis is the _____, but in land plants, the cell that

   divides by meiosis is the _____

EXERCISE 29   STUDENT NAME _____

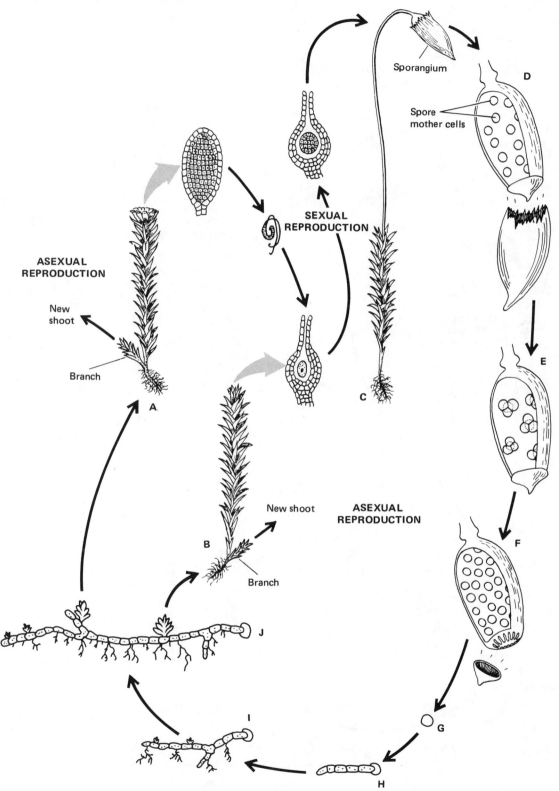

Figure 29.4   Life Cycle of a Moss (Polytrichum sp.) Showing Sexual and Asexual Methods of Reproduction

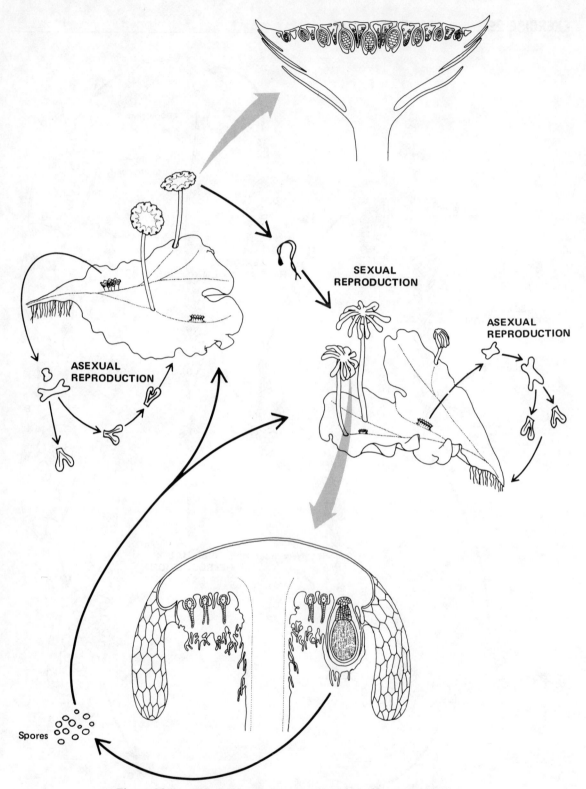

Spores

Figure 29.5   Life Cycle of a Liverwort (*Marchantia polymorpha*)
Showing Sexual and Asexual Methods of Reproduction

Exercise
30

LOWER VASCULAR PLANTS I:
PSILOTUM, CLUB MOSSES,
AND HORSETAILS

Today's work is a study of the club mosses (division Lycopodophyta), horsetails and scouring rushes (division Arthrophyta), and the most primitive living vascular plant, *Psilotum*. They are the living relics of early land plants. All are small, insignificant, and herbaceous, but many of their extinct relatives of 300 million years ago were trees.

These plants are also known as *lower vascular plants* because they have vascular tissue (xylem and phloem) but they represent a lower level of evolution than the more complex vascular plants—the seed plants. They show many adaptations to land life but still retain primitive life cycle features which are a holdover from their aquatic algal ancestors (see Table 30.1).

The sporophyte generation is the conspicuous and dominant thallus of the life cycle. The gametophyte is represented by a very small plant, ranging from a few millimeters to a couple of centimeters in overall size. This size difference between sporophyte and gametophyte is the reverse of that found in the bryophytes.

Table 30.1   Summary of the Progressive and Primitive Features of the Living Club Mosses and Horsetails as They Relate to Successful Life on Land

Progressive Features over the Byrophytes	Primitive Features
1. The incorporation of photosynthesis and spore dispersal in one thallus (the sporophyte) is a good combination of functions. Both have some of the same environmental needs (sun, air, sturdy erect plant). The sporophyte thallus suits both functions equally well 2. The sporophyte is well adapted to life on land: has cuticle, lignin, vascular tissue, and numerous stomates 3. Vigorous rhizome growth produces new shoots. This quickly renews the sporophyte generation and maintains the population in an area. It bypasses the slow cycling through the gametophyte generation	1. Leaf is small and has only one vein. This may limit the efficiency of translocation of materials and in effect may limit growth 2. Water is still required for fertilization 3. Alternation of two independent-living generations (gametophyte and sporophyte) having different growth needs and somewhat different habitats 4. Gametophyte restricted to wet, shady places: lacks cuticle, lignin, and (in most cases) vascular tissue 5. Gametophyte generation is slow-growing, requiring many months to many years before producing a zygote 6. Spore is slow to germinate

I. GENERALIZED LIFE CYCLE

Study the diagram of Figure 30.1, which illustrates the general life cycle of the lower vascular plants (*Psilotum*, club mosses, and horsetails). Also **study** Table 30.2. Remember that the gametophyte phase consists of a very small plant, and most of your work will concern only the sporophyte.

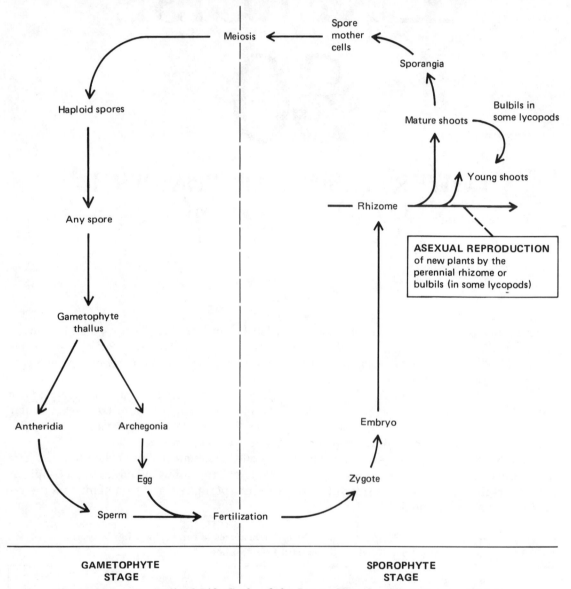

Figure 30.1 Generalized Life Cycle of the Lower Vascular Plants Lycopodophyta, Arthophyta, and Psilotum. (Note the similarity to the fern life cycle.)

II. *PSILOTUM*

► ACTIVITY 1 **Examine** living or preserved specimens or color transparencies of *Psilotum nudum*. *Psilotum* is the most primitive living vascular plant. It grows wild in Florida and Hawaii and other parts of the subtropics and tropics. It is also a fairly popular ornamental plant in greenhouses and homes farther north. For many years it has generally been considered a living relic of an extinct group known as the psilophytes. Some authorities now consider it a primitive fern, but not everyone is in agreement.

The simple plant you see is the sporophyte. The aerial (above ground) stem is a slender axis that has an equal forking type of branching known as DICHOTOMOUS BRANCHING. This

Table 30.2   Life Cycle Stages, Structures, and Cells Related to Methods of Species Survival

Survival Phase	Accomplished by
Multiplication	1. New shoots from the rhizome (asexual reproduction) 2. Spore production by sporophyte
Dispersal	1. Spores 2. Rhizome (to some extent) 3. Bulbils (in some species)
Assimilation	Shoots, leaves, and rhizome of the sporophyte thallus
Genetic variation	1. Meiosis by spore mother cells in the sporangium of the sporophyte 2. Sexual fusion in the gametophyte generation

branching pattern is found in many primitive plants. Are leaves present?   YES   NO. Are there small appendages that resemble scales?   YES   NO. If you have a living specimen, what color is the stem? _____. If you don't have a living specimen, yet you can see that there are no leaves, what color would you expect the stem to be? _____ The aerial shoot originates from an underground stem, the RHIZOME. You probably won't see the rhizome.

Look more closely at the branching axes and **find** the tiny, trilobed, globe-like SPORANGIA. Each sporangium occurs in the axil of a SCALE. SPORES are released from these sporangia and give rise to the GAMETOPHYTE stage as shown in the generalized life cycle of Figure 30.1.

III. DIVISION LYCOPODOPHYTA—CLUB MOSSES

Lycopods are called club mosses because of the moss-like appearance of the leafy stems and the club-shaped cones, although not all species have cones. *Lycopodium* and *Selaginella* are two representatives of the group.

A. *LYCOPODIUM*

▶ ACTIVITY 2   **EXAMINE** living, preserved, or pressed specimens of *Lycopodium*, such as *Lycopodium clavatum*; *L. lucidulum*; or *L. complanatum*. The leafy plant you see is the sporophyte. **Note** the way the stem branches. Is it an equal forking pattern?   YES   NO. What is this pattern called? _____. Is the stem axis naked as in *Psilotum*? YES   NO. Is the stem densely covered with leaves?   YES   NO. In those species which have cones, the parts of the axes that bear the cones seem almost naked because they have few leaves.

The leafy aerial (above-ground) shoots arise from a horizontal, leafy stem, the RHIZOME. The rhizome may grow below the soil's surface or just under the leaf litter of the forest floor. The rhizome forms a dichotomously branching network, and at its growing ends it produces NEW SHOOTS. The rhizome is perennial, and its production of new shoots (ASEXUAL REPRO-DUCTION) is the chief means of increasing the numbers of plants in an area.

Look for SPORANGIA. Depending on the species, these will be found either clustered to-gether in CONES or as solitary SPORANGIA occurring separately any place along the stem. In either case, the SPORANGIUM itself is always borne in the axil of a leaf, called the SPOROPHYLL.

Within each sporangium SPORE MOTHER CELLS divide by MEIOSIS to form HAP-LOID SPORES. These are freed from the sporangia, become dispersed by wind, and germinate into the GAMETOPHYTE THALLUS. A SPERM from an ANTHERIDIUM fertilizes an EGG in an ARCHEGONIUM to form a ZYGOTE that develops into the YOUNG SPOROPHYTE. This grows into the MATURE SPOROPHYTE PLANT.

The gametophyte stage of the life cycle can, and often does, require many months to many years for completion. It does introduce genetic variation, but because it requires such a long time for completion, it is not the primary means by which the population increases in numbers. Growth from the rhizome, as described, is the primary means.

Label the parts of the life cycle of *Lycopodium* (Figure 30.2), using the capitalized terms defined in the preceding description.

B. *SELAGINELLA*

Selaginella is a genus related to *Lycopodium*, but it has certain more advanced structural and reproductive features, and most species are creeping or prostrate in growth habit.

► ACTIVITY 3   **Examine** specimens of *Selaginella*, such as *Selaginella uncinata* (rainbow fern); *S. kraussiana brownii* (cushion moss); or *S. lepidophylla* (resurrection plant). **Note** the tiny LEAVES in four longitudinal rows. Do the branches show DICHOTOMOUS BRANCHING? YES   NO. **Look** to see if your specimen has any long unbranched, leafless axes coming off the leafy stem and hanging toward the ground. If so, this is a RHIZOPHORE. When it touches the ground, the tip will branch and develop underground rootlets.

Note the small CONES at the tips of the leafy stems. The cone of *Selaginella* has two kinds of sporangia, microsporangia that form microspores and megasporangia that form megaspores. The production of two kinds of spores is called *heterospory*. Each microspore produces a male gametophyte within the wall of the microspore. Each megaspore produces a female gametophyte within the wall of the megaspore. Microspores containing male gametophytes are blown by air currents to the vicinity of the female megaspore which is never expelled from the megasporangium.

Fertilization, embryo development, and growth of the young sporophyte all take place within the megaspore wall. This type of reproduction closely parallels features found among the seed plants.

Label the parts of the sporophyte of *Selaginella* shown in Figure 30.3, using the capitalized terms defined in the preceding description.

IV. DIVISION ARTHROPHYTA—HORSETAILS

► ACTIVITY 4   **Examine** living, preserved, or pressed specimens of *Equisetum*, such as *Equisetum hyemale* (scouring rush) or *E. arvense* (common horsetail). The surface of the stem has conspicuous longitudinal RIDGES or RIBS. Do the shoots have dichotomous branching? YES   NO. LEAVES are reduced to small scales that lie flat against the stem (APPRESSED LEAVES). These leaves occur as simple extensions of the ribs. They appear as a rather loose light-colored sheath found at the NODES (or joints) along the stem. Arthrophyte means "jointed stem." In *Equisetum arvense*, BRANCHES as well as LEAVES, occur at these nodes. A horizontal RHIZOME produces new shoots each year and is the primary means by which the population increases in numbers.

SPORANGIA are always clustered together in CONES located at the tips of shoots. When sporangia are young, the cones have a tight, smooth surface. When sporangia and spores are mature, the cone surface becomes more loose and open, exposing the sporangia to the air. Spores are released, dispersed, and each eventually gives rise to a gametophyte thallus which through the steps shown in Figure 30.1 completes the cycle back to the sporophyte stage.

Label the parts of the sporophytes of the two species of *Equisetum* shown in Figure 30.4, using the capitalized terms defined in the preceding description.

Complete Table 30.3.

V. FOSSIL FORMS

► ACTIVITY 5   **Examine** compressions or impressions of fossil stems and leaves of Lycopodophyta and Arthrophyta (if these are available).

Examine slides or color transparencies or acetate "peels" showing structural and anatomical detail of fossil vascular plants (if these are available).

EXERCISE 30   **STUDENT NAME** _____

QUESTIONS

1. List three tissues or special features of the sporophyte of the lower vascular plants which help sustain

   their life on land. (a) _____; (b) _____; (c) _____

   _____

2. What is the chief means of maintaining the populations of any of these plants studied? _____

   _____

3. Which division of plants is characterized by jointed stems? _____

4. If you found a plant with an equal-forking branching pattern, would you consider it to be a highly

   evolved plant? _____ What scientific term could you use to describe equal forking? _____

   _____

5. List all the haploid structures and cells in the generalized life cycle of Figure 30.1. _____

   _____

   _____

6. Name the cell in the life cycle that develops into the embryo. _____

7. Name the thallus on which the sex organs are located. _____

8. Which thallus of the life cycle is the chief, autotrophic, assimilative generation? _____

   _____

9. Are spore mother cells retained in the sporangium (of most species) or are they expelled? _____

   _____

10. Does meiosis occur within the sporangium? _____

11. Name two stages in the life cycle during which genes or groups of genes are either combined or re-

    arranged. (a) _____; (b) _____

12. Since populations are chiefly maintained by asexual reproduction through the growth of the rhizome,

    why then is the sexual phase retained? _____

    _____

13. Name one stage in the life cycle during which the chromosome number is (a) doubled

    _____; (b) halved _____

14. The ending "phyte" is from the Greek "phyta" meaning plant. Define the name "sporophyte": _____

_____

15. Define the name "gametophyte": _____

16. What structure provides the physical and nutritive support for the embryo? _____

_____

17. Name the structure(s) and/or cells that constitute the *dispersal* stage of the life cycle. _____

_____

18. Name the stages or events in the life cycle in which genetic variation is generated in the population.

_____

19. Is water necessary (a) for spore dispersal? _____; (b) for fertilization? _____

20. Coal is called a "fossil fuel." Explain: _____

_____

Table 30.3   Life Cycle Stages, Structures, and Cells Related to the Four Survival Phases in Club Mosses and Horsetails

Survival Phase	Accomplished by
Multiplication	
Dispersal	
Assimilation	
Genetic variation	

EXERCISE 30   **STUDENT NAME** _____

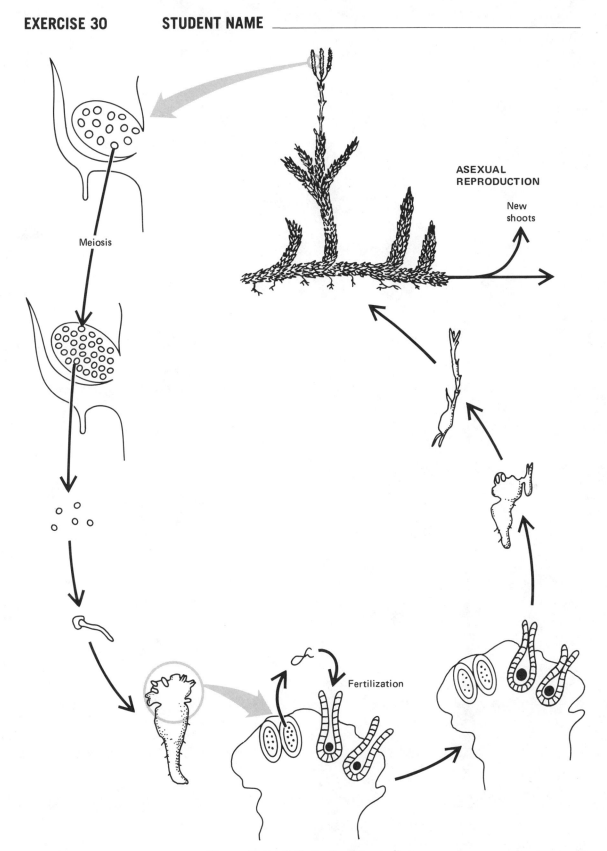

Figure 30.2   Life Cycle of *Lycopodium*

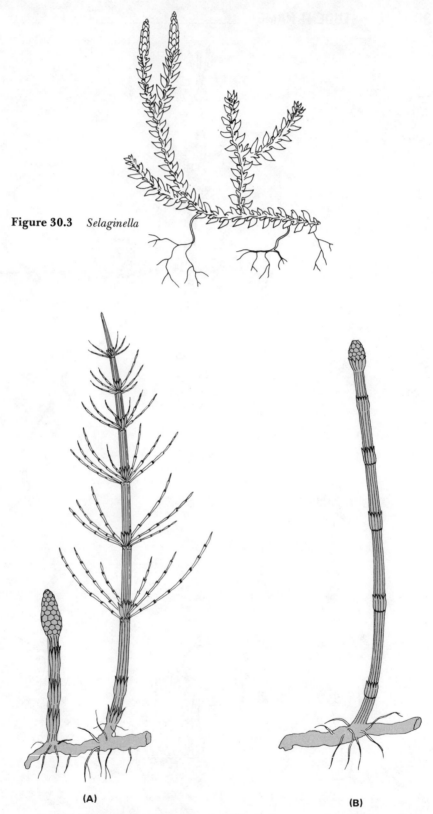

Figure 30.3   *Selaginella*

(A)

(B)

Figure 30.4   Two Common Species of Arthrophytes

EXERCISE
31

LOWER VASCULAR PLANTS II: FERNS

Ferns (division Pterophyta) are the most highly evolved of the lower vascular plants. In terms of variety and breadth of distribution, they are the most successful of the lower vascular plants. Primitive ferns and members of the divisions Lycopodophyta and Arthrophyta were conspicuous members of the world flora of past geological time. Many were trees.

Practically all of these plants are now extinct. Of the Arthrophyta, only one genus, *Equisetum* (25 species) survives. Of the Lycopodophyta, only four genera (about 800 species) exist. Ferns, however, have persisted with greater success.

We know of about 200 genera (10,000 species), some very ancient, some of more recent origin. It may be that because ferns have large, multiveined leaves, comparable to those of the seed plants, they have been more successful in "holding their own" in competition with the more advanced seed plants.

Like other lower vascular plants, living ferns have the same relatively primitive life cycle with two independent-living sporophyte and gametophyte generations (see Table 31.1). They differ from other lower vascular plants, as mentioned above, by their large (usually pinnate) leaves. Most ferns are herbaceous. A few tropical genera are tree-like.

Table 31.1 Summary of the Progressive and Primitive Features of Ferns as They Relate to Successful Life on Land

Progressive Features over Other Lower Vascular Plants	Primitive Features
Large leaves with numerous veins	Essentially the same as those of the other lower vascular plants: 1. Water still necessary for fertilization 2. Alternation of two independent-living generations 3. Gametophyte restricted to wet, shady places and generally lacking cuticle, lignin, and vascular tissue 4. Slow growth of gametophyte generation

I. MATURE SPOROPHYTE OF THE FERN

▶ ACTIVITY 1   **Review** the generalized life cycle of the fern in Figure 31.1 and refer to the pictorial cycle in Figure 31.2 as a guide in your study.

Examine living or pressed specimens of ferns such as Boston fern (*Nephrolepsis exaltata* v. *Bostoniensis*); Christmas fern (*Polystichum acrostichoides*); species of maidenhair fern (*Adiantum* sp.); and *Polypodium* sp.

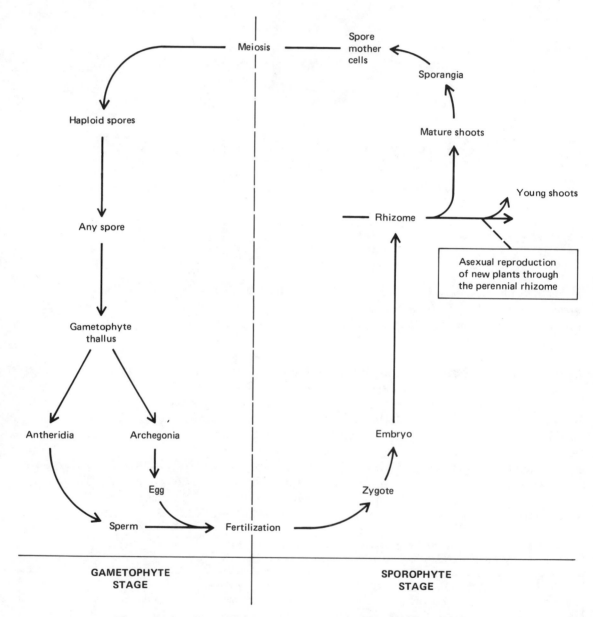

Figure 31.1   Generalized Life Cycle of Ferns (Division Pterophyta)

A. THE LEAF

Most of the body of the fern is made up of LEAVES. The leaves arise directly from the horizontal RHIZOME. There are no aerial (above-ground) stems. Each leaf is called a FROND. It consists of (1) a BLADE that is pinnately subdivided into smaller parts, called PINNAE (singular: pinna); and (2) a PETIOLE, a stalk that attaches the blade to the rhizome. Adventitious roots grow from the rhizome. **Label** the frond and its parts as shown in the life cycle of Figure 31.2, using the capitalized terms defined in the preceding description.

B. THE RHIZOME

Examine living specimens of ferns and note the horizontal stem, the RHIZOME, which may be at, or just below, the ground surface. The rhizome produces new leaves at its growing tip. Young leaves first appear in a tightly coiled mass, and, as growth goes on, the coil seemingly "unrolls." This coiled mass is called a "FIDDLE HEAD." Do you see any fiddle heads on your plants? YES   NO. Your chances of seeing them, of course, depends on the season.

Examine portions of rhizomes that have been dug up and find the ADVENTITIOUS ROOTS. As with all the other lower vascular plants, the primary means of increasing the population size is by growth from the rhizome.

Label the rhizome with its adventitious roots and a fiddle head as shown in the life cycle of Figure 31.2.

C. THE SORI AND SPORANGIA

Examine with a hand lens the undersurface of the pinnae and find brown or orange colored "spots." These are the SORI (singular: sorus). Sori are clusters of SPORANGIA. They are not always present at all times of the year. They also change in color and surface texture as they mature. Many persons mistake them for bugs.

In many genera (but not in all) the sori are protected by an epidermal flap of tissue, called the INDUSIUM. The shape of this flap of tissue differs in different genera. Does your specimen have sori?   YES   NO. Is there an indusium with each sorus?   YES   NO. What is the genus of fern that you are inspecting? _____.

If there is a specimen of *Adiantum* available, **examine** the underside of its pinnae. Here you will see that the edge of each pinna is folded under a little and forms a false indusium over the sori.

Examine leaves of *Polypodium* and determine whether or not the sori are covered by an indusium. Are they?   YES   NO.

D. THE ANNULUS—THE DISPERSAL MECHANISM
OF THE SPORANGIUM

► ACTIVITY 2   **Using** a dissecting needle, **scrape** the surface of one or two sori so that black specks (the sporangia) fall into a drop of water on a slide. Cover with a cover glass and examine under the microscope. Prepared slides may be substituted here if your ferns have no sori.

Look for golden-brown colored spheres. These are the SPORANGIA. The cells composing the walls of the sporangium are large and transparent so you should be able to see the SPORES within. These are formed by the meiotic division of DIPLOID SPORE MOTHER CELLS.

Release of the spores is an important function. It is accomplished by a special band of cells, called the ANNULUS. The annulus partly or completely encircles the sporangium. It forcibly shoots the spores for some distance. **Find** the annulus. You can recognize it by the thick inner walls and side walls of its cells. **Note** that the outer cell walls are thin. Two cells of the annulus have thin walls all around. These are the LIP CELLS. The annulus works as described in the following paragraph.

When the spores are mature, the annulus cells die, and as they lose water, the thin outer walls collapse and the thick side walls are pulled toward each other. Tensions are set up which, acting like a drawstring, straighten out the annulus. It breaks at its weakest point—the lip cells—and then flips backward, ripping open the wall of the sporangium. Almost immediately then it recoils and catapults the spores forward into the air.

► ACTIVITY 3   **Moisten** some sporangia and then let them dry out slowly. **Observe** the rupture of the annulus and the discharge of the spores.

Label the sporangium diagrams of the fern life cycle in Figure 31.2, using the capitalized terms defined in the preceding descripion.

II. FERN GAMETOPHYTE

Spores germinate into the gametophyte plant. This plant is a minute, green disk (a few millimeters in diameter), called the PROTHALLUS.

► ACTIVITY 4   **Examine** prepared slides of the prothallus. The prothallus is heart shaped, with one pointed end and one lobed end divided by a deep apical notch. It lies flat and on its underside are RHIZOIDS which form a dense, hairy mat. Scattered among the rhizoids are globular ANTHERIDIA and vase-shaped ARCHEGONIA. Antheridia occur more commonly near the margin of the prothallus, whereas the archegonia are usually near the apical notch. An ARCHEGONIUM consists of a venter and a neck. An EGG occupies the venter. Usually only the neck is visible because the venter is embedded in the tissue of the prothallus. SPERM are visible in antheridia.

FERTILIZATION requires water. The prostrate growth habit of the gametophyte with the sex organs on its lower wet side facilitates the swimming of the sperm to the vicinity of the archegonial necks. Fertilization occurs in the venter.

► ACTIVITY 5   **Examine** living fern prothalli. Usually they can be grown quite easily on the surface of clay pots kept moist or on agar plates. But their growth requires many months. **Use** a dissecting microscope and look for antheridia and archegonia. **Label** the parts of the prothallus as shown in the life cycle of Figure 31.2, using the capitalized terms defined in the preceding description.

III. YOUNG SPOROPHYTE OF THE FERN

► ACTIVITY 6   **Examine** prepared slides or living specimens of YOUNG SPOROPHYTES. These will still be embedded in the gametophyte thallus. While inside the archeogonium, the ZYGOTE develops into an EMBRYO. The embryo produces the rhizome, roots, and leaves. The young leaves grow upward through the apical notch. With continued growth, the rhizome produces more leaves. **Label** the young sporophyte growing from the gametophyte as shown in the life cycle of Figure 31.2. **Complete** the labeling of the life cycle, using the capitalized terms defined in the preceding description.

Complete Table 31.2.

EXERCISE 31   STUDENT NAME _____

QUESTIONS

1. In their life cycle, ferns are similar to what two other divisions of plants? (a) _____

_____; (b) _____

2. In their leaf structure, ferns are similar to what plants? _____

3. Is the sporophyte plant suited to life on land? _____

4. Barring some catastrophe, each year, stands of ferns in the temperate zone reappear or exhibit renewed

   life and growth. What produces these new fronds? _____

5. Identify the two stages or events in the life cycle where genetic variation is introduced. (a) _____

   _____; (b) _____

6. Fern rhizomes produce aerial (above-ground) stems from which fronds arise. Circle one: (a) true;

   (b) false.

7. From what structure does the embryo derive its initial support (nutritive and physical)? _____

8. Does the prothallus have true roots or rhizoids? _____

9. Does the rhizome have true roots or rhizoids? _____

10. A cluster of sporangia on a fern leaf is called a _____

11. What is the assimilative phase in the life cycle of the fern? _____

12. Is water necessary for fertilization in ferns? _____

13. Name the spore dispersal mechanism of the fern. _____

14. What are the two compatible functions that are carried out by the sporophyte plant?

   (a) _____; (b) _____

15. Would it make sense to have the gametophyte plant be the primary, photosynthesizing stage of the life

   cycle? _____ Explain: _____

   _____

16. Is the gametophyte prothallus dependent upon the sporophyte plant for its nourishment? _____

17. Ferns typically inhabit shady, sheltered areas. Is there any correlation between their habitat and their

life cycle? _____ Explain: _____

_____

18. Do ferns have vascular tissue? _____

19. Are sori on the upper or lower surface of frond pinnae? _____What is the advantage of this?

_____

20. What is the term commonly used to refer to the coiled mass of an immature frond? _____

**Table 31.2   Life Cycle Stages, Structures, and Cells Related
to the Four Survival Phases in Species of Ferns**

Survival Phase	Accomplished by
Multiplication	
Dispersal	
Assimilation	
Genetic variation	

EXERCISE 31   **STUDENT NAME** _____

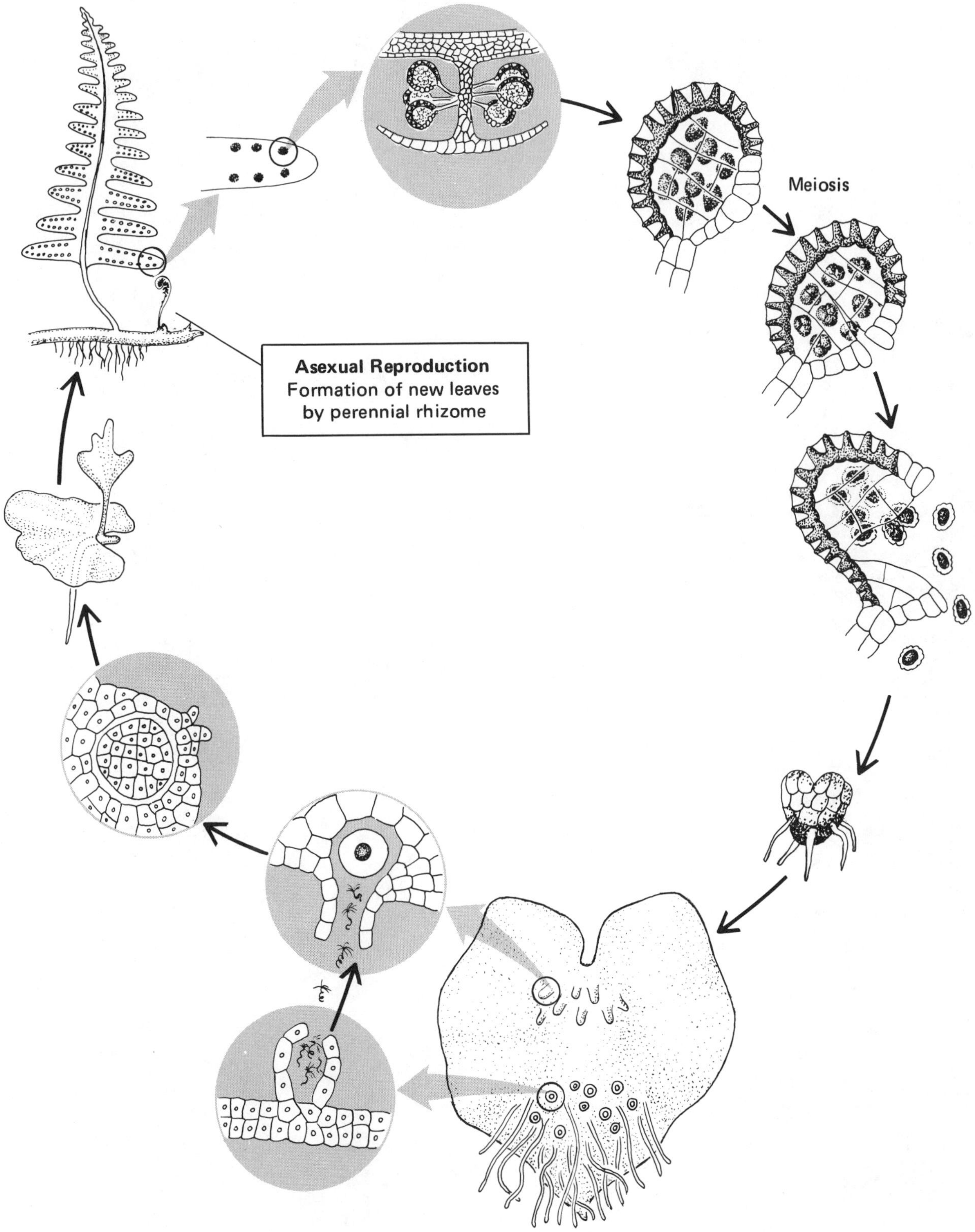

Meiosis

Asexual Reproduction
Formation of new leaves
by perennial rhizome

Figure 31.2   Life Cycle of a Fern

EXERCISE 32

CONE-BEARING SEED PLANTS

The evolution of the seed was a significant achievement for land plants. The seed contains an embryonic sporophyte supplied with nutrients for quick, vigorous growth. In the life cycle of higher vascular plants, the seed replaces the spore as the dispersal unit. It can survive adverse conditions better than a spore, and it grows and develops under conditions suitable for the mature sporophyte.

Many ancient land plants (now extinct) formed seeds. Living groups that produce seeds are the cone-bearing seed plants (gymnosperms) and the flowering seed plants (angiosperms). Gymnosperm seeds are described as naked because they are not enclosed in fruit as are those of the flowering plants (see Table 32.1).

Table 32.1   Summary of Progressive Features of Gymnosperms Giving Them Selective Advantage over Lower Vascular Plants

Progressive Features	Selective Advantage
1. The seed 2. Germination of spores inside the sporangia so that gametophytes are enclosed in sporangia 3. Reduction of the gametophyte plants to minature units that are tenants of the sporophyte plant 4. Sperm-carrying pollen tube 5. Wind-borne pollen 6. Tough, narrow, water-conserving leaves	1. As a dispersal unit, it has more survival assets than a spore; it protects and nourishes the young sporophyte in its embryonic and seedling stages 2. Change in world climate to drier conditions of Mesozoic Era favored selection of plants that retained their delicate spores and gametophytes inside the sporangia 3. Eliminates a vulnerable (free-living gametophyte) stage in life cycle 4. No longer dependent on water for fertilization; greater assurance of fertilization; less risk to sperm survival 5. Wider range of cross fertilization and greater potential for genetic variation 6. Survival advantage in dry climate

Conifers (division Coniferophyta) are the most numerous gymnosperms today. Examples familiar to you are pine, spruce, hemlock, fir, juniper, and cedar. Other divisions of gymnosperms, which include, for example, *Ginkgo* (Japanese fan tree) and cycads, will not be considered in this exercise.

Sporophytes of the Coniferophyta are trees or shrubs. The large, woody conifer, sporophyte cannot, however, be considered a significant evolutionary advance over the lower vascular plants. Many extinct lower vascular plants were of tree size. It is the seed which is the great difference.

The female gametophyte is a miniature kernel of cells. The male gametophyte is what we call the pollen grain. Both gametophytes are now totally dependent on the sporophyte and are borne in the cones of the sporophyte. No longer are there two independent-living generations.

The miniature female gametophyte thallus develops into the seed. The light, air-borne pollen grain, which carries the sperm, frees the plants from dependency on water for fertilization.

You have already studied the stems and leaves of the conifer sporophyte. The emphasis of this exercise, therefore, is on the gametophyte and the sporophyte cone structures in which it is found.

I. SPOROPHYTE AND GAMETOPHYTE

A. THE WOODY PLANT—THE SPOROPHYTE

► ACTIVITY 1   Weather permitting, the instructor will conduct you on a brief tour of campus to look at some examples of gymnosperms. The trees and shrubs that you will see represent the mature sporophyte plant. Herbarium specimens or cuttings from different gymnosperm plants can also be examined.

Note the generalized life cycle of pine (*Pinus* sp.) in Figure 32.1 and the pictorial life cycle in Figure 32.2 and refer to them as guides while you work through the exercise.

B. THE POLLEN (STAMINATE) CONE

► ACTIVITY 2   Examine POLLEN CONES of pine or some other conifer. If you have a portion of a twig, you will see that the cones occur in clusters. Examine a single cone. It is quite small. Measure or estimate its length. How long is it? _____mm.

Label the pollen cone in Figure 32.2.

Examine a prepared slide showing a longitudinal section of a staminate pine cone, *or* remove a single cone from the cluster and examine it with a hand lens and dissecting needle. It is composed of individual segments. Tease these apart with the needle. Each of these segments is a tiny modified leaf, called a MICROSPOROPHYLL. If your cone is mature, the microsporophyll will be papery. Place a few of these microsporophylls on a slide in a drop of water. Cover with a cover slip and examine under the dissecting microscope, or just examine them dry, using a hand lens. Can you find two elongate sacs embedded on one (the lower) surface of the microsporophyll?   YES   NO. These two sacs are the MICROSPORANGIA (pollen sacs). They produce MICROSPORES, which are not shed and dispersed as in lower vascular plants. Instead they remain within the sporangium where they germinate into male gametophytes.

Development of the Male Gametophyte

Examine either prepared slides of staminate cones or specimens of cones. Locate the sporangia on the undersurface of each microsporophyll. Sporangia are filled with many POLLEN GRAINS. These are the YOUNG MALE GAMETOPHYTES. They are oval bodies with two lateral, bladder-like WINGS.

The stages in the development of the pollen grain are shown in Figure 32.3. MICROSPORE MOTHER CELLS within the MICROSPORANGIUM (pollen sac) divide by MEIOSIS.

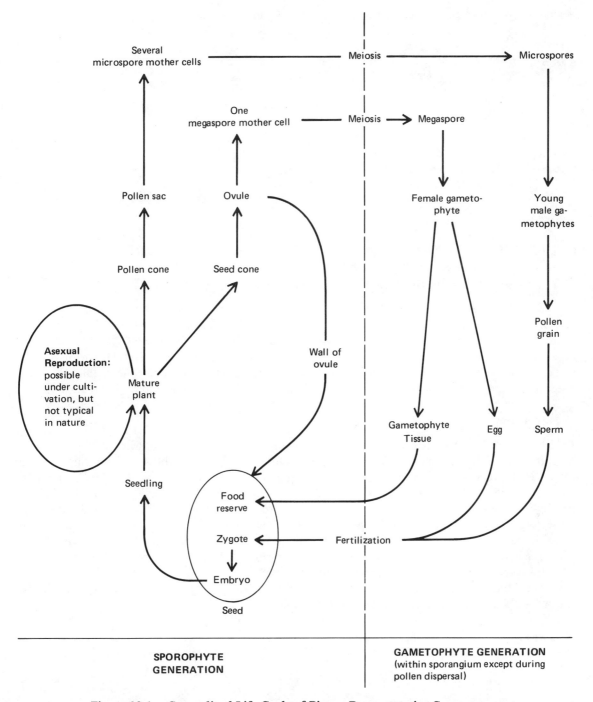

Figure 32.1   Generalized Life Cycle of Pine, a Representative Gymnosperm

Each produces four MICROSPORES. The microspore produces two PROTHALLIAL CELLS, which are remnants of the prothallus, and one ANTHERIDIAL CELL, the sole remnant of the antheridia. The two prothallial cells degenerate and the antheridial cell divides to form a GENERATIVE CELL and a TUBE CELL. In this two-celled stage, the pollen is shed.

   Label the various cells, structures, and stages in the development of the male gametophyte (pollen) in the life cycle of pine in Figure 32.2, using the capitalized terms defined in the preceding descriptions.

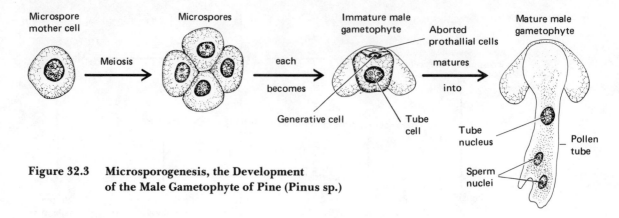

**Figure 32.3   Microsporogenesis, the Development
of the Male Gametophyte of Pine (Pinus sp.)**

C. THE SEED (OVULATE) CONE

Young Seed Cones in the First Spring of Growth

Examine preserved young (first-spring) cones or, if in season, pine twigs bearing first-spring cones. These cones are quite small (7 to 8 mm long). When alive, they are soft and green or reddish. The cone consists of modified branches, called OVULIFEROUS SCALES. Each scale bears on its upper surface two spore cases, each called a MEGASPORANGIUM. In this young stage, the scales are slightly separated and pollen grains sift down between them, falling on the megasporangia. This is pollination. It takes place in the first spring's growth of the ovulate cone.

Development of the Female Gametophyte

➤ ACTIVITY 3   **Examine** prepared slides showing longitudinal sections of ovulate cones of pine. **Note** the rather thick central axis of the cone. On either side of the axis you can see the OVULIFEROUS SCALES which have an oval bulge near their point of attachment to the axis. This oval bulge is the MEGASPORANGIUM or OVULE. Within it the FEMALE GAMETO-PHYTE develops. The stages in this development are shown in Figure 32.4. You will not see all these stages. The appearance of the cells in the ovule will vary depending on the age of the cone when it was sliced into sections to make the slide. You may be able to see the MEGASPORE MOTHER CELL, a large, light-colored, centrally located cell. You may also be able to see POL-LEN GRAINS lodged nearby, or growing into, the ovule.

Label the successive stages in the development of the female gametophyte in the diagram of the life cycle of Figure 32.2. **Refer** to Figure 32.4 as a guide.

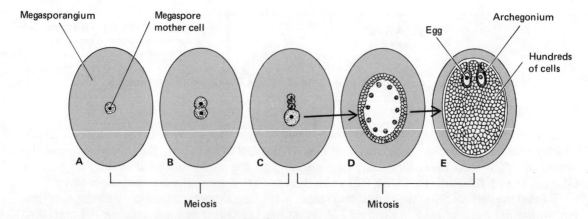

Figure 32.4   Megasporogenesis, the Development of the Female Gametophyte of Pine (Pinus sp.)

Young Seed Cones in the Second Spring of Growth

► ACTIVITY 4   **Examine** cones that are one year old. Are these cones as soft as the cones of the first spring?   YES   NO. Are the scales loose or closed together? (Circle one.) Sometime during the second spring the female gametophyte completes development. The pollen completes development also by forming a POLLEN TUBE that delivers the SPERM to the EGG. FERTILIZATION takes place in the second spring of growth of the ovulate cone.

Label the young one-year-old seed cone in Figure 32.2.

Mature Seed (Ovulate) Cones

► ACTIVITY 5   **Examine** mature SEED CONES which have shed or are shedding seeds. SEEDS are shed in the autumn that follows fertilization. In some species they are not shed for another year. Are the ovuliferous scales soft or woody? _____.
Is the cone closed or open? _____. Do you find any seeds on the surface of the scales?   YES   NO. How many on each scale? _____.

If you don't find the seeds, what can you see? _____
_____. Wouldn't it be better to have the seeds located on the lower side of each scale so they could drop out easily?   YES   NO. What is the advantage of having them on the upper side of the scale? _____.

Label the mature seed cone in Figure 32.2.

II. SEED

► ACTIVITY 6   **Dissect** a seed of piñon pine or nut pine (*Pinus cembroides* v. *edulis*). The seed is enclosed by a hard, stony protective coat (the remnant of the sporangium wall). **Break** open the seed coat. **Cut** the seed lengthwise. The white, oily material is the RESERVE FOOD (the remnant tissue of the female gametophyte). Embedded in the food material is the EMBRYO. It consists of an axis with one end, the EPICOTYL, surrounded by about eight finger-like COTYLEDONS. Below the cotyledons is the HYPOCOTYL at the very end of which is the RADICLE or root. The radicle is not visibly different from the hypocotyl.

Label the diagram of the dissected seed in Figure 32.2, using the capitalized terms defined in the preceding description. **Label** also the ovuliferous scale with the two seeds, each with one wing.

Examine the seeds and the cones for any features they might have to aid in seed dispersal. Is the seed propelled from the cone?   YES   NO. Is the cone or the seed attractive to animals by color?   YES   NO; by aroma?   YES   NO; by taste?   YES   NO. Do you think the seed itself can become readily air-borne?   YES   NO. Could a seed-eating bird easily break open the seed coat?   YES   NO. What's your conclusion? Do gymnosperms exhibit specialized dispersal mechanisms?   YES   NO. How does this affect the range of the populations of these plants? _____. What potential is there for gymnosperms exploiting new habitats, limited or unlimited? _____.

Complete the labeling of the diagram of the life cycle of the pine in Figure 32.2. Make certain that you have used all the capitalized terms defined in the preceding description.

EXERCISE 32   STUDENT NAME _____

QUESTIONS

1. Gametophytes of the lower vascular plants develop and live as independent plants. Give the specific

   place where the gametophytes of seed plants develop and live: _____

2. The spores of lower vascular plants are dispersed, and then they germinate someplace away from the

   parent sporophyte plant. Where do the spores of seed plants germinate? _____

   _____

3. The space limitations of the sporangium restrict the size of the gametophyte. How many cells make up

   the prothallus of the male gametophyte of pine? _____

4. Are there any antheridia in pine gametophytes? _____

   What represents the antheridia? _____

5. The seed of pine is housed in the remnants of what structure of the cone? _____

6. What tissue in the seed serves as the food supply for the embryo's growth? _____

7. What is the dispersal unit of gymnosperms? _____

8. Which has greater survival potential, a spore or a seed? _____

9. In seed plants, the male gametophyte plant has become extremely reduced and is referred to as the

   _____

10. What structure has eliminated the risk of survival to the sperm? _____

11. What is the functional advantage for the extreme reduction in size of the male gametophyte? _____

    _____

12. Are the gymnosperms the first land plants to have a woody growth habit? _____

13. Approximately how long (in months) is the interval between pollination and fertilization? _____

    _____

14. Give two ways that the seed is a more reliable dispersal unit than a spore. (a) _____

    _____; (b) _____

15. Since water is not used in fertilization in gymnosperms, by what means are sperm conducted to the egg

    over distances of a mile or more? _____

16. Name the two cell types in the life cycle that undergo meiosis? (a) _____

_____; (b) _____

17. Are the cells of the food reserve tissue in the seed haploid or diploid? _____

18. Are populations of gymnosperms likely to have their ranges extended as a result of the distribution of

their seeds? _____ Explain: _____

19. Give two reasons why you think gymnosperms are of little value as food crops to humans.

(a) _____;

(b) _____

20. The assimilative survival phase of the gymnosperm is the _____

21. Do gymnosperms have an asexual reproductive phase? _____

22. What is the only means by which they increase their number of individuals in a population? _____

_____

23. Spores of lower vascular plants grow into what? (a) the gametophyte or (b) the sporophyte? Circle (a)

or (b).

24. The seed germinates (a) into the gametophyte or (b) into the sporophyte? Circle (a) or (b).

25. Besides having a better food supply and being better protected all around, why is the seed a better dis-

persal unit than a spore when it comes to establishing the next generation? _____

_____

STUDENT NAME

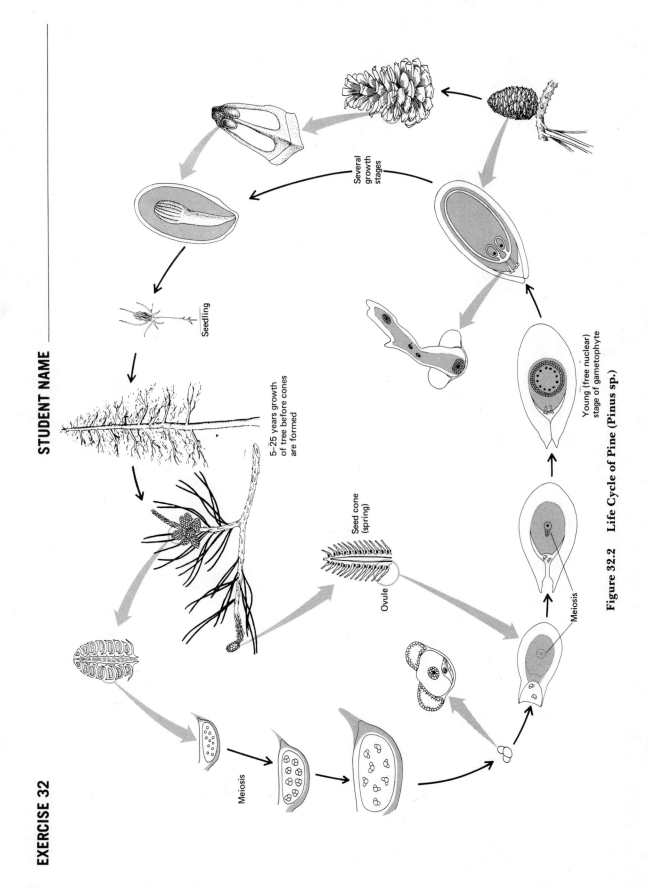

Several growth stages

Seedling

5–25 years growth of tree before cones are formed

Seed cone (spring)

Ovule

Meiosis

Meiosis

Young (free nuclear) stage of gametophyte

Figure 32.2   Life Cycle of Pine (Pinus sp.)

EXERCISE 33

FLOWERING SEED PLANTS

The flowering plants (division Anthophyta) are higher vascular plants whose seeds are enclosed in a fruit. Hence their traditional name "angiosperm" from the Greek *angeion*, meaning container, and *sperma*, meaning seed—a seed in a container (the fruit). Angiosperms are the peak of evolution in the plant kingdom. With few exceptions, they are the dominant flora of the land.

The 725 gymnosperm species are outnumbered 400 to 1 by the 350,000 or so known species of angiosperms. This success is due mainly to a combination of six features which give them tremendous competitive advantage over all other plant groups (see Table 33.1). These features are

1. Flower
2. Fruit
3. Endosperm
4. Herb growth form
5. Vessels and sieve tubes
6. Vegetative regeneration

Table 33.1   **Summary of Progessive Features of Angiosperms, Giving Them Selective Advantage Over Gymnosperms**

Progessive Feature	Selective Advantage
1. Flower	1. An organ that functions in precise pollen transfer and whose co-evolution and alliance with insects as agents of pollen transfer led to the origin of many new species adapted to wide varieties of habitats.
2. Fruit	2. An organ that protects the seed and disperses it with a high degree of efficiency.
3. Endosperm	3. A special triploid tissue in the seed with nutrients and growth-promoting hormones for the rapid and vigorous growth of the embryo.
4. Herb	4. A non-woody plant that matures quickly to reproduce itself in a matter of months (as compared to the several years required by most trees). The rapid life cycle results in more generations of plants per unit of time. This generates variation which allows for adaptations to a wide variety of habitats within any given time span.
5. Vessels and sieve tubes	5. Large water and food conducting tubes capable of rapid transport and mobilization of materials when needed for growth of buds, cambium, leaves, or developing fruit and seeds.
6. Vegetative propagation	6. The ability of vegetative organs (stems, roots, and leaves) to regenerate whole new plants; provides a rapid way to maintain and renew a population.

You have already studied all these features of the flowering plant in some detail. This exercise emphasizes the gametophyte generation of angiosperms. It has undergone even further reduction than in gymnosperms and has two new features associated with it—the endosperm and the fruit.

Note the diagrams of the generalized life cycle of the angiosperm shown in Figure 33.1 and the pictorial life cycle in Figure 33.2. **Refer** to them as a guide as you work through the exercise.

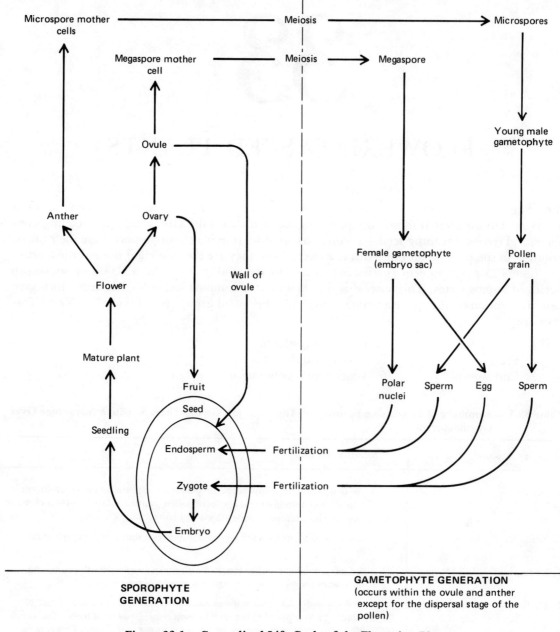

Figure 33.1   Generalized Life Cycle of the Flowering Plant

I. MALE GAMETOPHYTE

Remember from your study of the flower that the egg is in the ovule. But the ovule is enclosed in the ovary and is quite some distance from the stigma where the pollen grain is deposited. Hence the pollen grain must be able to produce a very long tube to reach the ovule and deliver the sperm to the female gametophyte. The evolution of the increase in pollen tube growth was probably closely correlated with the evolution of the enclosure of the female gametophyte.

▶ ACTIVITY 1   **Transfer** some pollen from the anthers of impatiens (*Impatiens* sp.) or some species of spiderwort (*Tradescantia* sp.) to a drop of sugar solution (10 percent cane sugar) on a slide. Include a few brush bristles or some such small bits of material before adding the cover slip to prevent the grains from being crushed. Put the slides in petri dishes containing several layers of filter paper saturated with water. Place the petri dishes in a warm place where the temperature can be kept at about 21 to 25°C. **Examine** the slides at intervals of 1 hour to see the development of the tube.

▶ ACTIVITY 2   **Examine** a stamen of a lily (*Lilium* sp.) or other flower. **Note** the stalk or FILA-MENT. It is a modified microsporophyll. At the tip of the filament is the ANTHER which is composed of four microsporangia fused together. Each microsporangium is a POLLEN SAC.

▶ ACTIVITY 3   **Deposit** some pollen into a drop of water on a slide and examine under the microscope. The coat (wall) of the pollen grain will have a particular surface pattern. This pattern varies and is useful in identifying species, genera, or families. It can be smooth or consist of spines, ridges, pores, plates, etc.

▶ ACTIVITY 4   **Examine** prepared slides of young lily (*Lilium* sp.) flower buds that show a cross section through the anther. The anther is four-parted. What are these four components? _____ _____. If the anther is very young, the entire anther mass will be filled in by tissue, some of which is sporogenous tissue. You can identify this tissue because it has numerous cells in stages of nuclear division. Some of these are MICROSPORE MOTHER CELLS. What kind of cell division is going on, mitosis or meiosis? Circle one.

If the anther on the slide is older, you will see a more loose assemblage of cells in the microsporangia. Usually you can find single cells and cells clustered in fours (tetrads). The single cells are MICROSPORE MOTHER CELLS. Each of the cells in the cluster is a MICROSPORE.

Label the diagram of the young anther as shown in Figure 33.2.

▶ ACTIVITY 5   **Examine** prepared slides of mature anthers of lily (*Lilium* sp.) or of lily pollen grains. If your slide shows anthers, the chambers of the TWO MICROSPORANGIA on either side of the central axis will be confluent and also open to the outside. POLLEN GRAINS at this stage consist of only two cells, a GENERATIVE CELL and a TUBE CELL. In this stage they are shed.

Label the mature anther and the young pollen grain just after its release as shown in Figure 33.2, using the capitalized terms defined in the preceding description.

After pollination, the generative cell forms two SPERM NUCLEI, and the tube cell elongates into the POLLEN TUBE. In this state, the pollen grain represents the MATURE MALE GAMETOPHYTE.

Label the mature male gametophyte as shown in Figure 33.2.

II. FEMALE GAMETOPHYTE

The female gametophyte develops within the OVULE (megasporangium) inside the OVARY. As with the male gametophyte, the female has undergone reduction.

Study Figure 33.3 which shows the stages of development of the female gametophyte. All these stages take place within the ovule. **Note** that, in comparison with gymnosperms (Figure 32.4), the gametophyte has undergone even further reduction in the number of cells that make up the final mature stage.

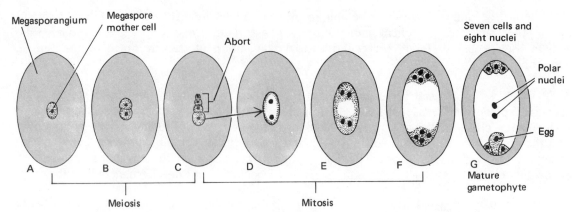

Figure 33.3   **Development of the female Gametophyte (embryo Sac) of Flowering Plants as it Occurs in the Majority of Species**

In the majority of flowering plants, the development of the female gametophyte proceeds as shown in Figure 33.3. A single MEGASPORE MOTHER CELL divides by meiosis to form four haploid MEGASPORES. Three of these degenerate (abort) and the remaining FUNCTIONAL MEGASPORE gives rise to the female gametophyte which is known as the EM-BRYO SAC. The functional megaspore enlarges greatly, its wall becoming the wall of the embryo sac. By a series of successive nuclear divisions, eight haploid nuclei are formed. Of these eight, three come to lie in the sac at the micropylar end of the ovule, three at the opposite end of the sac, and two in the center. The nuclei at the ends of the sac become invested by cell membranes. At the micropylar end is the EGG CELL with two SYNERGID CELLS. At the opposite end are three ANTIPODAL CELLS. The nuclei in the center are POLAR NUCLEI.

Figure 33.3 shows the most common type of embryo sac and of embryo sac formation. But there are other types. The differences are based (1) on the number of nuclear divisions occurring between the megaspore mother cell and the mature sac and (2) on the number and arrangement of of nuclei in the mature sac. An example is lily (*Lilium* sp.). Its antipodal nuclei and one of its polar nuclei are triploid instead of haploid.

Whatever the differences, the important new development in the flowering plant is the presence of the polar nuclei which are fertilized by a sperm nucleus to form an endosperm nucleus which has anywhere from three to five sets of chromosomes (depending on the species). This polyploid endosperm nucleus divides several times. Accompanying cell wall formation leads to the nutrient tissue known as the ENDOSPERM. As with most polyploid conditions (multiple chromosome sets), the polyploid endosperm is a vigorous tissue, supplying potent nutrients and growth-promoting hormones for the embryo. This has great selective advantage for the flowering plants because it provides guarantees for speedy germination and survival of the seed and seedling.

▶ ACTIVITY 6   **Examine** slides or preserved specimens that show transverse sections through a young ovary of lily (*Lilium* sp.). **Note** that the OVARY of lily is three-parted and has three chambers or LOCULES. Only two OVULES are visible in each locule on the side, but many more are actually present. The ovule with its supporting stalk is crook-shaped. If a MEGASPORE MOTHER CELL is present in an ovule, it appears as a large, ovoid, vacuolated, translucent cell surrounded by much smaller, more square-shaped cells.

Label the cross section of the lily ovary as shown in the cut-away view of the ovary in Figure 33.2, using the capitalized terms defined in the preceding description.

▶ ACTIVITY 7   **Examine** slides showing the ovules with MATURE FEMALE GAMETOPHYTES. **Locate** an OVULE and note the INNER and OUTER INTEGUMENTS enclosing it. The EMBRYO SAC (or mature female gametophyte) makes up the central portion of the ovule. It is difficult to find a sac showing all eight nuclei (**refer** again to Figure 33.3). This is because the sac is an oval, three-dimensional mass, and the nuclei do not all lie in the same plane. Hence it is impossible to cut a thin section in which all nuclei will appear.

Complete the labeling of the development of the female gametophyte as shown in the enlarged figures of the ovules in the life cycle diagram of Figure 33.2, using the capitalized terms defined in the preceding description.

III. FERTILIZATION

Study the illustration in Figure 33.2, which shows the pollen tube passing through the MICROPYLE, a small opening between the INTEGUMENTS at the base of the OVULE. This diagram and the next three that follow it illustrate FERTILIZATION. SPERM are shown in solid dots. Nuclei of the embryo sac are shown as open circles. The EGG is at the micropylar end of the embryo sac between the two, synergids.

One sperm nucleus fuses with the egg nucleus to form the diploid ZYGOTE. The second sperm nucleus fuses with the two polar nuclei to form the triploid ENDOSPERM NUCLEUS. This union of egg and sperm and of polar nuclei and sperm is called DOUBLE FERTILIZATION and occurs only in angiosperms.

Label the zygote and endosperm nucleus in Figure 33.2.

The endosperm nucleus divides many times and eventually the triploid ENDOSPERM tissue is formed, as noted earlier. The ZYGOTE cell divides many times and develops into the EMBRYO. During the maturation of the seed (remember: maturation and germination are two different processes) the embryo of most dicots digests all the endosperm and forms two nutrient-rich cotyledons. In some dicots not all the endosperm is digested. In the seeds of monocots, less of the endosperm is digested, the embryo forming only one cotyledon.

IV. SEED

ACTIVITY 8 **Examine** various seeds of flowering plants. **Note** that most of them tend to be small and lightweight. **Observe** the diversity of size, and surface features, such as spines, hairs, hooks, etc. The seeds of angiosperm trees, more so than herbs, are comparable to seeds of gymnosperms in size, weight, and ease of dispersal.

It is in the herbs that we see the tremendous modifications of both the seed and the fruit for dispersal. Seeds of herbs may be very small. Note though that seeds of some herbs used as food sources for humans may have been selected for millennia for large size, and these are, therefore, many times larger than their wild relatives.

V. FRUIT

ACTIVITY 9 **Examine** the various edible fruits made available to you and **note** the enclosure of the seeds. Good examples are found in produce that most people call vegetables, but which botanically are fruit. These are green pepper, cucumber, fresh peas in pods, or soybeans, melons, squash, etc.

Note that the seeds are attached to the wall of the FRUIT and in different ways depending on the species. The tissue of the fruit is the matured wall of the OVARY of the flower (and often includes other parts of the flower). As it matures, the seed produces growth-promoting hormones that stimulate the growth of the ovary and related parts. Remember (from Exercise 17) that the fruit isn't only fleshy but may also be papery or a dry container.

Examine also the fruits of trees, such as ash (*Fraxinus* sp.); oak (*Quercus* sp.); walnut (*Juglans* sp.); maple (*Acer sp.*); sweet gum (*Liquidambar*); elm (*Ulmus* sp.); buckeye (*Aesculus* sp.); etc.

Examine also the fruits of some of your local weeds such as thistle (*Cirsium* sp.); cocklebur (*Xanthium* sp.); or others as are available.

EXERCISE 33   **STUDENT NAME** _____

QUESTIONS

1. How many male gametophytes develop from one microspore mother cell? _____

2. How many female gametophytes develop from one megaspore mother cell? _____

3. Are archegonia present in the angiosperm gametophyte? _____; in the gymnosperm gametophyte?

   _____

4. How many cells make up the mature female gametophyte in angiosperms? _____; how many in the

   gymnosperms? _____

5. The home gardener or commercial producer of strawberries, white potatoes, or tulips never uses seeds

   for planting. How then is it possible that they can get a crop or a garden of flowers? _____

   _____

6. What does the angiosperm endosperm provide for the embryo? (a) _____;

    (b) _____

7. Is the fruit unique to the angiosperms? _____State the function of the fruit in three to four words.

   _____

8. Wind, animal fur and feathers, beaks and feet of birds are the usual dispersal agents for (a) trees;

   (b) fleshy fruit; (c) weeds. Circle the most appropriate.

9. List the six features that have given angiosperms superiority over all other plant groups.

   (a) _____; (b) _____

   (c) _____; (d) _____

   (e) _____; (f) _____

10. What are the two functions of the fruit? (a) _____ ; (b) _____

11. Why is the growth of an angiosperm seed usually so much faster than the growth of any other group of

    plants? _____

12. The gymnosperm embryo utilizes nutrients supplied by what tissue? _____

    _____Does this tissue have any special growth-promoting hormones? _____

13. The majority of angiosperm flowers are pollinated by (a) wind; (b) insects. Circle (a) or (b).

14. The rapid life-cycle turnover of herbs generates variation. New variations can only be kept and built up in a population if they are concentrated among a few individuals at the outset. This is done by precise pollen transfer among plants of the new mutant types. What pollen transfer agents are responsible for

    precise transfer between plants? _____

15. Is wind a very precise pollen transfer agent? _____ Does it indiscriminately blow pollen from one

    plant to any other plant? _____

16. Do you think the new mutant genes of a few individuals could become established in a population if the

    wind blew the pollen of these plants about? _____

17. Insects are pretty consistent in their activities. If an insect starts out the day visiting and picking up pollen from a purple-colored, mint-scented flower, it visits only that same type flower as long as it can find them. All these similar plants make up what is called a small gene pool. New mutants don't tend to become lost or "swamped out" in small gene pools. Is it possible for a new species to become established

    with the help of this constancy of insect visitation? _____

18. All gymnosperms and many angiosperm trees are wind pollinated. Most herbs are insect pollinated. In which group, or groups, therefore would you expect to find the greatest number of species?

    _____

19. There are only a few thousand species of angiosperm trees, but over two hundred thousand species of

    herbs. These figures should clue you that most herbs are pollinated by _____ and most trees

    are pollinated by _____.

20. Where would you expect to find the greater number of angiosperm species, in the tropics or in the

    cooler temperate zones? _____ Explain: _____

    _____

EXERCISE 33   STUDENT NAME _____

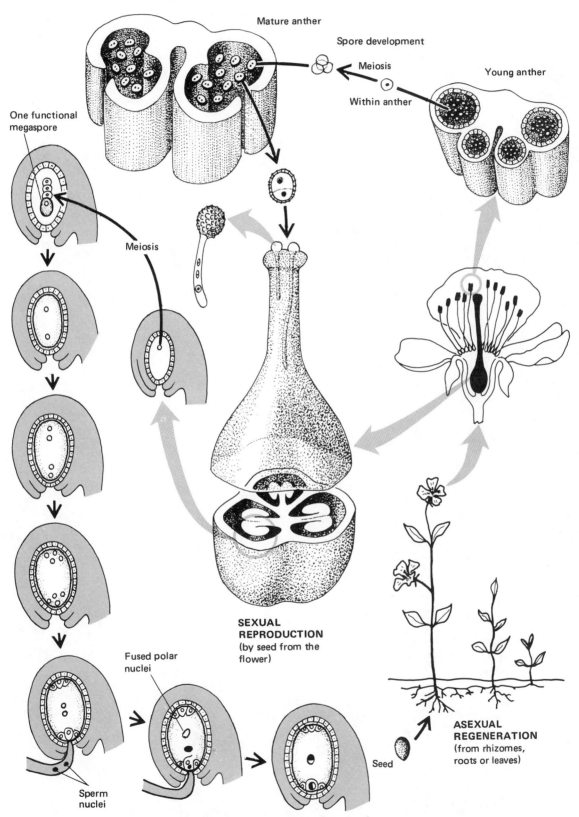

Mature anther

Spore development

Meiosis

Within anther

Young anther

One functional megaspore

Meiosis

SEXUAL REPRODUCTION (by seed from the flower)

Fused polar nuclei

Seed

ASEXUAL REGENERATION (from rhizomes, roots or leaves)

Sperm nuclei

Figure 33.2   Life Cycle of an Angiosperm

PART III

Genetics, Ecology, and Taxonomy

EXERCISE 34

MEIOSIS AND MONOHYBRID AND DIHYBRID CROSSES

In preceding exercises you have learned that *meiosis* is associated with sexual reproduction. Meiosis is a special kind of cell division which has three effects:

1. It ensures that eggs and sperms (gametes) have only half the number of chromosomes as the body cells (those of roots, leaves, etc.). This prevents continued multiplication of the chromosomes with each successive generation.
2. It maintains the identity of the species by keeping the chromosome number constant.
3. It scrambles the chromosomes and rearranges genes into different combinations so that new sets of chromosomes are passed into the gametes. Meiosis is analagous to shuffling and dealing cards in a card game. After each shuffle, the players are dealt new combinations of the same basic cards.

The sperm carries one set (the *haploid number*) of chromosomes that match the set of chromosomes carried by the egg. Their fusion creates a new individual with two sets of chromosomes (the *diploid number*).

The haploid sets match because for every chromosome in the sperm there is a matching chromosome in the egg. We call such chromosomes *homologous*. *Homologous chromosomes* have the same kinds of genes in the same arrangement.

I. TERMS AND DEFINITIONS USED IN GENETICS

ALLELE   One of two or more mutant (that is, chemically different) forms of a gene. Example: R = the normal or wild type allele; r = a mutant allele (form) of R.

CENTROMERE   The specialized part of a chromosome where the spindle fibers are attached during cell division.

CHROMOSOME	A thread of DNA (deoxyribonucleic acid) — the molecular basis of heredity.
DIHYBRID	A cross between parents differing with respect to two specified pairs of allelic genes. Examples: *RRTT* crossed with *rrtt*; or *rrTT* crossed with *RrTt*.
DOMINANT ALLELE	An allele that masks the expression of another allele of the same gene. Example: *R* masks (is dominant to) *r*.
F_1	The first filial generation; the offspring resulting from the first experimental crossing of plants or animals.
F_2	The offspring produced by self-fertilization of F_1 individuals.
GENE	A hereditary unit that occurs at a certain place (locus) on a chromosome. It can mutate to various allelic forms.
GENOTYPE	The genetic composition or formula for one or more genes (allelic pairs). Examples: formula (genotype) is *Rr, RR*, or *RRTt*, etc.
HOMOZYGOUS	Having a pair of like alleles for any one gene. Examples: *RR, rr*, or *tt*.
HETEROZYGOUS	Having a pair of unlike alleles for any one gene. Examples: *Rr* or *Tt*.
MONOHYBRID	A cross between parents differing with respect to one specified pair of alleles of a given gene. Examples: *RR* crossed with *rr*, or *Rr* crossed with *Rr*.
PHENOTYPE	The visible or detectable properties of an organism produced by the combined effect of the genotype and the environment.
PUREBRED	Having all specified alleles homozygous. Examples: *RR, rr, RRTT, rrtt*, or *rrTT*.
RECESSIVE ALLELE	An allele that is masked by another allele of the same gene. Example: *r* is masked by (is recessive to) *R*.
SISTER CHROMATIDS	Two genetically identical daughter strands of a replicated chromosome, joined by a single centromere.

II. COMPARISON BETWEEN MEIOSIS AND MITOSIS

Note Figure 34.1 which shows the essential differences between meiosis and mitosis. Mitosis involves *one* nuclear division; meiosis involves *two* nuclear divisions—referred to as MEIOSIS I and MEIOSIS II. Meiosis occurs during sporogenesis, the formation of haploid spores which (in plants) ultimately produce the gametes.

➤ ACTIVITY 1   **Look** at stage 3 (metaphase) in Figure 34.1. **Follow** along the description and **answer** these questions.

1. In meiosis the homologous chromosomes are side by side. Are they side by side in mitosis? YES NO.
2. In meiosis each centromere has a spindle fiber attached only to one side. In mitosis each centromere has a spindle fiber attached to _____ side(s).

Look at what has occurred between stages 3 and 4.

3. In mitosis, what has been separated, (a) homologous chromosomes; (b) sister chromatids? Circle (a) or (b).
4. In meiosis I, what has been separated, (a) sister chromatids; (b) homologous chromosomes? Circle (a) or (b).

In meiosis I you see illustrated *Mendel's law of segregation.* This law states that "the members of a pair of homologous chromosomes are separated (segregate) during meiosis." This law is not applicable in mitosis.

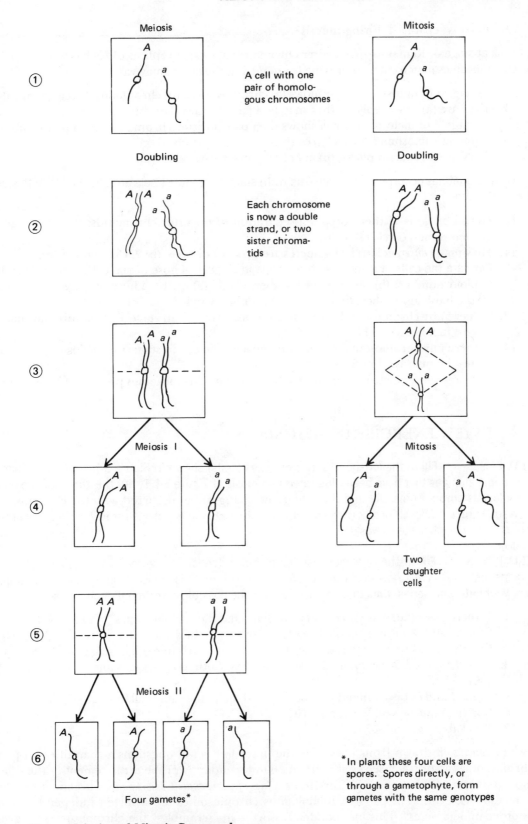

Figure 34.1   Meiosis and Mitosis Compared.
Only those stages showing essential changes are included. Nuclear membrane not shown.
Figures represent a pair of homologous chromosomes A and a. The letters can also be used
to represent genes on the chromosomes.

5. What is separated during mitosis? _____

Look at stage 4. Remember that sister chromatids are exact replicas of each other. Homologous chromosomes match each other gene for gene, but the alleles may vary.

6. In stage 4 of mitosis each cell shows (a) chromosomes; (b) chromatids. Circle (a) or (b).
7. They are (a) exact replicas; (b) a matched pair. Circle (a) or (b).
8. In stage 4 of meiosis each cell shows (a) a pair of sister chromatids; (b) a pair of homologous chromosomes. Circle (a) or (b).
9. They are (a) exact replicas; (b) matched pairs. Circle (a) or (b).

Look at what has occurred during meiosis II. In stage 5, spindle fibers are attached to both sides of the centromere.

10. During meiosis II there occurs a separation of (a) sister chromatids; (b) homologous chromosomes. Circle (a) or (b).
11. How many *different* kinds of gametes are produced? (a) 4; (b) 2. Circle (a) or (b).
12. **Compare** the cells of stage 6 with the cells in stage 1. Those of stage 1 represent (a) the haploid number; (b) the diploid number. Circle (a) or (b). Those of stage 6 represent (a) the haploid number; (b) the diploid number. Circle (a) or (b).
13. Is there any chromosomal difference between the cell in stage 1 of mitosis and that of stage 4 in mitosis?   YES   NO.
14. Is there any chromosomal difference between the cell in stage 1 of meiosis and that of stage 6 in meiosis?   YES   NO.
15. Does any one gamete in stage 6 have homologous chromosomes present?   YES   NO.

III. A REVIEW EXERCISE IN MEIOSIS

▶ ACTIVITY 2   **Fill** in the *left*-hand series of boxes (cells) shown in Figure 34.2. Show the same chromosome changes that have been illustrated in meiosis of Figure 34.1, but this time **use** two pairs of homologous chromosomes *Aa* and *Bb*. The same changes that occurred with the chromosome pair *Aa* in Figure 34.1 will occur with each of the chromosomes pairs *Aa* and *Bb* here. **Label** the chromosomes in each stage as was done in Figure 34.1.

▶ ACTIVITY 3   **Fill in** the *right*-hand series of boxes (cells) in Figure 34.2, but **answer** these questions first. We have a reason for asking you to do this exercise twice. **Look** at stage 3 (metaphase) of your left-hand series. **Note** that you have drawn one possible assortment of chromosomes.

1. Is there any other way you could have arranged the chromosomes?   YES   NO.
2. Can *AA* and *aa* be arranged so that they both go to the same cell in stage 4?   YES   NO.
3. Can *BB* and *bb* be arranged so that they both go to the same cell in stage 4?   YES   NO.
4. Can *AA* and *BB* be arranged so that they both go to the same cell in stage 4?   YES   NO.
5. Can *AA* and *bb* be arranged so that they both go to the same cell in stage 4?   YES   NO.
6. Can the same be said for *aa* and *BB*?   YES   NO.
7. What about *aa* and *bb*? _____

The *conclusion* drawn from answering the preceding set of questions is that different pairs of chromosomes assort themselves at metaphase independently of the arrangement of the other homologous pairs. This is the basis for *Mendel's law of independent assortment*. This law states that "the members of different pairs of homologous chromosomes are assorted independently of each other into gametes." This independent assortment "scrambles" the chromosomes. It gives new chromosome combinations to gametes and contributes to variation in the species.

Now **fill in** the right-hand series of boxes (cells) in Figure 34.2, this time showing an alternate arrangement of the chromosomes at metaphase (stage 3) from that of your left-hand series.

IV. MEIOSIS AND GAMETE FORMATION

Refer back to Figure 34.1. **Look** at stage 1 of meiosis. The genotype of the cell of stage 1 is *Aa*. The individual plant's genotype is also *Aa*. Tradiitonally, this is written as shown:

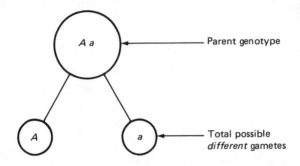

Refer back to Figure 34.2. Stage 1 represents two cells of one plant that undergo meiosis. The genotype of each cell is *AaBb*. The plant's genotype is also *AaBb*. We illustrate the gametes produced by this one plant as shown:

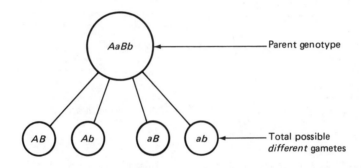

➤ ACTIVITY 4   **Practice** figuring the *possible* genotypes for the gametes produced by the following parent genotypes. If you think more gamete circles are needed, put them in.

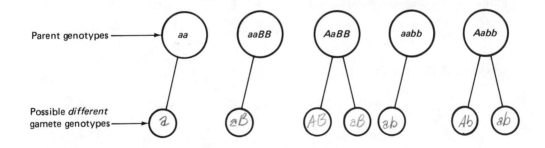

V. MONOHYBRID CROSSES

A. A MONOHYBRID CROSS WITH SIMPLE COMPLETE DOMINANCE

A plant that is homozygous or purebred for tall is crossed with one that is homozygous for dwarf. Let *T* represent the allele for tall and *t* the allele for dwarf.

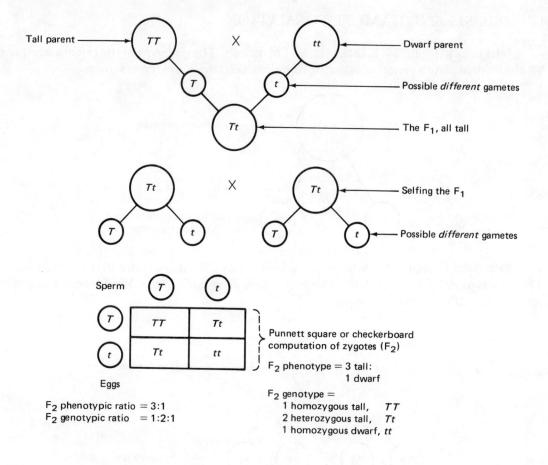

Tall parent — TT X tt — Dwarf parent

T t — Possible *different* gametes

Tt — The F_1, all tall

Tt X Tt — Selfing the F_1

T t T t — Possible *different* gametes

Sperm T t

Eggs		

T → | TT | Tt |
t → | Tt | tt |

} Punnett square or checkerboard computation of zygotes (F_2)

F_2 phenotype = 3 tall:
 1 dwarf

F_2 genotype =
 1 homozygous tall, TT
 2 heterozygous tall, Tt
 1 homozygous dwarf, tt

F_2 phenotypic ratio = 3:1
F_2 genotypic ratio = 1:2:1

B. THE MONOHYBRID BACKCROSS (OR TEST CROSS)

In crosses involving simple dominance, such as T and t of the preceding section, both TT and Tt appear tall. Often it is necessary to test which of the tall individuals is Tt and which is TT. This is determined by means of a BACKCROSS. The backcross (also called a test cross) involves crossing a homozygous recessive plant (tester stock) with plants showing the dominant trait and whose genotype is to be determined.

1. If

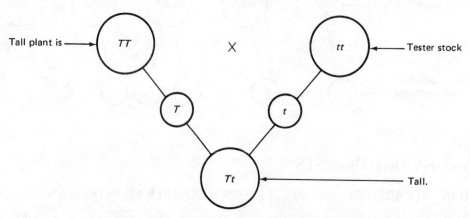

Tall plant is — TT X tt — Tester stock

T t

Tt — Tall.

If only tall plants are produced, then the tall parent can be assumed to be homozygous dominant for tall.

2. If

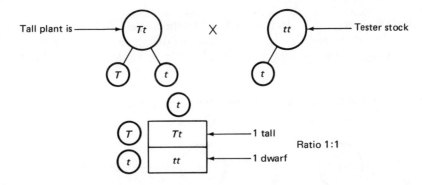

When half the offspring are tall and half are short, then the tall parent can be assumed to be heterozygous. This is a 1:1 ratio. Whenever data can be described in this ratio, you can assume that you are dealing with a backcross and that the individual showing the dominant trait is heterozygous and carries a recessive allele as well as the dominant one.

VI. MONOHYBRID OBSERVATIONS
AND ANALYSES

▶ ACTIVITY 5   **Examine** flats of F$_2$ generation seedlings which show color differences representing simple monohybrid inheritance. Suggested plants are corn (*Zea mays*) showing green and albino seedlings; tobacco (*Nicotiana tabacum*) showing green and albino seedlings; and sorghum (*Sorghum vulgare*) showing red and green seedlings.

    1. How many of the seedlings are green? _____ How many are nongreen (albino)? _____

    2. What is the ratio of green to albino seedlings? _____

    3. Are all the green seedlings homozygous?   YES   NO.

    4. Are all the green seedlings heterozygous?   YES   NO.

    5. Among these green seedlings you would expect the homozygous dominants to be (a) twice as numerous as the heterozygous dominants; (b) half as numerous as the heterozygous dominants? Circle (a) or (b).

    6. What proportion of these seedlings will breed true if self pollinated? _____

▶ ACTIVITY 6   **Obtain** an ear of corn (*Zea mays*) from the instructor and note the presence of both colored and white grains o the ear. These grains represent the F$_2$ generation of a certain cross. Which color is dominant? _____

    1. **Work** in teams and count the number of colored grains on the ear. **Count** only five rows. **Use** a toothpick or straight pin to mark your starting row as a reference point.

    2. **Count** and **record** the number of white grains. If counting only five rows, they must be the *same* rows as before.

    3. **Compile** the data for all teams on the blackboard and in Table 34.1. **Figure** the ratio and the parent genotypes and **record** in Table 34.1.

▶ ACTIVITY 7   **Obtain** another ear of corn from the instructor in which the grains are the result of a test cross (backcross). **Count** the number of colored grains and white grains in five rows. **Compile** and **record** the data in Table 34.1.

VII. DIHYBRID CROSSES

An example is shown below. In the garden pea, the allele *T* for tall is dominant to the allele *t* for dwarf. The allele *R* for red flowers is dominant to the allele *r* for white flowers. A purebred tall red plant is crossed with a purebred dwarf white.

► ACTIVITY 8   Fill in the blanks and circles where indicated.

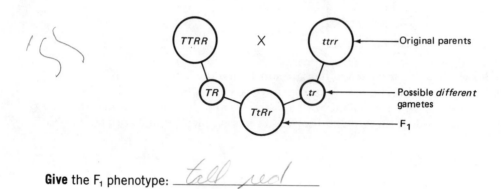

Give the F₁ phenotype: _tall red_

Fill in the genotypes of the gametes below.

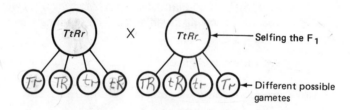

Fill in the following 16 squares with the zygote genotypes:

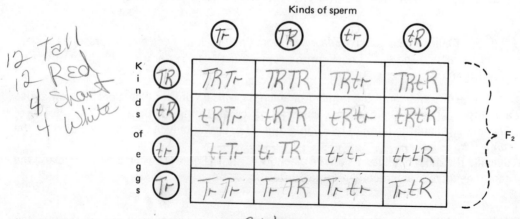

The phenotypic ratio of this F₂ is _____3:1_____

List the phenotypes from the most abundant to the least abundant and give the proportion of each:

Phenotype	Proportion
Tall Red	9
Short Red	3
Tall White	3
Short white	1

The cross shown above gives the classical F_2 ratio of 9:3:3:1. Whenever data of a given problem can be described in this ratio, you can assume that you are dealing with a dihybrid cross.

► ACTIVITY 9   **Complete** Table 34.2 which shows other dihybrid crosses. **Fill in** the genotypes as indicated.

VIII. DIHYBRID OBSERVATIONS AND ANALYSES

► ACTIVITY 10   **Obtain** from the instructor an ear of corn (*Zea mays*) in which the grains represent the F_2 generation of a certain cross. You will see four types of grains: purple full; purple wrinkled; white full; and white wrinkled. The full endosperm is starchy and the wrinkled endosperm is sugary.

Work in teams and **count** five rows of the ear of corn and tabulate the number of each of the four types of grains. **Use** a toothpick or straight pin to mark your starting row. **Count** the number of each of the four grain types found in these five rows. **Compile** and **record** all the figures from all individuals or teams in the class in Table 34.3 Use the letters *A* and *a* for the color and *S* and *s* for the endosperm.

Obtain another ear of corn (*Zea mays*) in which the grains represent the F_2 generation of a dihybrid test cross.

Work in teams and **count** five rows of the ear of corn. **Use** a toothpick or straight pin to mark your starting row. **Count** the number of each of the different grain types in these rows. **Compile** and **record** all the data in Table 34.3.

IX. SOME GENETICS PROBLEMS

For additional practice, you might want to work out the following problems or others provided by the instructor.

1. In summer squash, white fruit color dominates over yellow. A white-fruited squash plant, when crossed with a yellow-fruited one, produced 54 white- and 49 yellow-fruited plants.
   a. What are the genotypes of the parents?
   b. If the white-fruited parent is self-fertilized, what will be the fruit color of the offspring?

Table 34.3   Inheritance of Grain Color and Endosperm in Corn (*Zea mays*) **as Seen in the Phenotypes of an F$_2$ Generation and a Test Cross**

Student Team	F$_2$ Generation				Test Cross			
	purple, full	*purple, wrinkled*	*white, full*	*white, wrinkled*	*purple, full*	*purple, wrinkled*	*white, full*	*white, wrinkled*
Class totals								
Phenotypic ratio								
Parent genotypes								

    c. What is the genotypic ratio?

2. In squash, disk-shaped fruit type dominates over the recessive allele for sphere-shaped fruits. A squash plant with disk-shaped fruit is crossed with another squash plant. The F$_1$ seeds when planted give rise to 86 plants with disk-shaped fruits and 79 plants with sphere-shaped fruits.
   a. What are the genotypes of the parent plants?
   b. If two of the F$_1$ heterozygous plants for disk-shaped fruits are crossed and 464 seeds are produced, about how many will produce disk-shaped fruits; how many sphere-shaped fruits?
   c. How many will be homozygous for the character disk-shaped fruits?

3. In watermelon, fruit shape may be short or long; color may be green or striped. A homozygous long, green melon was crossed with a homozygous short, striped variety. the F$_1$ individuals were short, green melons. Work out the F$_2$ generation, and show the phenotypic ratio and the genotypic ratio.
   a. Which factors are dominant?
   b. Which are recessive?

EXERCISE 34   **STUDENT NAME** _____

QUESTIONS

1. Suppose the heterozygous condition produced a different trait than either the homozygous dominant or the homozygous recessive. Would this increase or decrease the number of phenotypes in the F_2 of a

    monohybrid cross? _____

2. What would be the F_1 phenotypic ratio if you crossed two such heterozygous individuals? _____

3. For each of the following species, the diploid number is given. In the space provided, give the number of homologous pairs this represents.

    Rye *(Secale cereale)* = 14;   _____ pairs

    Tobacco *(Nicotiana tabacum)* = 48;   _____ pairs

    Bean *(Phaseolus vulgaris)* = 22;   _____ pairs

    Corn *(Zea mays)* = 20;   _____ pairs

4. Suppose you examine the cells of a species of plant and found eight chromosomes. The homologous pairs looked as follows: one long straight pair, one short straight pair; one long bent pair, and one short bent pair. You then breed several generations of plants of this species. At the end of this time, would you expect to find

    a. Some plants with all the straight chromosomes and none of the bent ones? _____

    b. Some plants with all the long chromosomes and none of the short ones? _____Explain:

    _____

    _____

5. In plants like those described in Question 4, what proportion of the gametes would have

    a. Two straight chromosomes (one long and one short) and two bent chromosomes (one long and one

    short) _____

    b. Three straight chromosomes and one bent one? _____

    c. Four straight chromosomes? _____

6. An individual plant has two sets of chromosomes, a maternal set and a paternal set. When this plant produces gametes, do the chromosomes of the paternal set segregate from the chromosomes of the

    maternal set? _____

7. If the F_2 of a cross shows a 3:1 phenotypic ratio, how many *pairs* of alleles are involved? _____

8. What is the *purpose* of a test cross? _____

    _____

9. Plant traits such as vigor, yield, hardiness, and so on, are the result of the combined effect of numerous genes on different chromosomes. How does use of a tester stock help plant breeders develop lines with

desired combinations of these traits? _____

_____

_____

10. Does meiosis in plants give rise directly to eggs and sperm? _____

Table 34.1 Inheritance of Grain Color in Corn (*Zea mays*)

Student Team	Corn Ear with F$_2$ Generation Grains		Corn Ear with Grains from a Test Cross	
	colored	*white*	*colored*	*white*
Class totals				
Phenotypic ratio				
Parent genotypes				

EXERCISE 34   **STUDENT NAME** _____

Table 34.2   Dihybrid Crosses *

Parents	*TTrr* × *ttRR*		*TtRr* × *ttrr*		*ttRr* × *TTrr*	
Genotypes of all possi-ble *different* gametes						
Genotypes of offspring						

*The number of answer blanks does not necessarily reflect the number of possible genotypes.

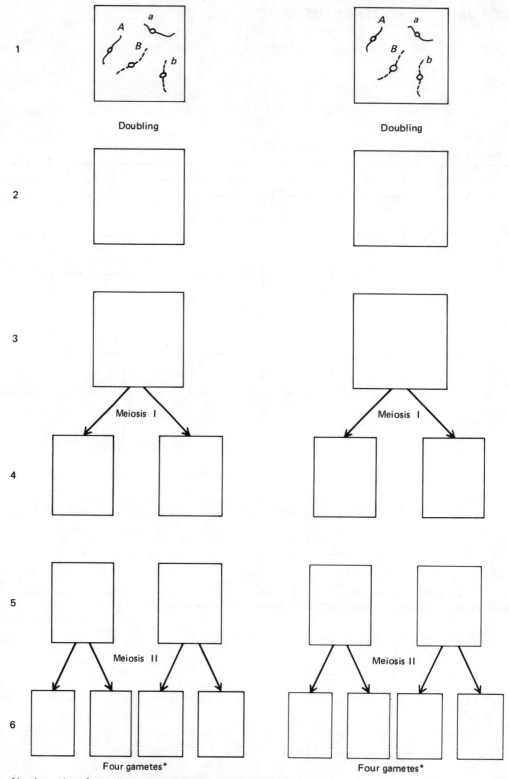

1

2

3

Meiosis I

4

5

Meiosis II

6

Four gametes*

Doubling

Doubling

Four gametes*

*In plants these four cells are spores. Spores directly, or through
a gametophyte, form gametes with the same genotypes.

Figure 34.2 **Meiosis Involving Two Homologous Pairs of Chromosomes Aa and Bb.**
Work out the sequence on the left first. Then work the sequence on the right, using an alternate arrangement of the chromosomes.

EXERCISE 35

PLANT ECOLOGY

Plant ecology deals with the interrelationships of plants and their environment. These relationships may be obvious, as in the case of species that will not survive freezing temperatures, or very complex, as the case where one forest type gradually gives way to another over the centuries. Since many of these principles must take the form of examples that will be explained to you, there will be only one laboratory investigation comparable to those of the previous exercises.

I. COMPETITION

Observe a demonstration showing a series of five or six sets of plants which have been growing for several weeks. The sets differ only in the number of plants that are in each pot. The number of seeds initially planted has been calculated (using expected germination percentages as you learned in Exercise 2) to result in eventual plant populations of 1, 2, 4, 8, and 16 (or 16 and 32) plants per pot. Suitable species for such a demonstration are corn (*Zea mays*); grain sorghum (*Sorghum* sp.); jimson weed (*Datura* sp.); or other fast-growing herbaceous plants.

► ACTIVITY 1   **Calculate** the average height of the plants in each pot and record this in Table 35.1. Weigh the total mass of plant material produced above the soil level in each pot. Or, if using peat cubes instead of pots of soil, weigh the entire mass of plants, including stems, leaves, roots, and peat cube. If these plants will not be needed for use by another class, the dry weight may be obtained by drying the plants (and peat cube) in an oven at about 70°C for 48 hours. The demonstration may have the weights and heights already calculated for you. In that case, just record them in Table 35.1.

The demonstration is intended to show the principle of competition, which is very important in ecology. In which pot(s) were the individual plants the largest? _____. In which were the plants the smallest? _____. Is this the result you expected?   YES   NO.

Table 35.1

Pot Code Letter	Number of Plants in Pot	Average Height (cm)	Total Weight (grams)	Weight per Plant (grams)
a	1			
b	2			
c	4			
d	8	.		
e	16			
f	32			

➤ ACTIVITY 2   Using the data in Table 35.1, **construct** a graph based on the weights per plant (Figure 35.1). Enter the values for weights of single plants and then the values for all the plants (or the entire pot). Connect the weights of single plants with a dashed line, and the weights of entire pots of plants with a solid line. (For this example it doesn't make any difference whether you weighed the tops or the whole plants, or whether they represent fresh weights or dried weights, so long as you are using weights of the same type, that is, all dried tops or all fresh weights of the whole peat cube with plants.) Also, if for some reason, you ended up with 14, rather than 16 plants or even six plants instead of eight, just enter on the figure the number you actually had. The important point is to remember to divide the total weight by 14, or 6 (rather than by 16 or 8), so that the value of weight per plant is correct. **Enter** all values obtained in Figure 35.1.

OBSERVATIONS   **Examine** the shape of the graph you have just made. How would you describe the trend in the weight of the average plant? _____
_____What about the trend in the total weights? Which pots had the greatest total weight of plants? _____. Was the pot with two plants twice the weight of the one with one?   YES   NO. How great was the difference? _____. Was the pot with 16 plants twice that of the one with eight?   YES   NO. Was it 16 times as heavy as the single plant? YES   NO. What was the actual difference? _____. The weights per plant generally decrease. Does this decrease show signs of leveling off? Is there a point where all plants are the same size, regardless of the number per pot?   YES   NO.

What about the trend in the total weight per pot? Generally, the maximum weight is attained with about eight plants. Was this true in your example?   YES   NO. Does the graph show a leveling off at about this point?   YES   NO. Does the graph level off sooner?   YES   NO. Where? _____. Does the total weight continue to increase throughout all the examples? YES   NO. If a leveling off, or plateau, is visible in the graph of total weights per pot, what does this represent in terms of the ecological principle of competition? _____
_____. Your laboratory instructor will discuss this with you at this time.

II. SUCCESSION

The seeds planted for the competition example were selected from a uniform lot of genetically identical (or very similar) characteristics. This is typical of the way that agricultural crops compete with each other. Plant breeders of vegetable, grain, and forest crops determine that a certain maximum density of plants will result in the greatest yield per plot (we used total weight to show this). Below this density, single plants are larger, and above this density, individual plants are small and stunted.

In nature the same basic principle applies. There is only a certain amount of plant growth that may be supported by a given amount of soil, moisture, and light. But in nature there are plants of thousands of different genetic backgrounds, belonging to dozens of species, in any one area. Some are bound to outcompete the others. They grow and flourish and reproduce themselves while the less successful species and individuals are lost from that spot.

Over the years, though, this competition causes changes in the successful species in an area. The vigorously germinating weedy species that may have been dominant when a bare spot first became available, may not be able to sustain themselves indefinitely in the same place. There are many reasons for this phenomenon, but the most common basis is that the originally successful species have, simply by growing successfully, so changed the environment that the new conditions of shade, moisture, and nutrients became better suited for the growth of other species.

▶ ACTIVITY 3   **Observe** photographs or a slide presentation or film showing the change in plant species over the years in a single area which is typical for your geographical location.

If the organization of your laboratory permits, the best way to examine these changes is through a field trip to a variety of nearby plant communities in various stages of succession. If you must rely on photographs or a film, be sure to ask the laboratory instructor for assistance whenever you have a question about any of the principles of succession that are demonstrated to you. Guest instructors or special assistants may also be in the classroom to help for this portion of the exercise.

OBSERVATIONS   What is the nature of the first colonizers on a patch of bare soil in your area? Are they large or small? (Circle one.) Are they annual or perennial? (Circle one.) Are they woody or herbaceous? (Circle one.) Do they typically grow in full sun, partial sun, full shade? (Circle one.) Do they (a) usually occur as mixed stands of dozens of species, or (b) are they usually in masses or clusters of two or three species only? Circle (a) or (b). Do they (a) start to grow first around the edges of the bare area, or (b) seem to start more or less all over the new habitat? Circle (a) or (b).

▶ ACTIVITY 4   **Fill** in Table 35.2. **Follow** the same line of inquiry, either from personal observation if you are able to go on a field trip, or from quizzing the laboratory instructor and the special assistants.

OBSERVATIONS   What sorts of changes can you note in this succession of stages? What change is there in the size of the plants? _____. Are the later stages characterized by woody plants? YES NO. Do the young seedlings of the dominant species germinate and grow well in the shade created by their parents? YES NO. What do you suppose would happen if you had a chance to observe a plant community where the seedlings of the dominant species grew and developed normally under the influence of their mature parents? Would there be a further change in the dominant species? YES NO. Explain why or why not:

_____.

The name given to an unchanging community is *climax*. Such a climax community perpetuates itself. Are there any examples of a climax area near your school? YES NO. Why or why not? Discuss the reasons with your instructor and special assistants.

III. PLANT GEOGRAPHY

As has been brought out in the laboratory discussion, climax areas are pretty scarce in most of North America. The settlers cut the forests, plowed the plains, and drained the swamps.

Nevertheless, examples still exist of most of the important natural plant formations of the continent. Even before you studied botany as a subject, you surely had some general idea of which of the major plant types was potentially dominant in your part of the country, such as deciduous forest, coniferous forest, or grassland. Why *are* there different dominant vegetations? As you studied the structure of the stem and leaf in previous exercises, several adaptations to adverse conditions were pointed out. These include mainly modifications that allow the plant to live in areas in which temperature and moisture are not ideal, especially the combination of high temperature and low moisture. Other adaptations are not so readily correlated to the appearance of the plant. Cold hardiness is one of these.

➤ ACTIVITY 5   **Observe** photographs or a slide presentation or a film that show examples of the major plant formations of North America, as mapped and named in Figure 35.2. This may be a continuation of the films or slides you observed before to see succession, or it may be a special class meeting or lecture just for this purpose.

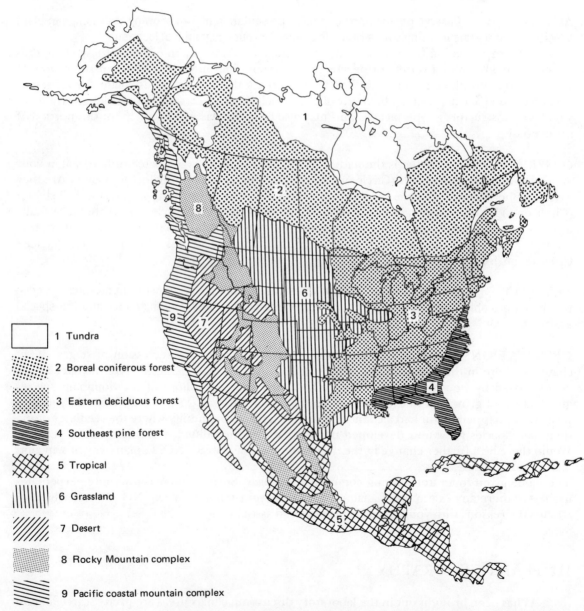

1 Tundra

2 Boreal coniferous forest

3 Eastern deciduous forest

4 Southeast pine forest

5 Tropical

6 Grassland

7 Desert

8 Rocky Mountain complex

9 Pacific coastal mountain complex

Figure 35.2   Vegetation Regions of North America

Locate your area on the map. Does the name of the potentially dominant vegetation region agree with your own observations?   YES   NO. If not, why not? _____ _____. Discuss this with your laboratory instructor or the special lecturer. Many terms will be used in these discussions, and they are defined in Table 35.3.

Table 35.3   **General Classification of Precipitation, Temperature, and Growing Seasons**

Precipitation		Temperature During Growing Season		Growing Season	
classified as	*(mm)*	*classified as*	*(° C)*	*classified as*	*(months)*
Low	75 to 500	Cold	2 to 10	Short	1 to 3
Medium	501 to 1750	Cool	11 to 20	Medium	4 to 8
High	1751 to 4000	Warm	15 to 25	Long	9 to 12
	or more	Hot	25 to 40		

▶ ACTIVITY 6   Taking the given information about existing North American plant formations (Table 35.4) combined with the definitions presented in Table 35.3, fill in the formation types represented in simplified fashion in Table 35.5. Not every possible set of conditions is represented here.

OBSERVATIONS   Are there any North American plant formations whose characteristics are not given in the table?   YES   NO. Which one, or ones? _____.
Fill in their combination(s) of conditions in line(s) seven or eight of Table 35.5. Can you dream up another theoretical set of conditions not given here? There are many possible sets not shown yet. **Enter** one or two more sets of hypothetical conditions in lines nine or ten. **Discuss** with your lab partners the overall effect of these conditions as they lead to stresses on the plants which may attempt to live in such an environment. Have you invented an environment in which nothing can survive? Most of the combinations not already given amount to just that—an area of little or no plant growth. North America has representatives of almost all typical biomes, even the tropical forests.

IV. HUMAN USE OF THE PLANT FORMATIONS

It is easy to think of some of the direct uses that we make of plant communities. The coniferous forests are cut for lumber and pulpwood and left to seed another generation of the same species. Are there any other formations which are used directly?   YES   NO. Which one(s) can you think of? (a) _____; (b) _____. It is normal for us to use the natural vegetation areas in one of four ways, direct use, as with the lumber from trees; agricultural adaptation of former plant communities; removal of vegetation for building sites for homes, industry, and transportation corridors; intrusion into natural plant formations for extraction of underground resources, such as minerals or oil. Each of these uses disturbs the natural successional processes, and many of them totally destroy the formation over a wide area.

▶ ACTIVITY 7   **Fill in** Table 35.6 showing the type of human use typically made in North America of each of the original plant formations. Where wide variation is known to you, indicate the range of use. Use text readings, outside reading assignments, class discussions, and evening "bull sessions" to assist in the formulation of these ideas. It is not intended that you be able to complete this table in the remainder of a single laboratory period.

Table 35.4   Characteristics of North American Plant Formations

Formation	Precipitation (mm)	Evaporation	Temperature (°C)	Growing Season (months)	Characteristic Species	Other Features
Tundra: arctic—high latitude alpine—high altitude	125 to 375	Low	Summer: 2 to 10 Winter: −40 to −20	1 to 2½, limited by cold	Grasses, dwarf shrubs, lichens	Treeless, permanently frozen ground
Boreal coniferous forest	375 to 900	Low	Summer: 10 to 20 Winter: −55 to −20	2½ to 3½	Spruce, fir, birch, aspen, larch	Many bogs and lakes, poorly drained, few tree species
Eastern deciduous forest	Eastern: 1200 to 1800 Western: 650 to 900	Medium	Summer: 15 to 25 Winter: −5 to +5	4 to 7	Oaks, hickories, maples, beech, basswoods, elms, buckeye	Many tree species, some shrubs and herbaceous plants
Southeast pine forest	1500 to 1800	Medium to high	Summer: 22 to 30 Winter: 5 to 10	5 to 8	Pines, cypress, oaks, hickories, magnolias	Pines successional and of high economic value
Grassland	Eastern: 650 to 900 Western: 300 to 500	High	Summer: 15 to 25 Winter: −15 to 10	North: 3 South: 8	Grasses and many broadleaved herbs	Most of precipitation in spring and early summer, fall droughts; tall grasses eastern part, mid and short grasses western part
Desert	75 to 250	High	Summer: 25 to 40 Winter: 2 to 10	1 to 3, limited by moisture	Cacti, many shrubs	Many annuals when rains come
Tropical forest	Western areas: 1800 to 4000 or more Dry areas: 750 to 1200	Medium to high	Summer: 25 to 33 Winter: 20 to 25	Wet: 12 Dry: 3 to 5	Mahogany, palms, figs, live oaks, grasses in some areas	Large number of species

**Table 35.5   The Types of Formations Associated with Particular
Combinations of Environmental Conditions**

Area	Precipitation	Evaporation	Temperature	Growing Season	Formation Type
1	Low	High	Hot	Short	
2	Low	Low	Cold	Short	
3	Medium	Medium	Warm	Medium	
4	High	Medium to high	Hot	Long	
5	Low to medium	Low	Cool	Short	
6	Low to medium	High	Warm	Short to medium	
7					
8					
9					
10					

EXERCISE 35   **STUDENT NAME** _____

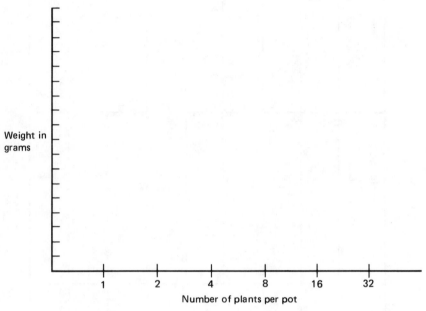

Figure 35.1

Table 35.2

Characteristic	Pioneer Stage	Second Stage	Third Stage	Oldest Stage
Size of plants				
Annual or perennial				
Woody or herbaceous				
Sunlight adaptation				
Mixed species or masses of one kind				
All over area or start from edges				

Table 35.6

Formation	Direct Use of Plant Formation if Any	Typical Agricultural Uses if Any	Other Human Intrusions Commonly Seen	Relative Abundance of Undisturbed Areas
Eastern deciduous forest				
Tundra and high latitude				
Boreal forest				
Southeastern coniferous forests				
Grasslands				
Desert				
Pacific coastal forests				
Tropical forests				

EXERCISE 36

KEY FOR IDENTIFICATION OF SOME COMMON TREE GENERA[1]

A dichotomous key is given here which will serve to identify several dozen of the most commonly planted ornamental and native trees. Obviously, not all genera can possibly be included. A very wide selection of trees has been made, and it includes most of the trees on most campuses in temperate North America. A few campuses in very mild climates may have a wide variety of other genera—so wide that it is considered beyond the scope of this manual to try to include them all. In a few places in the key, the phrase "frost-free" is used. The term is used rather loosely to describe species grown only in areas where hard killing frosts are infrequent. Examples might be the true cedar (*Cedrus*), palms, and *Araucaria*. **Compare** descriptions with illustrations in Figure 36.1.

A.   Leaves needle- or scale-like; plants without flowers and fruits; commonly bearing naked seeds in woody cones (except *Taxus*) ....**Division Coniferophyta** (Section II)

A′.   Leaves flat, broad, not needle-like; seeds never in woody cones ....B

B(A′).   Leaves fan-shaped, with parallel venation [Figure 36.1 (A)]; seed with bad-smelling fleshy outer layer, not enclosed in a fruit ....**Division Ginkgophyta** (Section I)

B′.   Leaves with netted or parallel venation; seeds always enclosed in a fruit which develops from the ovary of a flower ....**Division Anthophyta** (Section III)

---

[1]*Directions for Using Key*: This key consists of pairs or couplets of choices, 1–1′, 2–2′, and so on. Compare your plant specimen with the descriptions given in couplet 1–1′. Choose whichever statement (1 or 1′) is more applicable to your specimen and go on to the next statement as indicated by the number at the extreme right. Continue through the key, each time choosing between a pair of choices until you terminate at the name of a plant. In case you get lost and forget which choices you've previously made, the numbers written in parentheses indicate your previous position in the key.

I. DIVISION GINKGOPHYTA

1.   Flat, fan-shaped leaves [Figure 36.1 (A)]; contains the single species...**Ginkgo biloba**
             (maidenhair tree)

II. DIVISION CONIFEROPHYTA

1.   Leaves borne in clusters...2

1'.   Leaves borne singly along the stem 4

2(1).   Leaves more than five in a cluster 3

2'.   Leaves two to five in a cluster, persistent for several seasons...........................**Pinus**
             (the pines—at least forty species native and cultivated in the continental United
             States)

3(2).   Leaves deciduous, soft, linear, flattened ....**Larix**
             (larch or tamarack—native to Northern states and mountains)

3'.   Leaves persistent, stiff, quadrangular ....**Cedrus**
             (true cedar—cultivated in warmer areas without severe winters)

4(1').   Leaves (at least some of them) scale-like to triangular, opposite or whorled............5

4'.   Leaves needle-like or somewhat flattened, alternate or scattered along stem............7

5(4).   Branchlets flattened; cones longer than wide, brown and woody at maturity; all
             leaves scale-like, opposite...**Thuja**
             (arbor vitae)

5'.   Branchlets not flattened..6

6(5').   Leaves mostly in whorls of three, many of them needle-like; cones round, bluish, and
             berry-like at maturity ....**Juniperus**
             (juniper)

6'.   Leaves needle-like to triangular, growing in tight spirals and forming a rounded
             branch 2 to 8 cm in diameter; cones large and woody (frost-free areas).....**Araucaria**
             (monkey puzzle tree)

7(4').   Leaves soft, on green branchlets, the whole superficially similar to a compound leaf
             [Figure 36.1 (Z)]...8

7'.   Leaves rigid, on brownish branches, solitary 10

8(7).   Leaves dark green, persistent, the branchlets on a green stem, larger stem sometimes
             bearing solitary leaves ....**Sequoia sempervirens**
             (redwood)

8'.   Leaves light, yellow-green, deciduous together with the branchlets, larger stems usu-
             ally brownish 9

9(8').   Leaves up to 2.5 cm long, subopposite; leaf-bearing branchlets often over 10 cm
             long; youngest woody branches reddish brown with subopposite axillary buds 2 to 3
             mm long...**Metasequoia glyptostroboides**
             (dawn redwood)

9'.   Leaves not over 1.5 cm long, alternate; leaf-bearing branchlets 4 to 8 cm long;
             youngest woody branches yellowish tan with spirally arranged inconspicuous
             buds..**Taxodium**
             (bald cypress)

10(7').   Leaves four-sided in cross section; woody petiole persistent on stem.................**Picea**
             (spruce)

10'.   Leaves flattened in cross section 11

11(10').   Shrubs; leaves without lines of stomata on lower surface; seed at maturity enveloped
             by fleshy red tissue...**Taxus**
             (yew)

11'.	Trees; leaves with two lines of stomata on lower surface 12
12(11').	Leaves 8 to 20 mm long, round-tipped, two-ranked (forming two rows along stem); cones not more than 2.5 cm long ....**Tsuga** (hemlock)
12'.	Leaves 2 to 6 cm long with pointed tips, usually not two-ranked; cones larger.......13
13(12').	Leaves more or less straight, dark green; cones pendulous, falling of intact; bracts conspicuous, three-lobed ....**Pseudotsuga menziesii** (Douglas fir)
13'.	Leaves mostly strongly curved, light bluish green; cones upright, disintegrating on the tree; bracts hidden ....**Abies** (fir)

III. DIVISION ANTHOPHYTA

1.	Trees without branches, a single stem bears very large leaves directly (the palms)..... ....**Section D**
1'.	Trees with normal branching, several levels of branches and twigs bearing smaller leaves 2
2(1').	Branches (at least the younger ones) with spines..**Section A**
2'.	Branches without spines 3
3(2w).	Leaves simple [Figure 36.1 (J)]..4
3'.	Leaves compound [Figure 36.1 (K) through (M)]...7
4(3).	Leaves alternate 5
4'.	Leaves in whorls or opposite..**Section E**
5(4).	Leaf-margin entire [Figure 36.1(B)]...**Section B**
5'.	Leaf-margin toothed or lobed, not entire [Figure 36.1 (C) through (I)]...................6
6(5').	Margin toothed, often doubly toothed, or appearing shallowly lobed [Figure 36.1 (C) through (G)] ....**Section C**
6'.	Margin deeply lobed [Figure 36.1 (H) or (I)]..**Section D**
7(3').	Leaves alternate..**Section F**
7'.	Leaves opposite ....**Section G**

A. BRANCHES WITH SPINES

1.	Leaves simple 2
1'.	Leaves pinnately compound [Figure 36.1 (K) and (L)]...3
2(1).	Leaf margins entire..**Maclura pomifera** (Osage orange or hedge apple)
2'.	Leaf margins toothed and in some species shallowly lobed ....**Crataegus** (hawthorn)
3(1').	Spines modified stipules [Figure 36.1 (O)], up to 2 cm long; leaves once compound [Figure 36.1(K)] with 7 to 17 leaflets..**Robinia pseudoacacia** (black locust)
3'.	Spines modified branches [Figure 36.1(P)], up to 7 cm long, often forked (trunks often with clusters of very long, branched, adventitious spines); leaves once or twice compound, mostly with more than 18 leaflets............................**Gleditsia triacanthos** (honey locust)

B. BRANCHES WITHOUT SPINES; LEAVES SIMPLE, ALTERNATE, WITH ENTIRE MARGINS

1.	Leaves heart-shaped..**Cercis canadensis** (red bud)	
1'.	Leaves not heart-shaped 2	
2(1').	Leaves 20 cm or longer when fully grown 3	
2'.	Leaves not more than 18 cm long 4	
3(2).	Stipular scars encircling twigs at nodes; terminal bud large, with a single budscale covered with gray hairs [Figure 36.1(N)]................................**Magnolia soulangeana** (oriental magnolia)	
3'.	No stipular scars or stipules present; terminal bud small, thin, naked, covered with short brown hairs..**Asimina triloba** (pawpaw)	
4(2').	Leaves bristle-tipped, leathery; fruit an acorn...**Quercus** (oak—several oaks key out here, of which the more common are the shingle oak, *Q. imbricaria* and the evergreen live oak, *Q. virginiana*)	
4'.	Leaves not bristle-tipped, not leathery 5	
5(4').	Veins markedly incurving toward the tip of leaf [Figure 36.1(X)]; flowers perfect; usually shrubs..**Cornus alternifolia** (pagoda dogwood)	
5'.	Veins not incurving; flowers unisexual; usually trees 6	
6(5').	Leaves ovate with rounded or emarginate apex [Figure 36.1 (W)]................**Cotinus** (smoke-tree)	
6'.	Leaves ovate or oval; leaf apex acute or acuminate (pointed 7	
7(6').	Young twigs smooth; pith of twig chambered [Figure 36.1(Q)]; three bundle scars ....**Nyssa sylvatica** (sour gum)	
7'.	Young twigs hairy; pith of twig not chambered; one bundle scar ....**Diospyros virginiana** (persimmon)	

C. BRANCHES WITHOUT SPINES; LEAVES SIMPLE, ALTERNATE, WITH TOOTHED MARGINS

1.	Leaf margins singly toothed [Figure 36.1(C) and (F)]....................................2	
1'.	Leaf margins doubly toothed [Figure 36.1(D) and (E)]................................10	
2(1).	Sap milky...**Morus** (Mulberry)	
2'.	Sap not milky 3	
3(2').	Leaves triangular or heart-shaped, about as broad as long 4	
3'.	Leaves longer than broad, widest near the middle......................................5	
4(3).	Leaves asymmetrical with oblique base [Figure 36.1(V)], several (usually three) large veins arising from petiole ....**Tilia** (linden, or basswood)	
4'.	Leaves symmetrical, pinnately veined [Figure 36.1 (S)], with prominent midrib from petiole to apex of leaf...**Populus** (the many species of poplars, aspens, and cottonwoods belong to this genus)	
5(3').	Teeth of leaf margin coarse, two or fewer per centimeter [Figure 36.1(F)], or margin undulate [Figure 36.1 (G)]; buds mostly 15 to 25 mm long, pointed................**Fagus** (beech)	

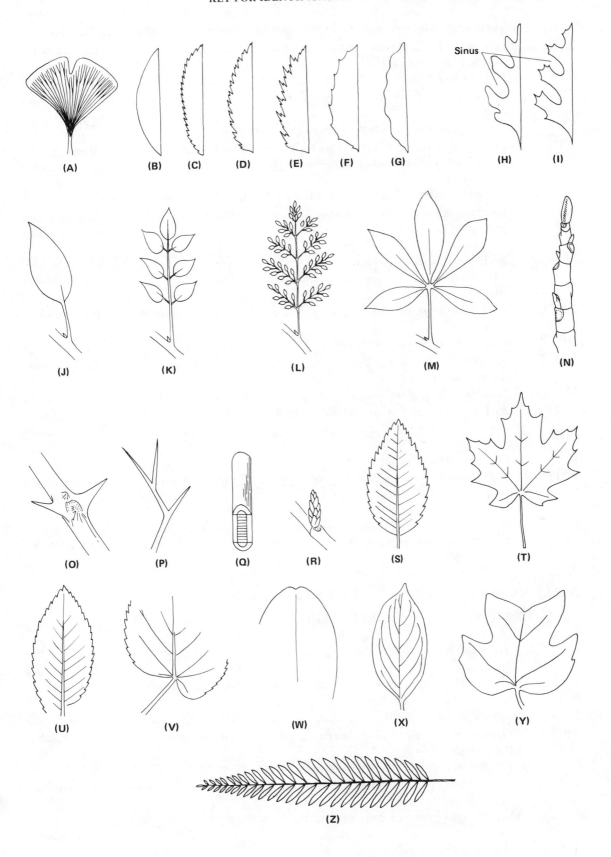

Figure 36.1   Illustration of Characters Used in the Identification of Trees

5'.   Teeth of leaf margin fine, three or more per centimeter; buds shorter 6

6(5').   Petioles with 1 to 4 small glands near leaf blade..**Prunus**
                 (plum, peach, cherry)

6'.   Petioles lack glands 7

7(6').   Leaves at least three times longer than broad ....**Salix**
                 (willow)

7'.   Leaves not more than two times longer than broad 8

8(7').   Leaves asymmetrical [Figure 36.1(V)], with 3 to 5 large veins arising from petiole;
                 pith of twig chambered [Figure 36.1 (Q)]...**Celtis**
                 (hackberry)

8'.   Leaves symmetrical, with one main midrib; pith of twig not chambered 9

9(8').   Buds elongate with many scales [Figure 36.1(R)]; flowers greenish; bark of trunk
                 scaling off in slender vertical strips..**Ostrya virginiana**
                 (hop-hornbeam)

9'.   Buds ovoid with few scales; flowers white to purple, showy; bark of trunk not scaling
                 off in strips..**Malus**
                 (apple)

10(1').   Leaf blade asymmetrical at base with one side smaller [Figure 36.1 (U)].........**Ulmus**
                 (elm)

10'.   Leaf blade symmetrical at base...11

11(10').   Petioles with 1 to 4 small glands near leaf blade..**Prunus**
                 (plum, cherry)

11'.   Petioles lacking glands 12

12(11').   Bark of trunk and larger branches chalky or silvery white, peeling horizontally in
                 thin papery strips ....**Betula**
                 (birch)

12'.   Bark brown, not peeling in horizontal strips 13

13(12').   Pistillate catkins conelike, woody, persistent until the second year; buds short-
                 stalked; two bud scales...**Alnus**
                 (alder)

13'.   Pistillate catkins with thin papery scales, not woody, soon deciduous, not persistent;
                 buds sessile with several scales...**Ostrya virginiana**
                 (hop-hornbeam)

D. BRANCHES WITHOUT SPINES; LEAVES SIMPLE, ALTERNATE, MARGIN DEEPLY LOBED

1.   Trees lacking branches, the single trunk bearing relatively few very large leaves; the
                 palms 2

1'.   Trees with definite branches bearing numerous leaves; not palm-like 7

2(1).   Leaves palmately divided...3

2'.   Leaves pinnately divided 5

3(2).   Petioles commonly lacking thorns or spines; usually not tree-like ....**Sabal**
                 (palmetto)

3'.   Petioles usually with thorns or spines; plants becoming tree-like.............................4

4(3').   Leaves glossy green; trunks marked with rings from old leaf bases...........**Livistonia**
                 (Australian fan palm)

4'.   Leaves light green or gray-green; trunks with persistent leaf bases and often dead
                 leaves...**Washingtonia**
                 (Washington fan palm)

5(2′).   Leaf bases and petioles forming a green sheath the top 1 to 2 m of the trunk ....
....**Roystonea**
(royal palm)

5′.   Leaf bases not forming a conspicuous green sheath...6

6(5′).   Leaf bases deciduous, forming a smooth trunk decidedly swollen at the lower end
and often leaning at a considerable angle from the vertical...............**Cocos nucifera**
(coconut)

6′.   A portion of the leaf bases persistent on the trunk; stem of constant diameter and
rarely leaning ....**Phoenix**
(date palm)

7(1′).   Leaves palmately veined, with more than one main vein arising from petiole [Figure
36.1(T)] 8

7′.   Leaves pinnately veined [Figure 36.1(S)] 10

8(7).   Sap milky; many of the leaves only toothed ....**Morus**
(mulberry)

8′.   Sap colorless...9

9(8′).   Leaves star-shaped, cleft into 5 to 7 wedge-shaped lobes with evenly toothed margins; terminal bud present; bark of trunk deeply furrowed, persistent; twig lacks encircling scars at nodes ....**Liquidambar styraciflua**
(sweet gum)

9′.   Leaves not star-shaped, 3 to 5 lobed, the margins not evenly toothed; terminal bud
absent; bark thin, more or less smooth, eventually peeling off in large pieces; twigs
with encircling scars at nodes [Figure 36.1(N)] ....**Platanus**
(sycamore, plane tree)

10(7′).   Branchlets yellowish green; leaves and bark aromatic....................**Sassafras albidum**
(sassafras)

10′.   Branchlets gray or grayish brown; leaves and bark not aromatic 11

11(10′).   Stipular scars encircling twigs at nodes [Figure 36.1 (N)]; leaf broadly notched at
apex, four lobed, about as broad as long [Figure 36.1 (Y)]...**Liriodendron tulipifera**
(tulip tree)

11′.   No encircling stipular scars present; leaf longer than broad with five or more margi-
nal lobes; *Quercus*—oaks 12

12(11′).   Lobes pointed [Figure 36.1(I)], bristle-tipped ....**Quercus**
(black oak group)

12′.   Lobes rounded [Figure 36.1(H) and (G)], not bristle-tipped ....**Quercus**
(white oak group)

E. BRANCHES WITHOUT SPINES; LEAVES
SIMPLE, IN WHORLS OR OPPOSITE

1.   Leaves three (sometimes two) at a node, ovate to ovate-oblong, with a long, pointed
tip, densely hairy beneath ....**Catalpa**
(catalpa)

1′.   Leaves opposite, sparingly pubescent or glabrous when fully grown 2

2(1′).   Leaves with entire margins and pinnate veins which are incurved toward apex
[Figure 36.1(X)]...**Cornus florida**
(flowering dogwood)

2′.   Leaves lobed, palmately veined [Figure 36.1(T)]...**Acer**
(maple)

F. BRANCHES WITHOUT SPINES; LEAVES COMPOUND, ALTERNATE

1.   Leaves twice compound [Figure 36.1(L)]..2
1′.   Leaves once compound [Figure 36.1(K)]..4
2(1).   Base of leaflets very asymmetric, midrib of leaflets decidedly off center........**Albizzia** (silk tree, "mimosa")
2′.   Base of leaflets symmetric, midrib in center of leaflet 3
3(2′).   Leaflets ovate, entire, with pointed apex ....**Gymnocladus dioica** (Kentucky coffee tree)
3′.   Leaflets oval or lanceolate, somewhat toothed, with rounded apex..............**Gleditsia** (honey locust)
4(1′).   Pith chambered [Figure 36.1(Q)]; fruit a large, nearly spherical drupe with indehiscent exocarp; leaflets 11 to 33...**Juglans** (walnut)
4′.   Pith not chambered; fruit not as above..5
5(4′).   Leaves 30 to 90 cm long with 13 to 41 leaflets, of unpleasant odor when crushed ....**Ailanthus altissima** (tree of heaven)
5′.   Leaves usually less than 30 cm long, no unpleasant odor..6
6(5′).   Rachis of leaves zigzag, the leaflets alternate, entire...........................**Cladrastic lutea** (yellowwood)
6′.   Rachis of leaves straight, the leaflets opposite...7
7(6′).   Leaflets less than 5 cm long with rounded or emarginate apex [Figure 36.1(W)], entire or inconspicuously toothed; fruit a legume...8
7′.   Leaflets more than 5 cm long with pointed apex, conspicuously toothed [Figure 36.1(C) through (E)]; fruit drupe-like...9
8(7).   Leaves with 7 to 17 leaflets, the terminal one usually present; stipules spiny [Figure 36.1(O)]...**Robinia pseudoacacia** (black locust)
8′.   Leaves mostly with more than 18 leaflets, the terminal one usually missing; stipules absent ....**Gleditsia** (honey locust)
9(7′).   Small trees or shrubs; sap milky; buds naked; drupes red, indehiscent, less than 1 cm in diameter, in pyramidal clusters ....**Rhus** (sumac)
9′.   Large trees; sap watery; buds with bud scales; drupes with dehiscent exocarp, more than 12 mm in diameter, singly or two to four in a cluster; *Carya*—hickories.........10
10(9′). Leaflets 9 to 17 per leaf; bark of trees furrowed but not shaggy; bud scales in pairs..**Carya illinoensis** (pecan)
10′.   Leaflets mostly 5 per leaf; bark of old trees separating into long shaggy strips; bud scales not in pairs...**Carya ovata** (shagbark hickory)

G. BRANCHES WITHOUT SPINES; LEAVES COMPOUND, OPPOSITE

1.   Leaves palmately compound [Figure 36.1(M)] 2
1′.   Leaves pinnately compound [Figure 36.1 (K)]...4
2(1).   Leaflets usually seven, coarsely doubly toothed [Figure 36.1(E)]; buds sticky ....**Aesculus hippocastanum** (horse chestnut)

2'.   Leaflets usually five, finely singly toothed [Figure 36.1(C)]; buds not sticky; buck-eyes..3

3(2').   Fruit smooth; petals dissimilar, the upper ones longer than the stamens; 25 m tall ...
         ....**Aesculus octandra**
         (sweet buckeye)

3'.   Fruit echinate (prickly); petals nearly equal, shorter than stamens; small tree, to 15 m
         tall ....**Aesculus glabra**
         (Ohio buckeye)

4(1').   Leaflets 3 to 5, coarsely and irregularly toothed and lobed..................**Acer negundo**
         (box elder)

4'.   Leaflets 5 to 11, entire or finely toothed; *Fraxinus*—ashes 5

5(4').   Young twigs four-angled..**Fraxinus quadrangulata**
         (blue ash)

5'.   Young twigs round in cross section..6

6(5').   Leaf scars concave at top; petioles glabrous or nearly so ....**Fraxinus americana**
         (white ash)

6'.   Leaf scars straight at top or nearly so; petioles velvety pubescent
         ....**Fraxinus pennsylvanica**
         (red ash)

APPENDIX

MATERIALS NEEDED, SOURCES OF MATERIALS, DIRECTIONS, AND GENERAL SCHEDULING

Depending on the individual school policy, the school may supply slides, forceps, and so on to the class, or each student may have to purchase his or her personal set. If the latter is the case, the following items are suggested as adequate for ordinary needs during one semester:

— 6 $1'' \times 3''$ glass slides
—12   cover slips
— 2   dissecting needles
— 2   single-edge razor blades
— 1   fine-point forceps
— 1   small 15 cm metric ruler
— 1   pipette (eye-dropper)
— 1   hand lens (6 – 10 ×)

No drawing paper is needed.

EXERCISE 1 — THE WHOLE PLANT

ACTIVITY 1

Potted plants for demonstration of vegetative parts. Plants not to be used up. One plant per four students. Two-month-old bean (*Phaseolus vulgaris*); two-month-old cucumber (*Cucmis sativus*); any dwarf variety of tomato (*Lycopersicon* sp.); coleus (*Coleus* sp.); ornamental pepper (*Capsicum an-*

nuum); Christmas cherry (*Solanum pseudocapsicum*); rubber plant (*Ficus elastica*) or any suitable species of house plant or potted woody dicot; or any leafy shrub such as lilac (*Syringa vulgaris*), privet (*Ligustrum* sp.) or forsythia (*Forsythia* sp.)

ACTIVITY 2

Plants bearing flowers for demonstration. Not to be used up. One plant per four students. Three month old bean plant *Phaseolus vulgaris*); house plants such as *Oxalis* sp., *Fuchsia* sp., flowering maple (*Abutilon ap.*); or any available garden plants with simple flowers such as *Petunia* sp., snapdragon (*Antirrhinum* sp.)

ACTIVITY 3

Plants bearing fruit or just samples of fruit for demonstration. Not to be used up. About one specimen per four students. Three-month-old bean plant (*Phaseolus vulgaris*); dwarf varieties of tomato (*Lycopersicon* sp.); ornamental pepper (*Capsicum annuum*); Christmas cherry (*Solanum pseudo-capsicum*) or any field, garden, or landscape plant seasonably available or dried, such as tulip (*Tulipa* sp.), milkweed (*Asclepias syriaca*), cherry (*Prunus* sp.), shepherd's purse (*Capsella bursa-pastoris*), oak (*Quercus* sp.), dogwood (*Cornus* sp.), mountain ash (*Sorbus* sp.), maple (*Acer*), catalpa (*Catalpa*) elm (*Ulmus* sp.), tree of heaven (*Ailanthus altissima*); peanut (*Arachis hypogaea*), ragweed (*Ambrosia* sp.); velvet leaf (*Abutilon theophrasti*).

ACTIVITY 4

One- or two-week-old seedlings grown in glass or plastic jars, tumblers, and so forth. They sould be old enough to show at least the first pair of true leaves. For demonstration. Not to be used up. Start them from seeds placed between the side of the jar and paper towelling. First line the jar with paper toweling. Stuff the center with crumpled up toweling. Moisten the papers, then insert two or three seeds between the side of the jar and the paper lining. Keep moist and in a sunny place. Suggested plants: bean (*Phaseolus vulgaris*); cucumber (*Cucumis sativus*); tomato (*Lycopersicon* sp.) One plant per four students.

ACTIVITY 5

Monocot plants for demonstration of vegetative parts. Not to be used up. Some should be in flower for use in flower examination in Activity 6. About one plant per four students. Wandering Jew (*Zebrina pendula*); spiderwort (*Tradescantia virginiana*); purple heart (*Setcreasea pallida*); day flower (*Commelina* sp.); *Philodendron* sp.; dumbcane (*Dieffenbachia* sp.); water hyacinth (*Eichornia* sp.); any lawn or pasture grass; any seedling or mature plant of corn (*Zea mays*); or oats (*Avena sativa*).

ACTIVITY 6

Monocot plants with flowers. For demonstration. Not to be used up. One plant per four students.

ACTIVITY 7

Seedlings of monocots that are one to two weeks old (old enough to show at least one or two leaves). Grow these in glass or plastic tumblers or jars. Follow same directions as for dicot seedlings, in Activity 4. One plant per four students.

EXERCISE 2 — SEED STRUCTURE AND GERMINATION

ACTIVITY 1

Seeds of the pinto, kidney, or lima bean (*Phaseolus* sp.). Soak them in water overnight or 4–6 hours before class. Enough seeds for at least two per student for dissection.

ACTIVITY 2

Three or four beakers or plates containing dry seeds of bean, pea, or other dicots—just for display.

Another display of seeds that have been kept moist for several days and have imbibed water. These seeds can be prepared in several ways. Here are three suggestions. (a) Place some seeds on moist filter paper (or paper toweling) in petri dishes. Do not flood the dish. (b) Put seeds in a beaker or container with moistened vermiculite. (c) Make up a "rag doll"—this is a wet paper towel rolled into a cylinder and with the seeds wrapped in the layers of the roll. Tie each end of the rag doll with string or twist tape. Store in plastic bag or carton to keep moist.

Mold is always a problem. Treat seeds with a fungicide such as captan, thiram, botran, ferbam, zineb, benomyl. Put a little of the liquid fungicide in an 8 or 16 oz. margarine tub and swish seeds around in it. Paper toweling for rag doll can be wet with water which has 1 part commercial bleach for every 9 parts water. Fungicides are available from garden supply centers, seed stores, or any of the seed, nursery, or horticultural supply catalog firms listed in this appendix.

ACTIVITY 3

This is a demonstration of the pressure exerted by the swelling of seeds that have imbibed water. Mix some plaster of Paris with enough water to make a mixture the consistency of pancake batter. Pour mixture into paper cups. Plant bean seeds in the mixture while it's still liquid. Allow plaster to set up hard. Tear off the paper cup. Set the hardened plaster cup in a saucer of water. Seeds inside the plaster will imbibe the water and swell up. The pressure exerted by the swelling seed will crack open the plaster of Paris cup. All this should be prepared several days before class. The cracked plaster of Paris is put on display for the class.

ACTIVITY 4

Plant bean seeds in sand (preferably) or soil. Stagger your plantings so that three age categories of seedlings (3-day-old; 6–8-day-old; and 10-day-old) are available to the class. Have enough seedlings for one of each age category per two students. Students will dig (lift) the seedlings from the growing medium. All seedlings will, therefore, be used up in each class. An easy way to 'dig' the seedling is to spread apart the fingers, plunge them in the medium under the seedling, then lift the seedling out. Gently shake off excess soil or sand. Prepare enough seedlings for all classes.

ACTIVITY 5

Prepare seeds of pea (*Pisum sativum*) the same as described for bean in Activity 4. Prepare enough seedlings for all classes.

ACTIVITY 6

The seed coat of honey locust (*Gleditsia triacanthos*) is impermeable to water unless it has been nicked or partially corroded away after several months exposure to the elements of weather. If you have seeds with intact coats, you can scarify (scratch or cut) the coat with a metal file. Scarify until you see a small white hole in the coat. This job is time consuming. It's best to get several files and let the students do it several days prior to the class. If you have a grinding wheel and use pliers to hold the seed against the wheel, the job can be speeded up. If seeds are not available locally, they can be purchased from Herbst Brothers Seedsmen Inc., 1000 N. Main St., Brewster, N.Y. 10509. Soak the seeds on moist paper in Petri dishes or in wet rag dolls for 2–3 days before class.

ACTIVITY 7

Soak grains of corn (*Zea mays*) overnight.

ACTIVITY 8

Prepare grains of corn (*Zea mays*). Follow the same directions as for bean in Activity 4.

EXERCISE 3 — FACTORS THAT INFLUENCE SEED GERMINATION

ACTIVITY 1

Students will work in groups of four. Each group will need the following: (a) 100 seeds of one kind of plant; (b) 1 beaker to soak these seeds; (c) enough bleach type disinfectant for soaking the seeds. Make up 1 part commerical bleach (5 percent sodium hypochlorite) with 9 parts water; (d) 2 paper towels per student; (e) string or twist tape.
Suggested seeds: lettuce (*Lactuca sativa*); radish (*Raphanus sativa*); spinach (*Spinacia oleracea*); corn (*Zea mays*); bean (*Phaseolus vulgaris*); pot marigold (*Calendula officinalis*); celosia (*Celosia argentea* v. *cristata*); dahlia (*Dahlia pinnata*); nasturtium (*Tropaeolum majus*); or morning glory (*Ipomoea purpurea*).

ACTIVITY 2

Staining with tetrazolium dye is facilitated when seeds are softened in water beforehand. Place seeds (corn, bean, and so forth) on top of or between moist paper towels overnight. Alternatively, corn can be put in a beaker of water for 3–4 hours at about 30° C. Don't soak beans this way, however.

Seeds of small-seeded legumes and some other kinds do not need softening, namely, bean, pea, soybean, and so on.

The pericarp of grasses (corn, sorghum, and so on) is not permeable to tetrazolium, so the seed should be bisected with a razor blade before testing. Bean and soybean do not need to be sectioned. Use a 1.0 percent tetrazolium solution for legumes (beans, soybeans, and so on) and a 0.1 percent to 0.25 percent solution for grasses and cereals (corn) that are bisected.

Corn ideally requires ½ to 1 hour staining time at 35° C but often results satisfactory for classroom accuracy are faster, even at room temperature (20° C). Soybean and bean ideally require 3–4 hours at 35° C but results (for classroom purposes) are obtainable at less than 1 hour.

To prepare a 1.0 percent solution, dissolve 1 gram of tetrazolium powder in 100 ml distilled or tap water. To prepare a 0.1 percent solution, mix 1 part of 1.0 percent solution with 9 parts water or dissolve 1 gram tetrazolium powder in 1000 ml water.

Store the solutions in the dark or in an amber colored bottle. Solutions may be kept for several months at room temperature. The solution used in a test should be discarded after each test.

The tetrazolium powder (2, 3, 5-triphenyl tetrazolium chloride or TTC) is available from several chemical supply houses. Two such supply houses are: Nutritional Biochemical Corporation, 26201 Miles Road, Cleveland, Ohio 44128, and Fisher Scientific Company, 1458 North Lamon Avenue, Chicago, Illinois 60651.

ACTIVITY 4

Seeds of honey locust (*Gleditsia triacanthos*) are needed—about 10–20 per student. Paper towels for each student to make a rag doll. Commercial bleach and fungicide as described in Activity 2 of Exercise 2 for wetting paper towels and sterilizing seeds. Twist tape or string.

ACTIVITY 5

Have students work in pairs. Supply about 10 seeds of corn (*Zea mays*) per student. Paper towels—one per student. Two small jars (at least one with a screw cap). Use baby- or junior-food jars. Sterilize or have students sterilize the seeds with fungicide as described in Activity 2, Exercise 2. Also, use dilute bleach water as described in Activity 2, Exercise 2 for the water required in this germination test.

ACTIVITY 6

Enough seeds of any one type for each student to have 20–25 seeds. One paper towel per student to make rag doll tester. Twist tape. Seeds of peas, turnips, cabbage, cucumber, green pepper, eggplant and sweet corn. Plastic bags for storing about 10 rag doll testers. Or use freezer cartons. Refrigerator and germination chamber (if possible).

EXERCISE 4 — USE OF THE MICROSCOPE FOR STUDY OF PLANT CELLS

ACTIVITY 1

For each student, one compound microscope, at least with two objectives, $10\times$ and $40\times$ and a $10\times$ ocular.

ACTIVITY 2

Prepared slides of commerical fibers—cotton, silk, and so on. One slide per student, or per two students. Available from most biological supply houses. See list of these.

ACTIVITY 3

Living elodea (*Anacharis canadensis*) plants—available from biological supply houses or locally from pet supply stores. One clean microscope slide, cover slip, and eye dropper per student.*

ACTIVITY 5

Onions—one or two for entire class and two to three bottles of IKI (iodine, potassium iodide). For each student*: 1 slide, 1 cover slip, 1 forceps.

*(Part of recommended student lab kit)

EXERCISE 5 — CELL DIVISION

ACTIVITY 1

For each student, one prepared slide of onion (*Allium cepa*) root mitosis. Available from all biological supply houses.

ACTIVITY 5

For each student*: 1 slide, 1 cover slip, 1 razor blade, 1 pipette. Also paper towel or filter paper.

For every three to four students: 1 Syracuse glass; 1-100 ml beaker with tap water; one onion bulb with roots. One week before class, set onions on top of jars or containers of about the same diameter as the onion. Put enough water in the jar so that the root end of the onion is just below the water surface. The onion can rest on the rim of the jar or you can put three toothpicks in it to bridge it over the top if the jar is too wide for the onion. Many roots will develop within one week.

A 1M solution of hydrochloric acid, if you do the aceto-carmine or aceto-orcein. A 3M solution of hydrochloric acid, if you do the Magneta xx stain. About 2 mm of acid per student should be enough. So, figure your total quantity based on this.

Aceto-carmine stain is prepared as follows: Boil 0.5 gm of carmine in 100 ml of 45 percent acetic acid for about three minutes. Cool and filter the solution. Keep this as your stock solution. Don't make up more than will be used during one semester. For class use, dilute 1 part of the stock solution with 2 parts of 45 percent acetic acid and put in dropper bottles for distribution.

To prepare aceto-orcein stain, heat 45 ml of acetic acid until hot but not boiling. Then add 2 gm of orcein. Allow the solution to cool, then dilute with 55 ml of distilled water. Put this in dropper bottles for class distribution.

For the aceto-carmine stain, purchase Carmine Alum Lake dry stain available from most biological supply houses.

For the aceto-orcein stain, purchase synthetic orcein dry stain available from most biological supply houses.

Magenta xx (also know as calcozine Magenta xx) is available from Pylam Products Inc., 95-10 218th Street, Queens Village, New York 11429.

EXERCISE 6 — ROOTS I: TYPES OF ROOT SYSTEMS AND GROWTH OF ROOT TIP

ACTIVITY 1

Collect plants for demonstration of different types of root systems. Not to be used up. Dandelion (*Taraxacum officinale*); any lawn grass or grassy weed; a mature (or at least 2-month-old) corn plant (*Zea mays*) and/or barley plant (*Hordeum vulgare*); turnip (*Brassica rapa*); beet (*Beta vulgaris*); carrot (*Daucus carota* v. *sativa*); bean (*Phaseolus vulgaris*); *Coleus blumei*; English ivy (*Hedera helix*); nasturtium (*Tropaeolum majus*); African marigold (*Tagetes erecta*); *Rhoeo* sp; wandering Jew (*Zebrina pendula*); *Kalanchoe* sp.

About 4–6 weeks before class, start cuttings in water, sand or peat cubes, pellets or blocks. The peat products are produced under various trade names, for example, Jiffy-7, Kys-Kube, Solo-Gro, Fertl-Cubes, and so forth. Most are available locally in garden supply centers, but can

*(Part of recommended student lab kit)

be purchased more cheaply in large quantities from the horticultural supply houses and some biological supply houses. See list of houses in appendix.

Suggested plants for cuttings: geranium (*Pelargonium* sp.); *Coleus blumei*; *Kalanchoe* sp.; bloodleaf (*Iresine* sp); *Impatiens* sp.; *Peperomia* sp.; African violet (*Saintpaulia ionantha*); snake plant (*Sansevieria* sp.). Plants are for demonstration and not to be used up.

ACTIVITY 2

Water hyacinth (*Eichornia crassipes*), water lettuce (*Pistia* sp.) to show root caps. Plants are for demonstration and not to be used up.

If green alder (*Alnus crispa*) is to be used, you will probably have to grow young seedlings from seed. One source of seed is Herbst Brothers Seedsmen, Inc., 1000 N. Main St., Brewster, N.Y. 10509. *Alnus* seed must be stratified for 60 days in a moist medium at 1° – 5° C to overcome dormancy (unless this has already been done by the seed supplier). Germination requires 30 to 40 days, preferably with 8-hour days of 30° C and 16-hour nights of 20° C.

Once you have these young plants available they are well worth the work. The red root caps are striking.

ACTIVITY 3

Prepared slides of longitudinal section of root tip of onion (*Allium cepa*). One slide per student. Slides available from biological supply houses. See list. One compound microscope per student.

ACTIVITY 4

Germinated seeds of bean (*Phaseolus vulgaris*) or pea (*Pisum sativum*), grown in sand or vermiculite, and whose primary roots are 3–4 cm long. Enough seeds for one to two per two students. Each pair of students to have available: paper toweling, 12 inches or so of thread, India ink, a metric ruler, a straight (common) pin, a small bottle and a cork to fit the bottle. A small hole in the cork is recommended, although not absolutely necessary (see text).

ACTIVITY 5

Radish seedlings about 4 days old germinating on moist filter paper in petri dishes. One dish per four students.

Additional material suggested but not absolutely necessary: other plants or stem cuttings rooted in peat cubes. In many ways these are to be preferred over the radish seedlings because of ease of handling by the students without disturbing the root hairs. Cuttings should be started a month before the class.

ACTIVITY 6

Prepared slides of orchid roots showing mycorrhiza. Available from biological supply houses. One slide per one or two students. One compound microscope per student.

EXERCISE 7 — ROOTS II: INTERNAL ANATOMY OF ROOTS

ACTIVITY 1

Prepared slide of cross section of young dicot root such as buttercup (*Ranunculus* sp.). One slide per student. Available from biological supply houses. Compound microscope—one per student.

ACTIVITY 2

Prepared slide of young willow (*Salix* sp.) showing origin of branch root. One slide per student. Available from biological supply houses. Compound microscope—one per student.

ACTIVITY 3

Prepared slide of a cross section of a monocot root. Slides available from biological supply houses are wheat (*Triticum* sp.); *Iris* sp.; corn (*Zea mays*); or greenbrier (*Smilax* sp.) One slide per student. Compound microscope—one per student.

ACTIVITY 4—OPTIONAL

Prepared slide of cross section of a woody root, such as basswood (*Tilia* sp.) or tulip tree (*Liriodendron tulipifera*).

EXERCISE 8 — HERBACEOUS STEMS I: TERMINAL MERISTEM GROWTH

ACTIVITY 1

Representative dicot herbaceous plants for students to examine. Not to be used up. One plant per two to four students. Suggested plants: *Coleus blumei*; geranium (*Pelargonium* sp.); any 1-month-old plant of bean (*Phaseolus vulgaris*), cucumber (*Cucumis sativus*), or squash (*Cucurbita* sp.) all of which you can start from seed.

ACTIVITY 2

For display purposes only, have a few dicot plants (preferably the same kinds as used in Activity 1) showing the effect of 2,4-D injury. Two to four days beforehand spray the plants with 2,4-D or related phenoxy herbicide such as Silvex or 2,4,5-T. These herbicides are widely available in garden supply centers or from horticultural supply houses. See list.

ACTIVITY 3

Prepared slides of longitudinal section of stem tip of *Coleus blumei*. One slide per student. Available from biological supply houses. See list. Compound microscope—one per student.

ACTIVITY 4

Monocot plants for examination (with some plants to be stripped of some leaves by students). Students are to see how the leaf and stem arrangement differs from that of dicots. Suggested plants: young (preferably 1-month-old) corn plants (*Zea mays*) about 1 per student for dissection; also for dissection or demonstration any of the common house plants, such as wandering Jew (*Zebrina pendula*); spiderwort (*Tradescantia* sp.); purple heart (*Setcreasea pallida*); *Rhoeo* sp.; dumbcane (*Dieffenbachia* sp.) or *Philodendron* sp.

EXERCISE 9 — HERBACEOUS STEMS II: STEM ANATOMY OF HER-BACEOUS DICOTS AND MONOCOTS

ACTIVITY 1

Prepared slides of cross section of stem of a dicot herb such as sunflower (*Helianthus* sp.) or geranium (*Pelargonium* sp.). One slide per student. Compound microscope—one per student.

ACTIVITY 2

Samples of twine, rope, bags, or cloth made from jute, hemp, or linen fibers. Available from arts and crafts shops, hardware stores or your grandmother's linen closet. For display only, not to be used up.

ACTIVITY 3

Prepare two sets of seedlings of bean or corn so that they are about three weeks old for this class. About a day or two (depending on conditions) do not water one set so that the plants show wilting.

OPTIONAL

Expose one of the well watered plants to a 150 watt incandescent light. Use a photoflood reflector or a type PAR bulb with built-in reflector so as to intensify the light and heat.

Have a variety of vegetables in different stages of wilting. Have pans or beakers for reviving some of these in water.

ACTIVITY 4

Prepared slides of cross section of older stems of dicot herbs such as geranium (*Pelargonium* sp.) or sunflower (*Helianthus* sp.). Prepared slides of 1-year-old twig of dicot woody plant such as species of *Tilia*, *Platanus* or *Liriodendron*. One slide of each per student. Compound microscope—one per student.

ACTIVITY 5

Prepared slides of cross section of stem of a monocot such as corn (*Zea mays*) or lily (*Lilium* sp.). One slide per student. Compound microscope—one per student.

ACTIVITY 6

Samples of products made from sisal, for example, twine, rope, shopping bags, hammocks, or from Manila hemp (abaca), for example rope, or any textiles products, like table mats, sandals or cloth. For display only.

EXERCISE 10 — WOODY STEMS I: EXTERNAL FEATURES OF YOUNG TWIGS

ACTIVITY 1

Collect leafless twigs (with at least 3–4 years growth) of woody plants in mid winter when the buds are fully developed. Store these in some dry place and use for this exercise. One twig per student is adequate. If the class can have available more than one species, it would be preferable. Recommended plants: horse chestnut or Ohio buckeye (*Aesculus* sp.); walnut (*Juglans* sp.); tree of heaven (*Ailanthus altissima*); ash (*Fraxinus* sp.); and tulip tree (*Liriodendron tulipifera*).

ACTIVITY 2

In mid winter collect twigs of plants with thorns or spines. Suggested plants: honey locust (*Gleditsia triacanthos*); hawthorn (*Crataegus* sp.); black locust (*Robinia pseudoacacia*); and crabapple (*Malus* sp.).

ACTIVITY 5 — OPTIONAL

Bring in fairly large branches for the students to practice pruning. Supply 1 pair of pruning shears per four students. These are widely available in hardware stores, department stores, garden supply shops and from horticultural supply houses.
  Collect branches that show the pollard effect from poor pruning.
  Have slides or photos to show poorly pruned trees.

EXERCISE 11 — WOODY STEMS II: INTRODUCTION TO WOODY STEM ANATOMY AND THE GYMNOSPERM STEM

ACTIVITY 1

Prepared slides of macerated wood of pine (*Pinus* sp.) or some other conifer. One slide per student. Available from biological supply houses. Compound microscope—one per student.

ACTIVITY 2

Prepared slides of cross section of three-year-old (or older) stem of pine (*Pinus* sp.). One slide per student. Available from biological supply houses. Compound microscope—one per student.

EXERCISE 12 — WOODY STEMS III: ANATOMY OF THE ADVANCED WOODY STEM—THE ANGIOSPERM STEM

ACTIVITY 1

Prepared slides of macerated angiosperm wood. One per student. Available from biological supply houses. Oak (*Quercus* sp.) is most commonly used for this purpose. Compound microscope—one per student.

ACTIVITY 2

Prepared slides of oak (*Quercus* sp.) showing three sections of cut: cross, radial, and tangential. One per student. Slides with three square sections of mature wood are preferable to those which use a 6–8 mm twig cross-section to represent the cross section. Compound microscope—one per student.

ACTIVITY 3

Prepared slides showing cross section of woody angiosperm stem. Many genera available from biological supply houses, for example, *Tilia*; *Liriodendron*; *Platanus*; and *Magnolia*. In this manual, Figure 12.4 and text description is based on *Tilia*.

EXERCISE 13 — WOODY STEMS IV: WOOD GRAIN AND TREE-RING DATING

ACTIVITY 1

Blocks of wood measuring about 5 × 10 cm and cut so they show radial, tangential, and cross section planes. One block per one to two students. Prepare these yourself or obtain from biological supply houses. Either gymnosperm or angiosperm wood is adequate. Suggested material: Red cedar (*Juniperus* sp.); almost any pine (*Pinus* sp.); Douglas fir (*Pseudotsuga menziesii*); white oak (*Quercus alba*); hickory (*Carya* sp.).

If you can, obtain blocks or rounds of wood cut from larger limbs; these are helpful for display purposes.

ACTIVITY 2

The same material may be used as in Activity 1.

ACTIVITY 3

"Hough" slides are no longer available from biological supply houses. They are thin cut sections of wood mounted between glass. They are suggested here because some schools may still own a set. In lieu of Hough slides, you could have available for the students (and your own reference) the following publications. (1) *What Wood is That* by Herbert L. Edlin, Viking Press, N.Y. (1969). This contains thin sections of wood glued to the pages; (2) *Wood Handbook*, (1974) U.S. Department of Agriculture Handbook No. 72, Forest Products Laboratory. Available from the Superintendent of Documents, Washington, D.C. 20402 (3) *Textbook of Wood Technology*, Vol. I by A. V. Panshin, Carl DeZeeuw, and H. P. Brown (2nd ed.) McGraw Hill Book Co. This has a key for wood.

ACTIVITY 4

Pieces of douglas fir or pine plywood about 10 cm × 10 cm should be cut from larger sheets so as to have one piece per two students. Grades A-C, A-D and CDX are all adequate. Veneered panels are interesting to the students, but be certain that genuine hardwoods are used, and not printed vinyl.

ACTIVITY 5

Wood "cores"—one set per one to two students. These are thin, long, rectangular slices of wood. Essentially for this introductory study, they show the same features as cores that are obtained with a wood coring instrument, but are sturdier and less easily broken.

They have been made available especially for this exercise by two companies—Carolina Biological Supply Co. and Ward's Natural Science Establishment, Inc.

EXERCISE 14 — LEAVES I: EXTERNAL FEATURES

ACTIVITY 1

Twigs or branches of most any conifer such as juniper (*Juniperus* sp.); pine (*Pinus* sp.); spruce (*Picea* sp.); fir (*Abies* sp.); hemlock (*Tsuga* sp.); or any small-leaved plant native to dry habitats such as sagebrush (*Artemisia tridentata*); creosote bush (*Larrea divariata*). Enough specimens for each student or pair of students. Not to be used up. Try to have a variety of specimens.

ACTIVITY 2

Specimens of cacti; or spurges such as crown of thorns (*Euphorbia splendens*) or redbird cactus (*Pedilanthus tithymaloides*), both common house plants. Also any succulents (the stonecrop family) commonly available in plant shops viz., species of *Sedum*: *Escheveria*; *Crassula*; *Kalanchoe*; and *Haworthia*. None to be used up.

ACTIVITY 3

These specimens are also all common house plants and can be bought or rented (for nominal amount) from plant shops. Obtain rubber plant (*Ficus elastica*); fiddle-leaf fig (*Ficus lyrata*); dumbcane (*Dieffenbachia* sp.); caladium (*Caladium* sp.) rex begonia (*Begonia rex*); banana (*Musa* sp.)—a leaf of this would suffice rather than the entire big plant; staghorn fern (*Platycerium* sp.). None of these to be used up.

ACTIVITY 4

Have boxes of dried and pressed leaves of a variety of angiosperm trees. Collect the leaves, place flat between folds of newspaper to dry. Preferably dry them in a standard plant press as available from all biological supply houses. Collect simple and compound leaf types—enough for several per student. Not to be used up.

ACTIVITY 5

Representative monocot plants to show leaf types. For display and examination only. Many of these are common house plants available from plant shops. Examples: dumbcane (*Dieffenbachia* sp.); spiderwort (*Tradescantia* sp.); wandering Jew (*Zebrina* sp.); *Sansevieria* sp.; *Dracaena* sp.; *Rhoeo discolor*; pineapple (*Ananas comosus*); gladiolus (*Gladiolus* sp.); lily (*Lilium* sp.); tulip (*Tulipa* sp.); or any grass, sedge or orchid (*Orchis* sp.). Leaves of grasses and iris may be pressed and dried as for Activity 4.

EXERCISE 15 — LEAVES II: INTERNAL STRUCTURE

ACTIVITY 1

Prepared slide of cross section of dicot leaf such as lilac (*Syringa* sp.) or privet (*Ligustrum* sp.). One per student. Compound microscope—one per student.

ACTIVITY 2

Leaves of plants for students to peel off the epidermis. Suggested plants: wandering Jew (*Zebrina* sp.), any *Sedum* species; German ivy (*Senecio mikanioides*) or corn (*Zea mays*). Razor blades,* microscope slides* and cover slips,* forceps,* pipette.* Enough living plants are needed to supply one living leaf for every one or two students. These will be used up. Compound microscope— one per student.

ACTIVITY 4—OPTIONAL

Zebrina plants for display. One plant (or set of plants) that's been grown in a well lighted place. Another plant (or set of plants) that's been grown for a month or more in a shadier place. The new leaves (if formed after the plant was placed in the shady place) will be the ones to examine. One set of plants per class.

ACTIVITY 5

Prepared slides of cross section of corn leaf (*Zea mays*). One slide per one to two students. Available from biological supply houses. Compound microscope—one per student.

ACTIVITY 6

Prepared slides of cross section of leaf (needle) of pine (*Pinus* sp.). One slide per one to two students. Available from biological supply houses. Compound microscope—one per student.

ACTIVITY 7

For display only. Living plants of rubber plant (*Ficus elastica*); fiddle-leaf fig (*Ficus lyrata*); and *Philodendron*—any of the large leaved varieties. These are all common house plants.

EXERCISE 16 — FLOWERS

ACTIVITY 1

Simple, single, regular flowers—to be used up. One per one to two students. Suggested: petunia (*Petunia hybrida*); tulip (*Tulipa* sp.); gladiolus (*Gladiolus* sp.); lily (*Lilium* sp.); primrose (*Primula* sp.); poppy (*Papaver* sp. or *Eschscholtzia*); *Browallia* sp.; *Amaryllis* sp.; *Hoya* sp.; *Clivia* sp.; *Azalea* sp. or *Magnolia* sp.

*Part of recommended student lab kit.

ACTIVITY 3

Flowers (male and female specimens) of wax begonia (*Begonia semperflorens*), a fairly common houseplant and tender perennial garden plant. For examination only. Not to be used up, although some will be damaged by handling. Enough for each one to four students.

ACTIVITY 4

Flowers of snapdragon (*Antirrhinum* sp.); garden pea (*Pisum sativum*) or sweet pea (*Lathyrus* sp.). One flower per student. To be used up. Available from florists, or grow your own plants from seed started 3–4 months earlier.

ACTIVITY 5

Flowers of composites—to be used up. Dandelion (*Taraxacum officinale*); aster (*Aster* sp. or *Cineraria* sp.); daisy (*Chrysanthemum leucanthemum*); sunflower (*Helianthus* sp.). Available from fields if in season and some from florists. Specimens to be used up. One head per four students.

ACTIVITY 6

Inflorescence of oats (*Avena sativa*). One floret per student. May or may not be used up, depending on whether or not the students dissect it. Allow about 15–20 weeks to flower from seed if you are growing your own.

EXERCISE 17 — FRUITS

DEMONSTRATION

This demonstration requires a series of four to six preserved specimens showing the stages of development of a fruit from pollination (the ovary) to maturation. A few are occasionally available from biological suppliers, often called the "life history" of the plant. Some of these series are embedded in plastic for permanent display.

Many examples, such as *Malus, Fragaria, Pisum, Phaseolus* and *Capsicum*, are easy to collect in quantity at the proper season of the year and preserve for class use. The usual preservative is FAA (50 percent ethanol; 2 percent glacial acetic acid; 10 percent formalin; 40 percent water), though materials may be transferred to 50 percent alcohol after a few days in the FAA.

ACTIVITY 1

Any number of examples of the five types given may be used. Most may be purchased in the grocery store or from a garden seed supplier. Some, such as the locusts (*Robinia* and *Gleditsia*), are best collected locally in the fall or early winter.

Allow one fruit of each category per every four to six students.

ACTIVITY 2

As for Activity 1, many of the fruit types may be found in the grocery store. A good selection requires that examples of fruits of many garden flowers and vegetables and field weeds be locally collected, so exact species used depend on availability.

ACTIVITY 3

All species suggested are common weeds or native trees, and local collection is normally used. Some biological supply houses do offer displays or sets showing these and other species under the heading of "seed dispersal" or a similar title.

EXERCISE 18 — WATER RELATIONS: OSMOSIS AND DIFFUSION

ACTIVITY 1

Either prepare before class or have the students prepare cellulose acetate tubing cut so as to make up 6 inch long tubes ("sausages") tied at both ends with a string, and filled with 10 percent sugar solution dyed with a tiny amount of eosin or methylene blue (0.005 percent).

Prepare enough of a 20 percent sugar solution to partly fill beakers into which the "sausages" will be put.

Need two beakers per each group of students. Students work in groups of one to four. One beaker has tap water. One has the 20 percent sugar solution. One triple-beam balance (scales) per 10 students. Paper toweling to dry "sausages."

ACTIVITY 2

Potatoes, turnips, or rutabagas—one of any plant per four students.

"French-fry" cutters—one per 20 students, or cork borers, 6–10 mm diameter, about one per four students.

One triple-beam balance (scales) per 10 students. Two beakers (100 ml or so) per pair of students. Tap water in one, strong (10 percent) salt (NaCl) solution in the other—enough to cover "plug" with liquid.

ACTIVITY 3

Set up the apparatus as shown in Figure 18.1. Fill one "sausage" with a glucose solution and the other with a starch "solution." Weigh and record each sausage, then insert the glass tubing into each and insert in tap water in the beakers as shown in Figure 18.1.
—Two test tubes
—Bottle of IKI (iodine-potassium-iodide)
—Bottle of Benedict's solution—allow about 10 ml per class for demonstrations. Biological supply houses have dry concentrate or solution for sale.
—Roll of Tes-Tape, a glucose test paper available from biological supply house or pharmacy
—Bunsen burner, tripod, asbestos mats or asbestos wire gauze mats, 500 ml beaker, tap water

ACTIVITY 4

Leaves of plants such as elodea (*Anacharis Canadensis*); *Rhoeo* sp.; or *Zebrina* sp.

One microscope slide* and cover glass* per student. Small dropping bottle with 5 percent salt (Nacl) solution—one per four to eight students. Paper toweling. Compound microscope— one per student.

*Part of recommended student lab kit

ACTIVITY 5

Take stem cuttings of the following plants and place them in water of different salt concentrations, for example, no salt; 0.5 and 3.0 percent. Or use some garden or lawn fertilizer and make up different concentrations in water. Since most fertilizers, as sold, are partly insoluble, use about 3.0 percent and 6.0 percent fertilizer by weight. Use test tubes for setting the cuttings in the solutions. Set in test tube racks or blocks.

Suggested plants: almost any non-woody plant, such as geranium (*Pelargonium* sp.); tomato (*Lycopersicon* sp.); coleus (*Coleus blumei*) and so on.

ACTIVITY 6

—Bottle of crystals of potassium permanganate ($KMnO_4$)
—Petri dishes—one per two to four students
—Non-nutrient agar or water in petri dishes
—Dishes with tap water (if petri dishes are unavailable)
—Forceps*—one per two to four students

ACTIVITY 7

Two–three hours before class put stalks of celery (*Apium graveolens*) or stems of white flowered carnation (*Dianthus* sp.) or *Chrysanthemum* sp. in beakers containing solutions of colored food dyes or methylene blue. Make sharp clean cuts so the dye will easily enter the plant.

ACTIVITY 8

—Two glass tubes each about 59 to 70 cm long and 5 cm in diameter
—Four rubber or cork stoppers to fit these tubes
—Cotton
—Dropper bottle with 1 N ammonium hydroxide (NH_4OH)
—Dropper bottle with 1N hydrochloric acid (HC1)

ACTIVITY 9

Set up demonstration overnight or 4-5 hours before class.
—Three bell jars per classroom
—Three glass plates to set jars on
—Plant in pot with soil covered over with plastic wrap or aluminum foil to seal escape of moisture from it. Or use plant growing in a peat cube and wrap the cube in plastic wrap or aluminum foil to prevent escape of moisture
—Wet sponge

ACTIVITY 10

For each group of (2–4) students: assemble materials needed for apparatus shown in Figure 18.3.
—Shallow pan
—A micropipette (capacity 0.1 ml or 0.5 ml, graduated in 0.01 ml is ideal) or a section of glass capillary tube about 15 cm long
—Glass or plastic elbow
—Rubber tubing to fit elbow to pipette or tube

*Part of suggested student lab kit.

—Rubber tubing to fit elbow to plant stem
—Razor blade
—Metric ruler 15 cm, reading in mm, if capillary tubes are used
—Plants for taking leafy stem sections. Geranium (*Pelargonium hortorum*); tomato (*Lycopersicon* sp.) or potato (*Solanum tuberosum*)

ACTIVITY 11

Set up plants under bell jars and water the soil heavily so as to induce guttation. Suggested plants: Two-week-old seedlings of barley (*Hordeum vulgare*); wheat (*Triticum* sp.); or tomato (*Lycopersicon* sp.)

ACTIVITY 12

Dry woody material and seeds to be put in water by the students to observe imbibition of water. Use large seeds, such as beans, mature cones of any pine, spruce, hemlock, and so on, or pieces of wood veneer.

EXERCISE 19 — PHOTOSYNTHESIS

ACTIVITY 1

—2 test tubes per student
—2 strips of filter paper per student to fit lengthwise in test tubes
—1–2 bottles of water soluble ink for class. Any brand name black writing ink is satisfactory, but office supply businesses are probably the only common source today. Ink may be applied with toothpicks; pens are not necessary
—Test tube racks or blocks
—Mortar and pestle or electric blender
—Fresh leaves (use spinach or geranium leaves if outdoor plant material is not available)
—Enough ether-acetone solvent to grind up the leaves and have a sufficient volume for class use (few ml per student). Make up solvent with 95 parts ether and 5 parts acetone. Distribute leaf extract in dropper bottles—one per 6–12 students

ACTIVITY 3

—Enclosed-element electric hot plates—one per four to eight students OR
  Bunsen burner, tripod, asbestos pad—one of each per four to eight students
—600 ml glass beakers—one per four to eight students
—90–95 percent alcohol (either ethanol or methanol)—enough for about 100–200 ml for four to eight students
—Small beaker (100 or 250 ml)—one per two students
—Bottle of IKI (iodine-potassium-iodide)—one per four to eight students
—Petri dish—one per two students
—Plant material to supply each one to two students with one leaf. Leaves must be variegated. Suggested: *Coleus blumei*; Algerian ivy (*Hedera canariensis* v. *variegata*); geranium (*Pelargonium hortorum* v. *marginatum*)

ACTIVITY 4

Cut 4 cm² (2 cm × 2 cm) squares of heavy black paper and attach (by using "sprung" or weakened paper clips) to leaves of plants of geranium (*Pelargonium* sp.) or bean (*Phaseolus vulgaris*). Store plants in darkness for 72 hours. Then, expose plants to full sunlight for 8–12 hours just before class. Enough plants to supply one leaf per one to two students.

—Enclosed-element electric plate or bunsen burner, tripod and asbestos pad—one per two to four students (just as used in Activity 3)

—Petri dish—one per two students

—Bottle of IKI—one per eight students

—90–95 percent alcohol—enough for about 50 ml per two students

ACTIVITY 5

—Boil some tap water (enough for 3 test tubes). Let it cool in a covered pan and pour into the tubes. Label tubes 1, 2, and 3, and have available for class

—One dropping bottle with phenol red indicator

—A soda straw or comparable glass tubing

—A sprig or two of elodea (*Anacharis canadensis*) or parrot's-feather (*Myriophyllum brasiliense*). These are usually available through aquarium suppliers

—Some lighting to supply bright light to illuminate the test tubes

—Set up two beakers with plants, funnels and tap water as shown in Figure 19.4

—Add several drops of 1 percent sodium bicarbonate ($NaHCO_3$) to the water in the beakers or blow your breath into it with a straw or glass tube

—Use same kind of plant material as in Activity 5

—Wood splinters (available from supply houses)

ACTIVITY 7

—White potatoes—about one per four to eight students

—Scalpel—one per four students; a paring knife will be adequate

—Microscope slide* and cover slip*—one per student. Compound microscope—one per student

—Dropping bottle of IKI—one per four to eight students

—Sample batches of flour made from rice (*Oryza sativa*); oats (*Avena sativa*); sweet potato (*Ipomoea batatas*); soybean (*Glycine max*), and so on. One sample (about 250g) of any one of the above is sufficient for entire class

EXERCISE 20 —DIGESTION AND RESPIRATION

ACTIVITY 1

—Corn grains which have been kept wet for 48 hours at about 20–22° C with good aeration. Cut in half lengthwise on the flat side so as to produce right and left halves. These germinating grains are placed, cut face down, on the surface of a fresh sterile plate of starch agar, poured about 2–3 mm deep in a 10 cm petri dish. Let the plates sit, covered, in the dark for about 36–48 hours before class. Needed as a demonstration, one plate per class

Starch agar may be purchased as a dry concentrate from biological supply houses, or may be made up by adding about *1* gm soluble starch (or ordinary cornstarch) to each 100 ml of water in a standard 1.5 percent *non*-nutrient agar recipe

—Bottle of dilute IKI for each class—about 5 ml will be used in each demonstration

ACTIVITY 2

—About 250g of wheat should be germinated for each class. This is best done by placing about 1 kg of grain in a 4 or 5 liter (or 1 gallon) jar and covering with water for 12 hours. Then pour

off the water while shaking the jar vigorously. Half fill with more lukewarm tap water and shake again. Pour off all liquid water, and let jar stand either upright or on its side with the cap very loosely in place. Repeat the washing and shaking every day for about 4–5 days before using as described
—An electric blender is needed for each class. The types which can use household canning jars are preferable, as a supply of clean jars can be made ready each day
—A 500 ml flask or bottle is needed for each class to hold the extract after filtering
—A large (100 mm) funnel with a cheesecloth pad for filtering the extract
—Bottle of Benedict's solution—about 10 ml will be used in the demonstrations
—Hot water bath on hot plate or bunsen burner for Benedict's tests
—Bottle of IKI solution—about 15 ml will be used in the demonstrations

ACTIVITY 3

— One 125 ml Erlenmeyer flask for each working group of two to four students
—About 50 ml of 1 percent starch suspension for each two to four student group
—At least 2 test (or culture) tubes for each student group. Sizes 10×75 mm to 15×125 mm are satisfactory
—Dropper bottles of IKI—one 30 ml bottle for each student group
—Dropper bottles of Benedict's solution—one 30 ml bottle for each student group
—Hot water bath on hot plate or bunsen burner for each four student group
—Test tube rack for each two-student group

ACTIVITY 4

This may be either a demonstration or may be performed by a group of two to eight students if time permits.

—5 test (or culture) tubes per group—any size from 10×75 through 15×125 will work
—Dropper bottles of 1 percent starch suspension—one per group
—Dropper bottles of amylase (available from ICN Pharmaceutical, Inc. The amylase derived from bacteria is satisfactory and by far the least expensive)
—Hot water bath and test tube rack as for Activity 3

ACTIVITY 5

This demonstration is started by soaking about 100g of dry seeds such as corn or beans for about 12 hours. The imbibed seeds are then tied loosely in cheesecloth and suspended in a tightly sealed jar. A 250–400 ml wide-mouth specimen bottle is useful, and the cheesecloth bag is usually suspended over (but not touching) about 10 mm of water simply by catching the string tying the bag under the bottle cap or stopper. One of the jars should contain a smaller jar or several open vials filled to the 20 mm level with a 5 percent solution of potassium pyrogallate or pyrogallol. Do not open the jars after setup. Allow 4–6 days to show suppression of germination.

ACTIVITY 6

Use any easy-to-root species, such as coleus (*Coleus blumei*); tomato (*Lycopersicon* sp.); wax begonia (*Begonia semperflorens*); geranium (*Pelargonium hortorum*); wandering Jew (*Zebrina pendula*) or purple heart (*Setoreasea purpurea*). If the season of the year is suitable, that is, late winter through late spring, many species of willow (*Salix*) or poplar (*Populus*) show this effect markedly. Use any tall, narrow container such as a graduated cylinder (500 or 1000 ml) or a 32×300 mm culture tube. Any small vibrator-type aquarium aerator pump will work, so long as the tube is kept at the bottom of the rooting container.

ACTIVITY 7

Set up the demonstration apparatus as shown in Figure 20.4. The bottles may be any type of specimen bottles from 200 to 500 ml (or 6 to 10 oz.) which may be tightly stoppered. The sizes of the stoppers and tubing illustrated should be altered to match the bottles utilized. Several extra bottles numbered 3 and 4 should be available for the lots of seeds held at the 3 different temperatures. It may be desirable to have different sets of seeds at each temperature if different class sections are close together, and time for evolution of CO_2 is short.

Common wheat, barley, beans, or corn (maize) are preferable for this demonstration, as their temperature sensitivity is clear. Their general treatment should be as outlined for Activity 2, above.

The illustration shows a small rubber squeeze-bulb positive displacement hand pump for air transfer. With this pump, squeezes may be counted and gas flow standardized easily. Almost any hand-operated tire or air-mattress pump could be adapted to this system. If none of these are available, an aquarium aerator pump might be used, with its time of operation measured exactly. In this last case, a time of about 15–30 seconds is about right, depending on the volume of the pump.

The NaOH or KOH in Bottle 1 is about 7 normal, and should be adequate for an entire day's classes. Add more fresh solution if any precipitate begins to collect in Bottle 2. The $Ba(OH)_2$ in Bottles 2 and 4 is about .5 normal, and several liters should be prepared at once, in advance. About 300–500 ml will be used for each class, depending on the size of the bottles used.

A funnel (about 75 mm) is needed for each of the 3 sample temperatures used. Circles of medium-coarse filter paper are needed to remove the $BaCO_3$ precipitate from each Bottle 4. Number the paper in pencil before filtering with the temperature (5°, 20°, 30°) of the germinating seeds tested. Weigh another circle of filter paper to determine the weight for subtracting from the weight of dried precipitate plus paper.

A drying oven set at 80° C and a good balance are needed. One oven for all classes is adequate. If an oven is not available, the wet filter papers and precipitate can be dried directly on a hot plate set to about 50° C. If a sensitive balance for each class is not available, a student could be sent to a central prep room where an assistant is stationed to weigh the precipitate on a torsion or electronic balance.

ACTIVITY 8

Prepare about 2 hours before each class period enough yeast suspension to allow about 20 ml per student. To make the suspension, mix one cake of compressed yeast or 2 packages of dry yeast with one liter of 25° C water and about 50g of sucrose (or glucose). After 1–2 hours, bubbles of CO_2 should be evident. After 4–6 hours, the sugar will be used up. More may be added, or a new culture started.

—Fermentation tubes, one for every one or two students. The type graduated in ml is preferable, but the ungraduated type may be used if a 15 cm metric ruler* is available
—Dropper bottles of phenol red, one per every four to eight students
—Foam plastic ice bucket with crushed ice or ice water, capacity approximately 2 l (½ gal.). One per 8–12 students
—Access to a refrigerator

EXERCISE 21 — PLANT MOVEMENT AND GROWTH RESPONSES TO STIMULI

ACTIVITY 1

—Pots of seedlings of oats (*Avena sativa*) or barley (*Hordeum vulgare*) containing 4–8 seedlings

*Part of recommended laboratory kit

each. Use 1½–2 inch plastic or clay pots or compressed organic media such as Jiffy-7, Kys-Kube, Fertl-Cube or Solo-Gro, all of which are available through horticultural supply houses. The seeds should be started in the dark 5–10 days before they are needed. Some experimentation may be needed to determine the exact time span desirable under your temperature regime. A coleoptile length of 2–4 cm is good, but seedlings are too old if the first leaf breaks through the coleoptile. One pot will be used by every group of two to four students
—Boxes of ordinary toothpicks. Each student will use three or four
—Small jars of 1:10,000 indole acetic acid (IAA) mixed with lanolin. IAA is available from biological supply houses. One 50–100g jar or cup per four to eight students
—A DARK box or cabinet for in-class coleoptile development. It is a help if the temperature is at least 25° C, but not over about 32° C inside the box

ACTIVITY 2

A demonstration should be prepared about 36–96 hours before class of several susceptible dicot plants such as tomato (*Lycopersicon* sp.), garden bean (*Phaseolus* sp.), dandelion (*Taraxacum officinale*) or jimson weed (*Datura* sp.). Any lawn or field weed could also be used by transplanting them to pots either before or after treatment with the herbicide. Use any phenoxy herbicide such as 2, 4-D; silvex (2,4,5-TP); or 2,4,5-T mixed according to label directions. Wet the foliage of the test plants, but do not soak them as if watering the plant. Do not spray in the greenhouse, or even return the sprayed pots to the greenhouse at any time, as many of the other greenhouse plants are very susceptible to residual fumes of the herbicide.

ACTIVITY 3

The radish seedlings should be planted about 4–8 per 1½–2″ pot or compressed organic cube as were the seedlings for Activity 1, above. The pots should be started about 4–7 days before needed, and should be kept the entire time in their assigned box, with monochromatic light if specified. Only 2 monochromatic filters are needed, corresponding to wavelengths in the blue (450 nm) and red (650 nm) portions of the spectrum. They are available from at least one biological supply house.

ACTIVITY 4

This demonstration calls for two 8–10cm (3–4″) pots planted with any fast-growing herbaceous plants. Seedlings of oats (*Avena sativa*), barley (*Hordeum vulgare*), radish (*Raphanus sativus*), and cucumber (*Cucumis sativus*) about 3 to 6 cm tall are the most successfully used. Potted geraniums (*Pelargonium hortorum*) or *Coleus blumei* may also be used, but will require 5–10 days to show effects. The seedlings will demonstrate phototropism in 36–48 hours if placed near any unidirectional source of light.

The clinostat, itself, is available as an item of scientific apparatus designed especially for this experiment. It revolves at a rate of 2 revolutions per hour (rph). Any clockwork or motor-driven turntable may be used, and any speed of rotation from 1 revolution per minute (rpm) to 1 rph is satisfactory. Speeds much faster than 2–4 rpm lead to centrifugal effects in auxin distribution, hence unmodified record turntables (33 rpm) are not satisfactory.

ACTIVITY 5

This demonstration requires a mature (10–15 day old) *asexual* culture of *Phycomyces blakesleeanus* growing on potato-dextrose agar in the bottom of a 500–1000 ml Erlenmeyer flask or any bottle of similar size. The culture has been allowed to develop with the flask or bottle covered with aluminum foil or black construction paper to exclude light. About 72 hours before class, or when asexual sporangiophores have begun to develop, open a small (2–3 cm) "window" in the covering about one-half or two-thirds of the way up the side of the flask. Place the culture so that moderate light may enter through the opening. No especially strong spot lights are needed, and a north window is ideal as to intensity.

ACTIVITY 6

This demonstration requires placing imbibed corn grains on the bottom of a petri dish, embryo side down to the glass. Four to six grains are adequate per dish. Cover the grains with a circle of heavy filter paper; then add more filter paper or a crumpled paper towel to fill the dish a little beyond the top rim. Now place the cover on the petri plate and secure with one or two heavy rubber bands. The seeds are then allowed to germinate as the plate is held vertically, either in a special rack or by embedding one edge in a lump of modelling clay. The paper towel is kept moist, but not dripping wet. Allow 4–5 days of development before required for class.

ACTIVITY 7

Any healthy, fast-growing potted plant may be used for this demonstration. As suggested in the Activity, specimens growing in compressed organic media such as Jiffy-7, Kys-Kubes, and so on, show exposed root tips more easily. If growing a plant on its side for several days, provide an aluminum foil cover for the root mass to retain moisture and exclude light except when being examined by the class.

ACTIVITY 8

This activity requires that several healthy potted *Mimosa pudica* plants be available in each classroom for observation of the sensitive reaction. They require about 3–6 months to attain a suitable size from seed, but may be propagated by cuttings after this. Seed is available from major seed houses. If the plants are healthy and well-watered, they will "recover" from the sensitive reaction in 15–30 minutes.

Insectivorous plants may be purchased from major biological supply houses and, occasionally, through sellers of house plants. All types are best kept in a bog terrarium of at least 35 l (10 gallon) size, and maintained according to the instructions supplied with the plants.

The demonstration slides of grass leaf cross sections are available from all major biological supply houses. Some supply corn (*Zea mays*) for this purpose; some stock only *Poa*; and some supply an unnamed xerophytic grass. Any of these are satisfactory. One demonstration set-up, consisting of two paired microscopes, should be available for every 8–10 students.

EXERCISE 22 — BACTERIA

ACTIVITY 1

—Prepared slides of bacteria illustrating the three cell shapes. One slide per student. Available from biological supply houses
—Compound microscope—one per student

ACTIVITY 2

—Compound microscope—one per student
—Agar slant cultures of *Bacillus megaterium* and *Micrococcus luteus*. One culture of each organism per each four students. Available from biological supply houses

—Dropping bottle with crystal violet dye; one bottle per 10 students. Crystal violet is available from biological supply houses either already made up in 1:1000 solution or you can buy the dry stain at much less cost and make up your own solution. Use 10–15 gm crystal violet stain dissolved in 100 ml of 95 percent ethanol, then make up to 1000 ml with distilled water
—A couple of paper towels for each student
—1–2 cartons of ordinary household scouring powder
—Bunsen or alcohol burners, one per four students
—Microscope slides*—three per student
—Bacteriological inoculating loop—one per two students. Available from biological supply houses
—250 ml beaker—one per two students. Optional—see #3 under Staining the Slide in Activity 2

ACTIVITY 3

—Broth culture of *Rhodospirillum rubrum*—one per four students.
—Bacteriological inoculating loop—one per student
—Jar of vaseline. One or two per class
—Microscope slide* and cover glass*
—Compound microscope—one per student
—Bottle of oil for oil immersion lens. One or two per class—only if microscopes have oil immersion lenses

ACTIVITY 4

—Broth cultures of *Bacillus megaterium* and of *Escherichia coli*. One of each per four students. Available from biological supply houses
—One sterile agar petri dish per two students
—One forceps* per two students
—Grease pencils. One per four students
—Cotton swab or Q-Tip. One per each student
—Paper disks impregnated with a solution containing some bacteriostatic agent. Agents to use are iodine; methiolate; vinegar; Listerine or other mouth wash; Lysol concentrate; any liquid bleach (sodium hypochlorite); 20 percent solution of table salt (sodium chloride—NaCl)
—Disks can be punched out of filter paper using an ordinary paper punch. They are then coded (using ordinary pencil for the marker) and soaked in any one of the suggested solutions. They should be dry, and each type stocked in a marked beaker for class use
—Disks impregnated with antibiotics are available from Ward's Natural Scientific Establishment, Inc., or if you have a source for antibiotics, you can make up aqueous solutions of them and soak filter paper disks in them as described above

ACTIVITY 5

—Plants of soybeans (*Gycine max*); bean (*Phaseolus vulgaris*) or other legume. You can grow these in sand (preferably) or soil and pull them up to show the roots to the class. Class will examine the nodules on the roots.
—Microscope slide* and cover glass. One per student
—Compound microscope—one per student

*Part of suggested student lab kit.

EXERCISE 23 — FUNGI I: INTRODUCTION TO FUNGI, PRIMITIVE FUNGI, AND SLIME MOLDS

ACTIVITY 1

—Water cultures of living *Saprolegnia* or *Achlya* available from biological supply houses
—Compound microscope—one per student
—Microscope slide* and cover glass.* One per student

ACTIVITY 2

—Living culture of bread mold (*Rhizopus stolonifera*) available from biological supply houses
—For best results, culture the mold as follows: Make up potato-dextrose agar, autoclave it, and then pour into sterile petri dishes. Cut squares or rounds of cellulose acetate paper to fit the petri dish, and lay on top of the solidified agar. Inoculate the agar with the *Rhizopus* organism. Store covered in a warm 20–25° C, (preferably) dark place. In about 4–7 days the mycelium and sporangia will be abundant. For class distribution, let each student cut out a 1 cm² of the acetate, using a razor blade. If you use commercially made bread to culture the mold, you will have to use brands that do not contain fungus preservative (usually it's calcium propionate).

ACTIVITY 3

Make up potato-dextrose agar petri dishes as for Activity 2. Inoculate one-half of each dish with positive strain *Phycomyces blakesleeanus*, and the other half with the negative strain of *P. blakesleeanus*. Set in 20–25° C dark (or relatively so) place. Zygospores should be well formed in about 10–14 days. *P. blakesleeanus* and agar are available from biological supply houses.

Set up 2–3 plates on dissecting microscopes for student examination.

ACTIVITY 4

Culture the slime mold *Physarum polycephalum* on non-nutrient agar or on moist filter paper in petri dishes. The organism is available from biological supply houses. Grind up some rolled oats in a mortar and sprinkle over the culture periodically. Avoid overfeeding. Every second day rinse the agar surface with tap water. This is essential to reduce bacterial growth. Add grain of oatmeal when old ones have been consumed.

Keep culture in low light.

One plate per four or more students.

EXERCISE 24 — FUNGI II: THE HIGHER FUNGI

ACTIVITY 1

—Dried or preserved specimens of *Peziza* (cup fungus). Available preserved from biological supply houses. Or collect and dry them yourself, if you have a local source. One specimen for examination per four students.
—Prepared slides of cross section of the Ascocarp of *Peziza* sp. One per one to two students. Available from biological houses
—One compound microscope per student

*Part of suggested student lab kit.

ACTIVITY 2

Dried or preserved specimens of morels (*Morchella esculenta*). One per four students. Preserved material available from biological supply houses. Collect and dry them yourself, if you have a local source.

ACTIVITY 3

—Dried leaves of lilac (*Syringa* sp.) infected with powdery mildew. You can collect these leaves during late summer from lilac bushes. Dry them flat in newspaper (in plant press if available) and store in cardboard boxes. Have enough leaves (for example, one per two students) for student to scrape off bits of the powder onto a drop of water on a slide.
—One microscope slide,* cover glass,* dissecting needle (or something comparable for scraping).

ACTIVITY 4

Prepare cultures of *Sordaria fimicola* (an ascomycete) on non-nutrient agar in petri dishes and have available for class—1 dish per four to eight students. Culture by autoclaving 6–10 flakes of oatmeal in the petri plate; then pouring sterile non-nutrient agar *over* the sterile oats. Innoculate with the living material when agar cools—or—follow special instructions packed with the culture.

    Sordaria is available from biological supply houses. Depending on conditions, if the culture is grown at 20–25°, it should be mature and have perithecia ready for class use, if you start the culture two weeks before class, and innoculate the new plates with a generous (5×5mm) chunk of the old culture.

ACTIVITY 6

—Dissolve one yeast cake or two 7-gm packages of dry yeast and 50-100 gm of sucrose in 1000 ml of water. Let stand for 2-3 hours to assure budding formation by the time students examine it
—One compound microscope, one slide,* one cover glass* per student

ACTIVITY 7

—Prepare cultures of *Penicillium* and/or *Aspergillus* on potato-dextrose agar. Follow same directions as for culturing *Rhizopus* for Activity 2 of Exercise 23. Start the cultures 4–5 days before class use. One culture plate per four to eight students
—30 ml dropper bottle of 5 percent potassium hydroxide (KOH). One per four students
—30 ml dropper bottle of 70 percent ethanol
    One bottle per four students
—One compound microscope, 2 slides,* 2 cover glasses* per student

ACTIVITY 8

Mushrooms—*Agaricus campestris* or *A. bisporus*—available fresh from grocery, or canned, or available preserved from biological supply houses. About one per two students.

ACTIVITY 9—OPTIONAL
—Fresh specimens of mushrooms in the spore shedding stage
—White paper or index cards

*Part of recommended student lab kit.

—Finger bowls or glass beaker, and so on, to cover mushrooms
    One or two demonstrations per class.
    Set up material 3–5 days (depending on the maturity of the mushroom) before class.
    Don't let mushroom deliquesce.

ACTIVITY 10

—Prepared slides of cross section of cap of gill type mushroom such as *Coprinus* sp. Available
    from biological supply houses. One per student.
—One compound microscope per student.

ACTIVITY 11

Demonstration material of wheat rust (*Puccinia graminis*) and grain smut (*Ustilago* sp.)—available
preserved or dried from biological supply houses. Or collect your own specimens during the
summer.

EXERCISE 25 — ALGAE I: GREEN ALGAE AND EUGLENOIDS

ACTIVITY 1

—Cultures of *Chlamydomonas* (or *Carteria*) are needed—sufficient for each student. All major
    biological supply houses now supply living cultures of *Chlamydomonas*. Each student will use
    about 2 ml of the culture
—Standard glass microscope slide* and cover slip,* one of each per student
—Dropper bottles of methyl cellulose—one per four to eight students. This is available from all
    biological supply houses either in solution or as a dry powder. It is usually mixed with water to
    10 percent (by weight) and added to the drop on the slide
—Dropper bottles of iodine
—Compound microscope—one per student

ACTIVITY 2—OPTIONAL

—Biological supply houses may have these strains of *Chlamydomonas* listed as + and − or as "mat-
    ing type" kits. They are not always available.
—Standard glass microscope slide* and cover slip*—one of each per student
—Compound microscope—one per student

ACTIVITY 3

Prepared slides of *Pandorina* are available from all major biological supply houses. Living cul-
tures are sometimes available as well.
—Standard glass microscope slide* and cover slip*—one of each per student
—Compound microscope—one per student

ACTIVITY 4

Cultures of *Volvox* are available from all major biological supply houses, as are prepared slides.
—Standard glass microscope slide* and cover slip*—one of each per student.
—Compound microscope—one per student

*Part of recommended student lab kit.

ACTIVITY 5

Prepared slide of *Ulothrix* are available from all biological supply houses. Living cultures are frequently available as well. One slide per student.

ACTIVITY 6

Prepared slides of *Oedogonium* are available from all biological supply houses. The activity describes, and Figure 25.5 is based upon, the macrandrous species of this genus. Most unspecified examples of *Oedogonium* are macrandrous, whether slides, living cultures, or preserved material are offered. Do not use specimens described as, or found to be, nanandrous. One slide per student.

ACTIVITY 7

The same slides or specimens as used in Activity 6 are to be used here as well.

ACTIVITY 8—OPTIONAL

Preserved material of *Ulva* is available from all major biological supply houses. Allow one thallus for every four to eight students for gross observation only. Not to be used up.

ACTIVITY 9
—Living cultures of *Euglena* are available from all major biological suppliers, as are prepared slides
—One microscope slide* and cover slip* per student (or one prepared slide per student)
—Dropper bottle of methyl cellulose, as used for Activity 1, above
—Compound microscope—one per student

EXERCISE 26 — ALGAE II:
BLUE-GREEN ALGAE AND GOLDEN-BROWN ALGAE

ACTIVITY 1
—All major biological suppliers have cultures of *Gloeocapsa* and prepared slides, as well. It is also a common moist soil alga, and you may be able to collect your own from rocks, tree bark, flower pots, bricks and greenhouse walks.
—One microscope slide* and cover slip* per student (or one prepared slide per student)
—One compound microscope per student
—One dropper bottle of methylene blue stain per four to six students
—Small pieces (1 × 2 cm) of paper towel—one per student

ACTIVITY 2

Oscillatoria is available from biological suppliers as living cultures, preserved cultures, and prepared slides. It is also one of the most common naturally-occurring algae of stagnant water.
  One microscope slide* and one cover slip* per student (or one prepared slide per student)
  One compound microscope per student

*Part of recommended student lab kit

ACTIVITY 3

Anabaena is commercially available from supply houses as preserved material or prepared slides. It is also a common aquatic alga in more or less permanent ponds. If present in a pond at all, it is likely to be very abundant.
—One microscope slide* and cover slip* per student (or one prepared slide per student)
—One compound microscope per student

ACTIVITY 4

Nostoc is available from biological suppliers as living cultures, preserved cultures and prepared slides. It may also be found in colonies in all types of fresh water as well as on moist soil and in moist moss mats.
—One microscope slide* and cover slip* per student (or one prepared slide per student)
—One compound microscope per student
Note: If prepared slides of *Oscillatoria, Anabaena* and *Nostoc* are purchased for class use, it is necessary to have only enough of each genus for one third of the expected students. The class may then work simultaneously on all three types, with the students sharing slides.

ACTIVITY 5

Soil diatoms may be collected from any moist, undisturbed soil. They are also sold by biological suppliers as living cultures, preserved specimens, or prepared slides.
—One microscope slide* and cover slip* per student (or one prepared slide per student)
—One compound microscope per student
Diatomaceous earth and preserved cultures of marine diatoms are available from suppliers, as are prepared slides. Especially selected prepared mounts are often available for demonstration purposes showing large, unbroken centric diatoms, but they are too costly for general class use.
—One microscope slide* and cover slip* per student (or one prepared slide per student)
—One compound microscope per student

EXERCISE 27 — ALGAE III: IDENTIFICATION OF SOME COMMON FRESH WATER ALGAE

—Make up cultures containing a variety of genera of fresh water algae. Genera should correspond to those in the Algae Key of this exercise. All genera are available from biological supply houses either as single cultures or mixtures of genera. Distribute in beakers or test tubes about one per four students. Pipettes or droppers with each beaker or test tube
—Or, if possible, collect water cultures from lakes, ponds, reservoirs, or rivers
—One compound microscope per student
—Several slides* and cover glasses* per student

EXERCISE 28 — ALGAE IV: BROWN ALGAE, RED ALGAE AND LICHENS

ACTIVITY 1

Living or preserved specimens of *Fucus*. Not to be used up. One per two students. Biological

*Part of recommended student lab kit.

supply houses offer both living and preserved material and plastic mounts. Collect your own, if you are near the ocean.

ACTIVITY 2

Living or preserved specimens of *Laminaria* and/or *Nereocystis*. Available from biological supply houses as living or preserved material, or in plastic mounts. One or two specimens for entire class. Not to be used up.

ACTIVITY 3

Living or preserved specimens of *Sargassum*. One sprig or so per each student. Not to be used up. Available from biological supply houses as living or preserved material, or in plastic mounts. If you live in the southern Gulf States, you may know where to collect your own material.

ACTIVITY 4

Various specimens of red algae available from biological supply houses as living or preserved material. Suggested specimens: *Gelidium* sp.; *Rhodymenia* sp.; and *Stenogramma* sp.

ACTIVITY 5

Dried or fresh specimens of lichens representing the three growth forms: crustose, foliose, and fruticose. Available from biological supply houses or collect them yourself. For general demonstration for class.

ACTIVITY 6

Prepared slides of *Physcia* showing median section of thallus and apothecia. One per student. Available from Ward's Natural Science Establishment, Inc.
—One compound microscope per student

EXERCISE 29 — MOSSES AND LIVERWORTS
(AND INTRODUCTION TO LAND PLANTS)

ACTIVITY 1

Living and preserved material of *Mnium*, *Polytrichum* and similar genera are offered for sale by biological suppliers. The Activity is based upon the availability of a recently-collected living moss mat. Allow one plant per student.
—One microscope slide* and one cover slip* per student
—One compound microscope per student

ACTIVITY 2

All biological suppliers offer prepared slides showing a longitudinal section through the antheridial head of a moss. The Activity and Figure 29.4 assume a dioecious species of *Mnium* or a similar genus.
—One prepared slide per student
—One compound microscope per student

*Part of recommended student lab kit.

ACTIVITY 3—OPTIONAL

This Activity requires living male gametophytes of *Mnium, Polytrichum,* or a similar type of moss. Allow one male gametophyte per two students. Local collection is about the only way to acquire living material at the right stage, and success is, therefore, dependent on the season of the year and your locality.
—One microscope slide,* one cover slip,* and one dissecting needle* per two students
—One compound microscope per two students

ACTIVITY 4

All biological suppliers offer prepared slides of a longitudinal section through the archeogonial head of a moss. As with Activity 2, above, a species of *Mnium* or a similar type is assumed. Note that slides may be described either as "with archegonia" or as "median section" (or similar wording) of the archegonium. The latter are far better in showing the desired features, but are *several dollars per slide* more expensive. A good compromise is to order one or two of the higher quality mounts per class for demonstration.
—One prepared slide per student
—One compound microscope per student

ACTIVITY 5

Both *Polytrichum* and *Mnium* may be purchased with mature sporophytes attached in the living or preserved state. Local collection of these is so simple in most areas at all times of the year that purchase of specimens should be only rarely required. Allow one sporophyte per student.
—One microscope slide,* one cover slip,* and one fine forceps* per student
—One petri dish containing 50–100 g of petroleum jelly per 6–10 students
—Toothpicks for transferring petroleum jelly—one per student
—Pieces of paper towel (about 2 × 4 or 3 × 6 cm)—one or two per student
—Dropper bottle of tap water—one per four students
—One compound microscope per student

ACTIVITY 6—OPTIONAL

This activity also requires mature capsules of *Polytrichum* or *Mnium*, and these may be purchased from biological supply houses or collected locally. Requires one capsule per student.
—One microscope slide* and one cover slip* per student
—One compound microscope per student

ACTIVITY 7

Moss protonemae are available as whole mounts on prepared slides from all major biological suppliers. They are occasionally offered with buds, which are older plants and are preferable for this Activity. Many species of mosses may be cultured on moist soil, moist clay pots, or agar fortified with mineral nutrients such as Knop's solution. Spores may be taken from locally collected species, or viable spores may be purchased from some supply houses.
—One prepared slide per student, or one living culture per four students. The living culture may be a demonstration. Allow about 45–60 days for buds to form on artificial cultures of protonemae.
—One compound microscope per student

*Part of recommended student lab kit.

ACTIVITY 8—OPTIONAL

Living and preserved *Sphagum* plants are available from all biological supply houses. Dried horticultural *Sphagnum* is sold by the cubic foot in bales or bags. One cubic foot is adequate for 50–100 students.
—Disposable plastic drinking glasses—about 6 oz to 8 oz capacity—one per two to four students—may be re-used in other classes
—Triple-beam balance—one per 8–12 students
This activity may be performed as a demonstration, as well.

ACTIVITY 9—OPTIONAL

Living and preserved *Sphagnum* plants are available from all biological supply houses. Individual preserved "leaves" are also sold for microscopic examination. Prepared slides of these "leaves" are also available from some suppliers. Allow one "leaf" per one or two students.
—One microscope slide* and one cover slip* per student or one prepared slide of *Sphagnum* "leaf" per two to four students
—One compound microscope per student

ACTIVITY 10

—Gametophytes of *Marchantia* and *Conocephalum* are sold as living or preserved material by all major biological supply houses. Living material may also be fairly easily cultured on non-calcareous stones or coarse gravel in an aquarium or moist terrarium under moderate light levels.
—Allow one piece of thallus about 1–2 × 4–8 cm per pair of students. This should be examined grossly only and will not be used up.
—Gametophytes of *Marchantia* bearing antheridial and archegonial structures are available primarily as preserved material. Some suppliers are able to provide living material in certain seasons (usually spring only) as well.
—Allow one piece of thallus bearing each type of structure per two to four students. They then work in pairs or groups of four sharing samples of both sexes. It should not be used up.
—One dissecting microscope per two to four students, or one 6× to 10× hand lens*

EXERCISE 30 — LOWER VASCULAR PLANTS I:
PSILOTUM, CLUB MOSSES AND HORSETAILS

ACTIVITY 1

Living or preserved specimens or color transparencies of *Psilotum nudum*. Living specimens available from Ward's Natural Science Establishment, Inc. Also available as preserved or on color transparencies from most biological supply houses.

Many greenhouses, commercial and/or university affiliated, often having living *Psilotum* or you can grow it as an ordinary pot plant indoors. It's fairly easy to maintain and will grow best if it's grown in the same pot with some other plant. One or two plants may be about all that you obtain for the entire class.

*Part of recommended student lab kit.

ACTIVITY 2

Living, preserved, or pressed specimens of *Lycopodium*. Available from most biological supply houses. Carolina Biological Supply Co. offers living material as well as preserved. Try to have species with cones and species without cones. *L. lucidulum* is a common coneless species. *L. clavatum* and *L. complanatum* have cones. You may be able to find *L. lucidulum* sold in plant shops as a terrarium plant. Its common names are ground pine and running pine. One plant per four students if possible. Not to be used up. Hand lens, 6–10×*—one per student

ACTIVITY 3

Living or preserved specimens of *Selaginella*. Sometimes available from plant shops, sold as a terrarium plant. Living material available from Carolina Biological Supply Co. All supply houses carry preserved and/or plastic mounted material. You can easily grow and maintain it yourself as a pot plant. One plant per four students if possible. Not to be used up. Hand lens, 6-10×*—one per student

ACTIVITY 4

Living or preserved specimens of *Equisetum hyemale* and *E. arvense*. *Equisetum* is much more common than *Lycopodium* or *Selaginella* so you may be able to collect your own specimens from old stream banks and roadsides, especially in areas of sandy or infertile soil. Otherwise, plants are available preserved or in plastic mounts from biological supply houses. Plants not to be used up. Hand lens, 6–10×*—one per student

ACTIVITY 5—OPTIONAL

Microscope slides showing anatomical detail of stems and leaves of fossil plants are offered by Carolina Biological Supply Co. and Ward's Natural Science Establishment, Inc.

Rock compressions or impressions of fossil leaves and/or stems of *Lycopodophyta* and *Arthrophyta* are also available from Ward's.

EXERCISE 31 — LOWER VASCULAR PLANTS II: FERNS

ACTIVITY 1

—Have two or three living fern plants for general class demonstration. You can often rent ferns (and other foliage plants) from local plant shops if you don't want to purchase these.
—If possible, depending on the season, make available for class observation ferns with new young fiddle-head shoots
—If possible, obtain fern material that can be dug to show its rhizome with roots. Several classes can examine the same material. Suggested ferns: Boston fern (*Nephrolepsis exaltata* var. *Bostoniensis*); Christmas fern (*Polystichum acrostichoides*); maidenhair fern (*Adiantum* sp.); and *Polypodium* sp.
—Collect fern leaves sometime before or during the school session and press and dry flat in newspapers in plant press (if available). Collect only leaves with mature sori. Sori will be dark or brown if sporangia are mature. Store leaves in boxes. Usually a student will use up one pinna of a leaf when examining it, so store enough leaves to suit the size and number of your classes
—Hand lens*—ideally, one per student

*Part of recommended student lab kit.

ACTIVITY 2
—One compound microscope, one slide,* one cover glass,* one dissecting needle* per student
—Enough fern leaf material for each 1–2 students to work with one pinna

ACTIVITY 4

—Prepared slides of fern prothallus available from biological supply houses. One per student
—Compound microscope—one per student
—Living fern prothalli
     Available from Carolina Biological Supply Co. Or you can start your own cultures, but you need to do this about a month or more before class.
     Soak a Jiffy-7 peat pellet in water. After it has expanded, dust it with fern spores which you can obtain from your own plant material or purchase from biological supply houses. Place the peat cube in a saucer or petri dish and cover with a plastic drinking glass or beaker, and so forth. Set it in a cool (14–20° C) place with medium light and don't let it dry out. In a month, several prothalli will have formed. After about three months growth, young sporophytes will be formed.
     Other methods for culturing fern spores are described in Turtox Service Leaflet No. 44 and in *A Sourcebook for the Biological Sciences* by E. Morholt, P. F. Brandwein and A. Joseph.

ACTIVITY 6

—Prepared slides of fern prothalli bearing young sporophytes. One slide per student
—Or living young sporophytes which you can prepare yourself as described for Activity 5, or they can be purchased
—Compound microscope—one per student

EXERCISE 32 — CONE-BEARING SEED PLANTS

ACTIVITY 1

Cuttings of different gymnosperms made available for general demonstration to the class. Or take students on a tour to see different gymnosperm trees.

ACTIVITY 2

—Prepared slide showing longitudinal section of a staminate pine cone. One per student. Available from biological supply houses
—One compound microscope
—*Alternative*: Preserved staminate cones. Collect and preserve these yourself or order from biological supply house
—Hand lens* or dissecting microscope, one slide,* one cover slip,* and dissecting needle* per student

ACTIVITY 3

Young (first-spring) ovulate cones preserved at pollination time. Either collect and preserve yourself or order from biological supply houses.
—One per student. Not to be used up, but ultimately they are damaged

*Part of recommended student lab kit.

—Prepared slides of ovulate pine cone showing megaspore mother cell. Available from biological supply houses. One slide per student
—Compound microscope—one per student

ACTIVITY 4

One-year-old ovulate cones (collected the second summer). Available from biological supply houses, or collect and preserve yourself. One cone per one or two students.

ACTIVITY 5

Mature ovulate cones with woody scales and mature seeds. Available from biological supply houses or collect yourself. One cone per one or two students.

ACTIVITY 6

Seeds of pine (*Pinus* sp.) The best kind to use are Colorado Pinyon pine (*Pinus cembroides edulis*). They are large, easy to dissect and the embryo and other seed parts are readily identifiable. However, they are more expensive than other, smaller, seeds. You can get the most Pinyon pine seeds for your money from Herbst Brothers Seedsmen, Inc. Biological supply houses also carry them, but they are more costly.

    Other less expensive pine seeds (not identified as to species by the suppliers) are available from biological supply houses. Mellingers (a horticultural supply house) has several species of pine seed but doesn't always have Pinyon pine.

EXERCISE 33 — FLOWERING SEED PLANTS

ACTIVITY 1

—This activity and Activity 3 can be done together.
—Living specimens of impatiens (*Impatiens* sp.) or spiderwort (*Tradescantia* sp.) with flowers in the pollen-shedding stage. If these plants are not available almost any other kind will suffice as long as the flowers have anthers with mature pollen
—Students will transfer some pollen to a slide—so three or four plants per class should be adequate. Some of the same pollen material is to be used in Activity 3
—30 ml dropper bottles with 10 percent cane sugar solution. One per four students
—One slide,* one cover slip,* one dissecting needle* per student
—One petri dish lined with 3–4 layers of filter paper. One dish per two students
—A few brush bristles or comparable material to hold cover slip up off the slide

ACTIVITY 2

Stamens of lily (*Lilium* sp.) flower or other plant. Either preserved or fresh. One or more per student. For examination only. Not to be used up.

ACTIVITY 3

—This activity can be done simultaneously with Activity 1. The same plant material is used
—One slide,* one cover slip,* and one compound microscope per student

*Part of recommended student lab kit.

ACTIVITY 4

—Prepared slides of young flower buds of lily (*Lilium* sp.) showing a cross section through the young anthers. One slide per student
—Compound microscope—one per student
—Available from biological supply houses

ACTIVITY 5

—Prepared slides of cross sections of mature anthers of lily (*Lilium* sp.) or of lily pollen grains. One slide per student
—Compound microscope—one per student

ACTIVITY 6

—Prepared slides of cross sections of young ovary of lily (*Lilium* sp.) showing megasporocytes (megaspore mother cells). One slide per student. Available from biological supply houses
—Compound microscope—one per student

ACTIVITY 7

—Prepared slides of cross sections of ovary of lily (*Lilium* sp.) showing ovule (ovulary) with the mature female gametophyte. One slide per student. Available from biological supply houses
—Compound microscope—one per student

ACTIVITY 8

A collection of a wide variety of angiosperm seeds. Have the seeds of herbaceous plants separately grouped from those of trees and shrubs. Include seeds with spines, hairs, hooks, and so on, to show that in many plants the seed itself is the dispersal agent.

   For your own reference, see *Seed Identification Manual* by Alexander C. Martin and William D. Barkley (1961) University of California Press, 1414 So. Tenth St., Richmond, California 94804

ACTIVITY 9

A collection of a variety of edible fruits. Include many of the common vegetables that students know, such as green peppers (*Capsicum* sp.); tomato (*Lycopersicon* sp.); eggplant (*Solanum melongena*); melons, squashes, and pumpkins (*Cucurbita* sp.); okra (*Hibiscus esculentus*); grape (*Vitis* sp.); apple (*Malus* sp.); orange (*Citrus* sp.), and so on. Also fruits of trees such as ash (*Fraxinus* sp.); oak (*Quercus* sp.); walnut (*Juglans* sp.); maple (*Acer* sp.); sweet gum (*Liquidambar styraciflua*); elm (*Ulmus* sp.); and buckeye (*Aesculus* sp.) etc.

   Also fruits of weeds locally available such as thistle (*Cirsium* sp.); cocklebur (*Xanthium* sp.); dandelion (*Taraxacum officinale*); goatsbeard (*Tragopogon* sp.); jimson weed (*Datura* sp.); or any other species used in Exercise 17.

EXERCISE 34 — MEIOSIS AND MONOHYBRID AND DIHYBRID CROSSES

ACTIVITY 5

—Flats or pots of F_2 generation seedlings which show color differences representing simple monohybrid inheritance. Will not be used up

—Suggested plants: corn (*Zea mays*); tobacco (*Nicotiana tabacum*) and sorghum (*Sorghum vulgare*) Seeds for this are available from biological supply houses. Seeds should be started 6–10 days before class

ACTIVITY 6

—Ear of corn (*Zea mays*) showing colored and white grains in a 3:1 ratio. Available from biological supply houses
—One ear per two students (ideally)
—Supply of toothpicks or straight pins

ACTIVITY 7

—Ear of corn (*Zea mays*) showing the results of a testcross. Available from biological supply houses.
—One ear per two students (ideally)
—Supply of toothpicks or straight pins

ACTIVITY 10

—Ear of corn (*Zea mays*) in which the grains represent the F_2 generation of a cross involving two colors—white and purple—and two endosperm types—full (or starchy) and wrinkled (or sugary). Available from biological supply houses.
—One ear per two students (ideally)
—Supply of toothpicks or straight pins.

EXERCISE 35 — PLANT ECOLOGY

ACTIVITY 1

This Activity may be done either by groups of two to four students or by the instructor as a demonstration. In either case, sets of five or six identical pots of soil are planted with varying numbers of seeds. 3–4″ plastic pots are good, and use of a largely synthetic soil medium containing mostly perlite or vermiculate assists calculation of root mass. Use corn (*Zea mays*); grain sorghum (*Sorghum* sp.); jimson weed (*Datura* sp.); or any available fast-growing herbaceous plant with upright growth habit. They should be started 4–12 weeks before use, depending on the species and size of pot. During this period, water each pot of the sets with roughly equal quantities of water, and do not feed any fertilizer unless you are using a totally artificial potting soil. Be sure to plant extra seeds to make up for those which do not germinate, and also to pull up any extra seedlings at about 2–3 cm in height.
—One set of five or six pots is needed per group, or per demonstration.
—Either fresh or dry weights may be obtained. Dry weights are considered more exact, but mean that the plants must be harvested 48–72 hours before class, and dried for 40–48 hours at 70°C before weighing. The class will then see a display showing a set of living plants to be measured and a set of dried plants from similar pots with weights shown.
—One of the most satisfactory overall methods involves the use of prepared compressed media such as Jiffy-7 or Solo-Gro. To these small "pots," small seeds such as sorghum are better suited than are large corn grains. Some fertilizer is already present, so none need be added. The uniformity of the medium will allow entire masses of plants, roots, and media to be oven-dried easily. The weight of a dry, unused block or pellet may easily be subtracted from the eventual total weight to arrive at the plant biomass. With these smaller rooting masses, too, it may take only 20–30 days to complete maximum growth.
—Access is needed to a forced-draft drying oven if dry weights are to be obtained.

—Paper bags of the type sold for carrying lunches are desirable to hold plants while they are being dried.

—Access to a good electronic or torsion balance—although a triple-beam balance is adequately accurate if plants are large and the experiment has been run for 2½–3 months.

BIOLOGICAL SUPPLY HOUSES*

1) Carolina Biological Supply Company
Burlington, North Carolina 27215

2) Turtox/Cambosco
8200 S. Hoyne Avenue
Chicago, Illinois 60620

3) Ward's Natural Science Establishment
P.O. Box 1712
Rochester, New York 14603

or

P.O. Box 1749
Monterey, California 93940

4) ICN Pharmaceuticals, Inc.
26201 Miles Rd.
Cleveland, Ohio 44128

HORTICULTURAL SUPPLIERS
(INCLUDING SEEDS)

1) Mellinger's .... All gardening supplies;
2310 W. South Range   potting materials, tree seeds
North Lima,

2) Herbst Brothers Seedsmen, Inc. .... Seeds of trees, shrubs,
1000 N. Main St.   vegetables, flowers & grass
Brewster, New York 10509

3) Geo. W. Park Seed Co., Inc. .... Seeds of herbaceous plants;
Greenwood, S.C. 29647   bulbs, and some garden supplies

4) W. Atlee Burpee Co. Seed Growers ....Vegetable and flower seeds
Warminster, Pa. 18974

        or

Clinton, Iowa 52732

        or

Riverside, Calif. 92502

*Suppliers mentioned in this manual. Several other sources exist.